I

MODERN RADAR SYSTEM ANALYSIS

Recent Titles in the Artech House Radar Library

David K. Barton, Series Editor

For further information on these and other Artech House titles, including previously considered out-of-print books now available through our In-Print-Forever® (IPF®) program, contact:

Artech House
685 Canton Street
Norwood, MA 02062
Phon: 781-769-9750
Fax: 781-769-6334
e-mail: artech@artechhouse.com

Artech House
46 Gillingham Street
London SW1V 1AH UK
ePhone: +44 (0)171-973-8077
Fax: +44 (0)171-630-0166
e-mail: artech-uk@artechhouse.com

Find us on the World Wide Web at: www.artechhouse.com

MODERN RADAR SYSTEM ANALYSIS

DAVID K. BARTON

ARTECH HOUSE

Barton, David Knox. 1927-
    Modern radar system analysis.

    Includes bibliographies and index.
    1. Radar.   I. Title
TK6575.B364   1988   621.3848   88-6341
ISBN   0-89006-170-X

Contents

v

Preface

This book began as a new edition of *Radar System Analysis,* the manuscript for which was prepared originally in 1963. Published first by Prentice-Hall in 1964, and republished in the Artech House edition starting in 1976, that book had begun to lose its relevance to radar systems of the 1980s. Its twenty-five year life span was due to the fact that much of its material covered the basic theoretical aspects of radar which do not change with advancing technology. However, in using this book as the reference text for radar courses in The George Washington University program for continuing education, a great deal of supplementary material was needed to address issues that were not covered in the old text.

As the new manuscript advanced, it turned out that only a small fraction of the original material remained, and hence the title was changed to *Modern Radar System Analysis,* to reflect something other than a new edition of the earlier book. In Chapter 1, material has been included to cover radar range calculations in environments including clutter and jamming as well as the benign background of thermal noise. The basic notations and methods originated by Lamont Blake are still used, representing the nearest available approach to a standard radar range calculation procedure. The search radar equation, derived originally by Barlow and his colleagues at Sperry Gyroscope Company and presented in *Radar System Analysis,* is also presented here as a basis for later discussions of this type of radar. Chapter 2 now presents methods of determining detection performance on a broad class of signals, represented by the chi-square distribution with arbitrary number of degrees of freedom, which encompasses not only the steady target and the Swerling target cases, but other target models and diversity modes of radar operation. Material on constant-false-alarm-rate (CFAR) detection has been added. New data on clutter models are included in Chapter 3. Special emphasis has been placed on the characteristics of land clutter, as observed by surface radars, and on chaff as a source of clutter in military radar systems.

Chapters 4 and 5 discuss the key issues of radar resolution, as provided in angle by the antenna and in range and doppler coordinates by the waveform and signal processor. Basic antenna and array theory is summarized to the extent necessary for the radar user or system engineer to understand the issues, rather than as preparation for a career in antenna and microwave design. The ambiguity function is presented not in mathematical form, but rather as a process of measurement in which one or more targets are observed by the radar and indicated by a meter at the output of the receiver and processor. New material on pulsed doppler systems has been included.

The material on propagation, presented in Chapter 6, has been extended to the millimeter-wave bands, and a section on diffraction has been added. These extensions and the new process for modeling the multipath environment, developed over the past ten to twelve years by the author and his colleagues at Raytheon and in US Army organizations, provides a basis for evaluation of radar performance against low-altitude targets, so important in many of today's radar system applications.

The basic theory and modeling approaches developed in the first six chapters are applied in the second half of the book to specific types of radar. Chapter 7 discusses search radars: 2D and 3D air surveillance radars, horizon scanning radars for both air targets and marine navigation, and mapping radars using real and synthetic apertures. Examples are used to illustrate the performance of different classes of waveforms and processing, including MTI, pulsed doppler, and pulse compression. Material on search radar ECM and ECCM has been included, written from the viewpoint of the radar system engineer, rather than that of the electronic warfare specialist. Chapters 8 and 9 discuss basic radar measurement theory, applicable to both search and tracking radars, but most relevant to the dedicated tracker and the multifunction array radar. Chapter 10 covers several topics of special concern to designers and users of tracking radar: acquisition and loss of target tracks, tracking dynamics, application of phased arrays, and considerations of ECM and ECCM. Discussion of monopulse operation in the jamming environment has been made possible by prior publication of applicable material in the Soviet open literature, relieving earlier concerns about military secrecy. The final chapter on Radar Error Analysis contains new material on multipath error calculation, along with a more compact presentation of material from the original *Radar System Analysis*.

Wide availability of computers in radar analysis and simulation tasks has made possible detailed performance data covering different targets and environments. The material presented in Chapter 1 suggests that graphical approaches will remain useful, especially in providing the system engineer and radar user with insight into the effects of different factors in both radar design and environmental conditions. The graphical procedures suggested have not yet been implemented as a program for either personal computer or mainframe. However, certain key elements have been published, and are illustrated in the radar examples of this book. These include the basic range calculation, RGCALC [1.7], based on Blake's material, and determination of vertical coverage, VCCALC [6.20]. A program for calculation of the low-altitude propagation factor, designed by the author for the TI-59 calculator, is included as Appendix A. The author's multipath error program, modified by Bill Skillman and adapted for personal computer use, has been published [11.9]. Skillman has also published a program, SIGCLUT, for evaluation of clutter and jamming effects. All these tools, and undoubtedly others which will become available to the radar engineer during the lifetime of this book, can assist in performing analyses of radar systems.

The understanding and insight of the radar system engineer remains the basic tool for design, evaluation, and effective application of radar. *Modern Radar System Analysis* is addressed to the development and refinement of this ability.

DAVID K. BARTON
HARVARD, MASSACHUSETTS
APRIL 1988

ACKNOWLEDGMENTS

Much of the material in this book originated through collaborative efforts with my colleagues in Raytheon and at several laboratories and agencies of the Department of Defense. An attempt has been made to give references to prior publications of these contributions, but where this has not been possible, I want to make this general acknowledgment of their help. More specifically, I must single out the many useful discussions with Bill Shrader and Gregers Hansen on search radar subjects, with Pete Cornwell and Pete Kirkland on propagation and multipath phenomena, and with Hal Ward (who also provided perceptive review of the manuscript) on the entire range of radar system subjects.

Chapter 1
The Radar Range Equation

The development of radar during the first half of the twentieth century marked the greatest advance in methods of sensing remote objects since the telescope was invented in the year 1608. In spite of subsequent progress in infrared and optical systems, nothing challenges radar as a reliable sensor and measuring device for objects appearing unexpectedly in large volumes of airspace or on the surface of the earth. Radar's ability to scan rapidly over large angular sectors in a short time makes it uniquely suitable for warning of the approach of dangerous targets, in both civil and military applications. Providing its own source of illumination, radar makes possible not only detection but accurate measurement of radial distance and velocity of targets. The ability of radio waves to penetrate the atmosphere under all conditions of weather, and the absence of strong ambient illumination in the frequency bands used by radar, make possible much longer ranges and greater sensitivity than are obtained in the visible portion of the electromagnetic spectrum.

This chapter will discuss the basic approach to radar detection of target objects, and will derive the radar range equations that establish the coverage volume within which target detection can be obtained. Subsequent chapters will consider the characteristics of noise and other interference components entering the receiver, the required ratio of target signals to this interference, and the means by which this ratio can be obtained in the actual environments where radars operate.

1.1 RADAR FUNDAMENTALS

Basic Functions

The two most basic functions of radar are inherent in the word, the letters of which stand for *RA*dio *D*etection *A*nd *R*anging. Measurement of target angles has been included as a basic function of most radars, and doppler velocity is often measured directly as a fourth basic quantity. Resolution of the desired target from background noise and clutter is a prerequisite to detection and measurement, and resolution of surface features is essential to mapping and imaging radar. The radar resolution cell is a four-dimensional volume, bounded by the antenna beamwidths, the width of the processed pulse, and the bandwidth of the receiving filter. Within each resolution cell, a decision may be made as to the presence or absence of a target, and if a target is present its position may be interpolated to some fraction of the cell dimensions.

The block diagram of a typical pulsed radar is shown in Figure 1.1.1. The equipment has been divided arbitrarily into seven subsystems, corresponding to the usual design specialties within the radar engineering field. The radar operation is controlled by a master clock, located within the synchronizer or in the associated exciter block. The synchronizer generates a series of pulses which initiate the radar transmission, the receiver gates or sampling strobes, the signal processing functions, and the display sweeps. When called for by the synchronizer, the modulator supplies a pulse of high voltage to the RF amplifier, simultaneously with an RF drive signal from the exciter. The resulting high-power RF pulse is passed through the transmission line or waveguide to the duplexer, which connects the line to the antenna for radiation into space. The antenna forms a beam which is steered by mechanical or electrical means to a specific direction in front of the antenna aperture.

After reflection from a target, the echo signal reenters the antenna, which has been connected to the receiver by the duplexer. A local oscillator signal furnished by the exciter translates the echo frequency to intermediate frequency (IF), which is then amplified and filtered in the receiver prior to more refined signal processing. The processed signal is passed through an envelope detector, which recovers the pulse waveform and makes it available for display or further video processing. Data to control the antenna steering and to provide outputs to an associated computer are extracted from the time delay and modulation on the video signal.

There are many variations from the diagram Figure 1.1.1 which can be made in radars for specific applications. For example, the block diagram of a modern, computer-controlled, phased-array radar system is shown in Figure 1.1.2. The radar synchronizer is now a specialized portion of the digital hardware that accepts radar control messages from the computer scheduler, translating them into waveform selection and timing controls, beam switching controls, signal processor sampling strobes, and processor controls. The exciter and modulator, under digital control, synthesize the carrier frequency and waveform modulation. The beam-steering processor controls settings of the phase shifters that establish the direction of the

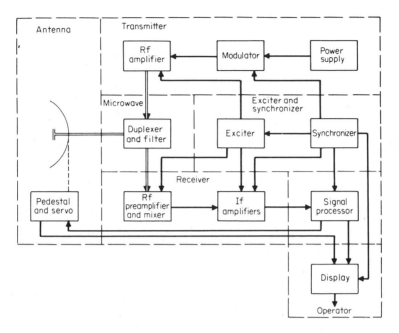

Figure 1.1.1 Block diagram of typical pulsed radar.

transmitting and receiving beams. Receiver paths and gain settings are also computer controlled to minimize the chance of overload by jamming or clutter. For each scheduled radar action, consisting of transmission and receiving of one or more pulses in a particular beam position, the waveform and receiver-processor configuration can be selected to optimize performance. Successful control of such a radar requires detailed knowledge of the environment and availability of radar resources both in terms of time and transmitter power or energy.

Radar Applications

A complete catalog of radar applications would extend for many pages, with new entries added each year. The major fields of application, however, remain as shown in Table 1.1. Other miscellaneous applications, not readily categorized, can also be mentioned: monitoring of bird migrations, ground vehicle control, rendezvous of space vehicles, *et cetera*. The basic principles of radar, to be discussed in this volume, apply to all of these systems, with suitable definition of target parameters and resolution or measurement requirements.

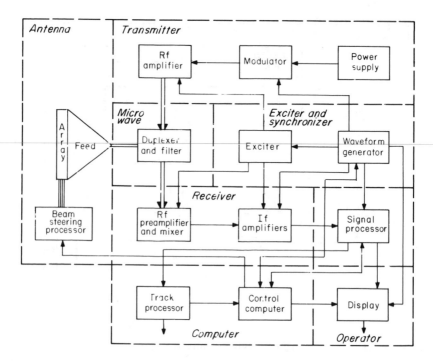

Figure 1.1.2 Block diagram of phased-array radar.

Radar Frequency Bands

Radar techniques can be used at any frequency from a few megahertz up to the optical and ultraviolet ($f > 3 \times 10^{15}$ Hz, $\lambda < 10^{-7}$m), but most equipment has been built for microwave bands between 1 and 40 GHz. The IEEE has adopted as a standard the band letter system, which has long been used in the engineering literature of radar and communications, shown in Table 1.2.

One reason for identifying these separate radar bands, rather than using the coarser ITU designations of VHF, UHF, SHF, and EHF, is that the propagation characteristics and applications of radar tend to change quite rapidly over the microwave regions. Attenuation in rain varies as f^2, when measured in decibels, and backscatter from rain and other small particles varies as f^4 over most of the microwave region. Ionospheric effects vary inversely with the square of frequency, and can be important at frequencies below about 3 GHz. The dimensions of the radar resolution cell vary inversely with frequency, with given antenna size and percentage bandwidth of the signal. These factors lead to the following general preferences and limitations in the use of the different bands:

Table 1.1 Radar Applications

Air Surveillance:	Long-range early warning (including airborne early warning); Ground-controlled intercept; Acquisition, for weapon system; Height-finding and 3D radar; Airport and air route surveillance.
Space and Missile Surveillance:	Ballistic missile warning; Missile acquisition; Satellite surveillance.
Surface Search and Battlefield Surveillance:	Sea search and navigation; Harbor and waterway control; Ground mapping; Intrusion detection; Mortar and artillery location; Airport taxiway control.
Weather Radar:	Observation and prediction of precipitation and winds; Weather avoidance (aircraft); Clear-air turbulence detection; Cloud visibility indicators.
Tracking and Guidance:	Antiaircraft fire control; Surface fire control; Missile guidance; Range instrumentation; Satellite instrumentation; Precision approach and landing.
Astronomy and Geodesy:	Planetary observation; Earth survey; Ionospheric sounding.

HF band: over-the-horizon radar, combining very long range with low spatial resolution and accuracy;
VHF and UHF bands: long-range, line-of-sight surveillance (200 to 500 km) with low to medium resolution and accuracy, and freedom from weather effects;
L-band: long-range surveillance with medium resolution and some weather effects;
S-band: short-range surveillance (100 to 200 km), long-range tracking (50

Table 1.2 Standard Radar-Frequency Letter-Band Nomenclature

Band Designation	Nominal Frequency Range	Specific Frequency Ranges for Radar Based on ITU Assignments for Region 2, see Note (1)
HF	3 MHz–30 MHz	Note (2)
VHF	30 MHz–300 MHz	138 MHz–144 MHz
		216 MHz–225 MHz
UHF	300 MHz–1000 MHz (Note 3)	420 MHz–450 MHz (Note 4)
		890 MHz–942 MHz (Note 5)
L	1000 MHz–2000 MHz	1215 MHz–1400 MHz
S	2000 MHz–4000 MHz	2300 MHz–2500 MHz
		2700 MHz–3700 MHz
C	4000 MHz–8000 MHz	5250 MHz–5925 MHz
X	8000 MHz–12 000 MHz	8500 MHz–10 680 MHz
K_u	12.0 GHz–18 GHz	13.4 GHz–14.0 GHz
		15.7 GHz–17.7 GHz
K	18 GHz–27 GHz	24.05 GHz–24.25 GHz
K_a	27 GHz–40 GHz	33.4 GHz–36.0 GHz
V	40 GHz–75 GHz	59 GHz–64 GHz
W	75 GHz–110 GHz	76 GHz–81 GHz
		92 GHz–100 GHz
mm (Note 6)	110 GHz–300 GHz	126 GHz–142 GHz
		144 GHz–149 GHz
		231 GHz–235 GHz
		238 GHz–248 GHz (Note 7)

Table 1.2 (cont'd)

NOTES: (1) These frequency assignments are based on the results of the World Administrative Radio Conference of 1979. The ITU defines no specific service for radar, and the assignments are derived from those radio services which use radar: radiolocation, radionavigation, meteorological aids, earth exploration satellite, and space research.

(2)There are no official ITU radiolocation bands at HF. So-called HF radars might operate anywhere from just above the broadcast band (1.605 MHz) to 40 MHz or higher.

(3) The official ITU designation for the ultra high frequency band extends to 3000 MHz. In radar practice, however, the upper limit is usually taken as 1000 MHz, L- and S-bands being used to describe the higher UHF region.

(4) Sometimes called P-band, but use is rare.

(5) Sometimes included in L-band.

(6) The designation mm is derived from *millimeter*-wave radar, and is also used to refer to V- and W-bands when general information relating to the region above 40 GHz is to be conveyed.

(7) The region from 300 GHz–3000 GHz is called the submillimeter band.

to 150 km) with medium accuracy, subject to significant weather effects in snow or heavy rain;

C-band: short-range surveillance, long-range tracking and guidance with high accuracy, subject to increased weather effects in snow or medium rain;

X-band: short-range surveillance in clear weather or light rain; long-range tracking and guidance with high accuracy in clear weather, reduced to medium or short range (25 to 50 km) in rain;

K_u- and K_a-bands: short-range tracking and guidance (10 to 25 km), used especially when antenna size is very limited and when all-weather operation is not required. Wider use in airborne systems at altitudes above most weather;

V-band: very short-range tracking (2 to 5 km) when long-range interception of signals must be avoided;

W-band: very short-range tracking and guidance (2 to 5 km);

mm-wave bands: very short-range tracking and guidance (2 to 5 km).

There are, of course, many cases in which band usage is stretched, such as to provide accurate tracking at L-band with very large antennas and with compensation for ionospheric propagation effects, or to search at C- or X-band with special doppler processing to reject rain clutter and with high power to overcome the atmospheric attenuation and limited aperture size. However, once a radar band and overall size and power are established, the potential for search and tracking functions is fairly well constrained.

Major Problem Areas

Radar has been described as a mature art because the basic scientific principles are well understood, and the problem areas are steadily yielding to an advancing technology. Most of these problems have been recognized from the early days of radar, and remain only partially solved today. A few of these, and approaches to their solution, are the following.

(a) *Adequate signal-to-noise ratio in free space.* The radar equations describe what is needed, and technologists are responding with higher average powers, larger antenna apertures, higher frequencies (for trackers), reduced receiver noise, and more efficient processing over longer periods of target illumination.

(b) *Clutter reduction.* In a given environment, clutter competing with potential targets can only be reduced by selecting the most appropriate radar frequency and reducing the size of the radar resolution cell in any of its four dimensions: two angular coordinates, range delay, and doppler frequency. Especially in search operations, where the required coverage volume is large, the design limitations and trade-offs offer limitless opportunities for further effort by systems engineers and designers.

(c) *Interference reduction.* The third source of background signals, competing with the target, comprises the active RF generators (friendly or hostile) that surround the radar. Many system designs are dominated by this factor, and the electronic environment changes as rapidly as the radar technology designed to overcome it.

(d) *Signal selection.* Given a limited ratio of signal to total background, or a background which varies in space and time over a wide dynamic range, signal processing techniques are required to select desired signals on the basis of some set of characteristics which distinguish them from the background. This is one of the areas in which modern digital signal processing has been most fruitful, and in which the future potential is high.

(e) *Measurement accuracy.* Once a signal has been detected and selected for use, it must generally be measured in spatial coordinates, and possibly in doppler and amplitude as well. The reduction of errors from internal radar sources, environment, and the target itself is another continuing field of radar work.

(f) *Size, weight, cost, and reliability.* Solutions to the preceding problems are useless if they involve equipment that is too cumbersome, expensive, or complex to be deployed and operated by the intended user. These practical constraints have consigned most developed radar systems to the museum or the junkyard before they were placed in production. The challenge to the radar system designer is to know how little can be incorporated into the equipment to obtain adequate performance, and that is the purpose of this book.

1.2 DERIVATION OF THE RANGE EQUATION

The radar equation is a basic relationship which permits the calculation of echo signal strength from the known parameters of the radar, the propagation path, and the target. The radar equation may be extended to express the ratio of signal power to noise power (SNR) at the output of a radar receiver, and this ratio may then be used to calculate the expected performance of the radar as a detection or measurement device. If exact values of the parameters and propagation factors used in the radar equation are known, the equation will yield an exact value of received signal and SNR. In practice, however, knowledge of the radar and target parameters is imperfect, and a series of loss factors is included to make the results of the equation agree with the actual performance. The calculation or estimation of these loss terms is an important step in evaluating the performance of a practical radar.

Received Signal Power

Assume first that the power P_t of the radar transmitter is radiated isotropically from the antenna. Since the surface area of a sphere at radius R from the radar is $4\pi R^2$, the uniform distribution of transmitted power over this surface will produce a power density given by

$$I_s = \frac{P_t}{4\pi R^2} \tag{1.2.1}$$

If a transmitting antenna with power gain G_t is now used in place of the isotropic antenna, the power density along the axis of its beam will be increased to

$$I_s = \frac{P_t G_t}{4\pi R^2} \tag{1.2.2}$$

where I_s is in W/m^2 for P_t in W and R in m. Imagine now a spherical target of radius a at range R on this beam axis. The target has a projected area $\sigma = \pi a^2$, which will intercept a fraction of the radar power given by

$$P_i = I_s \sigma = \frac{P_t G_t \sigma}{4\pi R^2} \tag{1.2.3}$$

This power will be reradiated isotropically by the sphere, producing at the radar a power density equivalent to that of an isotropic transmitter of power P_i located at the target:

$$I_r = \frac{P_i}{4\pi R^2} = \frac{P_t G_t \sigma}{(4\pi R^2)^2} \tag{1.2.4}$$

The radar antenna may be characterized by its effective aperture area A_r, which will capture from the reradiated wave a power:

$$S = I_r A_r = \frac{P_t G_t A_r \sigma}{(4\pi)^2 R^4} \tag{1.2.5}$$

This is the power available to be transferred to the radar receiver. If the receiving antenna is described by its gain G_r, we may use the following relationship:

$$A_r = G_r \lambda^2 / 4\pi \tag{1.2.6}$$

to arrive at the received power:

$$S = \frac{P_t G_t G_r \lambda^2 \sigma}{(4\pi)^3 R^4} \tag{1.2.7}$$

This is the equation for received signal power at the radar, for a spherical target of cross section σ. An actual target, viewed at a particular aspect angle, may be replaced by an equivalent spherical target that produces the same signal power. The target is then characterized by a radar cross section σ, which is not necessarily related to its physical area, but which describes its ability to scatter power toward the radar.

The preceding equation may be restated in the form of the product of three factors:

$$S = (P_t G_t) \frac{\sigma}{4\pi R^2} \frac{A_r}{4\pi R^2} \tag{1.2.8}$$

where the first factor is the effective radiated power (ERP) of the radar transmission in the direction of the target, the second is the fraction of this ERP intercepted by the effective spherical target cross section, and the third is the fraction of the resulting scattered power captured by the receiving aperture.

As an example, consider a radar transmitting $P_t = 100$ kW with a gain $G_t = 10^4$. The target is $\sigma = 1$ m^2 at a range $R = 90$ km, and the receiving aperture ($G_r = 10^4$ at $\lambda = 0.1$ m) is $A_r = 8$ m^2. The three terms and resulting received power are

$$S = (10^9)(10^{-11})(8 \times 10^{-11}) = 8 \times 10^{-13} \text{ W}$$

Bistatic Radar Equation

The signal power equation in the form of (1.2.8) can easily be modified to express the power received in a bistatic radar system, in which the receiving site is separated from the transmitter by a significant distance. The target cross section must now be stated as a bistatic value σ_b, expressing the ability of the real target to scatter energy incident from the direction of the transmitter into the direction of the receiver. The second term of the equation will contain the transmitter-to-target range R_1, while the third term will contain the target-to-receiver range R_2:

$$S = (P_tG_t)\frac{\sigma_b}{4\pi R_1^2}\frac{A_r}{4\pi R_2^2} \qquad (1.2.9)$$

Radar Beacon Equations

If the two-way radar equation is compared to the one-way transmission equations, applicable to the interrogation and response of a transponder beacon, to interception of the radar signal by a receiver on the target aircraft, or to jamming of the radar by a transmitter on the aircraft, the advantages of one-way operation are immediately apparent. For interrogation by or interception of the radar signal,

$$S_i = P_tG_t\frac{A_b}{4\pi R^2} = P_tG_tG_b\frac{\lambda^2}{(4\pi R)^2} \qquad (1.2.10)$$

and for transponder or jammer response,

$$S_r = P_bG_b\frac{A_r}{4\pi R^2} = P_bG_bG_r\frac{\lambda^2}{(4\pi R)^2} \qquad (1.2.11)$$

Even for a small airborne antenna, $G_b = \pi$, $A_b = \lambda^2/4$, there is only one large path-loss term in each equation, rather than two in cascade. The one-way advantage is

$$\frac{S_i}{S} = \frac{A_b}{A_r}\frac{4\pi R^2}{\sigma}, \quad \frac{S_r}{S} = \frac{P_bG_b}{P_tG_t}\frac{4\pi R^2}{\sigma} \qquad (1.2.12)$$

For our previous example, with a 100 W beacon,

$$S_i = 10^9(2.5 \times 10^{-14}) = 2.5 \times 10^{-5} \text{ W}$$

$$S_r = 314(8 \times 10^{-14}) = 2.5 \times 10^{-8} \text{ W}$$

The dominant term $(4\pi R^2/\sigma) = 10^{11}$ in each case overwhelms the radar's advantage in aperture area and ERP.

Receiver Noise

The ability of the radar to detect the target echo depends not only on the signal power S, but also on the competing noise present in the receiver. The sources of this noise lie both within the radar and in the outside environment, as shown in Figure 1.2.1.

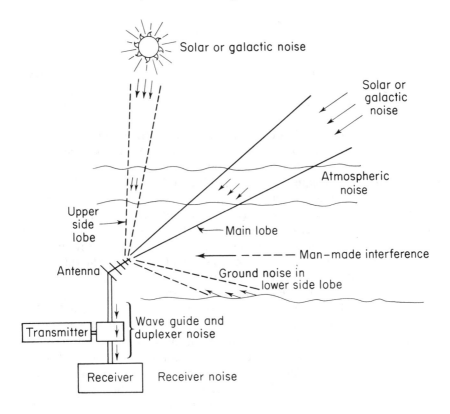

Figure 1.2.1 Sources of radar noise.

It is convenient to summarize the effect of these noise components in terms of a single source at the antenna output terminal, represented by an input termination resistor with a temperature T_s, which makes available to the receiver a noise with spectral density:

$$N_0 = kT_s \qquad (1.2.13)$$

where k is Boltzmann's constant, 1.38×10^{-23} W/(Hz-K), and T_s is in kelvins, giving N_0 in W/Hz. This system input noise temperature, using the method of Blake [1.2], may be divided into three major components:

$$T_s = T_a + T_r + L_r T_e \qquad (1.2.14)$$

where

$$T_a = (0.88 T_a' - 254)/L_a + 290 \qquad (1.2.15)$$

$$T_r = T_{tr} (L_r - 1) \qquad (1.2.16)$$

$$T_e = T_0 (F_n - 1) \qquad (1.2.17)$$

Here, the temperature T_a is the contribution from the antenna, T_r is that of the RF components connecting the antenna to the receiver, and T_e is that of the receiver itself. T_a' is the apparent temperature of the sky as viewed at the radar frequency, plotted in Figure 1.2.2, L_a is the dissipative loss within the antenna, T_{tr} and L_r are the physical temperature and loss of the input RF components, T_0 is the reference temperature of 290 K, and F_n is the noise factor of the receiver. For antennas pointing at an absorbing ground surface, $T_a = 290$ K, and $T_s = 290 L_r F_n$. However, for antennas which look into cool space, $T_s \rightarrow 290 (L_r F_n - 1)$. The difference in system performance can become appreciable when $L_r F_n \rightarrow 1$, and is 1 dB for $L_r F_n = 5.0$ (or 7 dB). Note that all terms in (1.2.14) to (1.2.17) are expressed in degrees or as power ratios, not in dB. Thus, a loss $L_r = 1.0$ dB must be expressed as a ratio 1.26 when used in any of these equations.

In effect, (1.2.13) to (1.2.17) represent the system noise in terms of an input termination resistor heated to a temperature T_s, followed by an ideal (noise-free) receiving system having the gain and bandpass characteristics of the actual system. This permits us to analyze the system without becoming involved in details of the receiver noise sources and gain structure, which are instead summarized in F_n and T_e. The output signal-to-noise ratio can be calculated by assuming that the input signal S competes with an effective input noise power:

$$N = N_0 B_n = k T_s B_n \qquad (1.2.18)$$

and that both are passed through the receiver having a noise bandwidth B_n defined as

$$B_n = \frac{1}{|H(f_0)|^2} \int_{-\infty}^{\infty} |H(f)|^2 df \qquad (1.2.19)$$

where $H(f)$ is the frequency response of the receiving filter and f_0 is the center frequency of that filter (Figure 1.2.3). The noise bandwidth B_n is usually close to the half-power bandwidth, but for single-tuned filters, B_n is greater by a factor $\pi/2$. As will be shown below, the output SNR can be calculated by combining (1.2.18) and (1.2.19) only for the case of a wideband filter, $B_n\tau \gg 1$, where τ is the transmitted pulsewidth. For matched and narrower bandwidth filters, a different procedure must be followed.

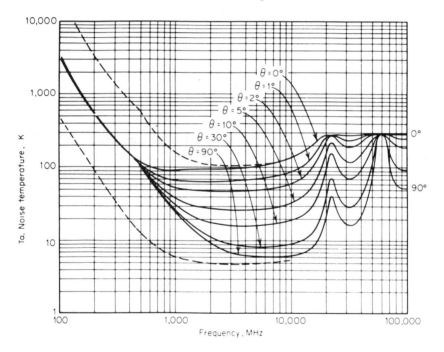

Figure 1.2.2 Noise temperature of an idealized antenna (lossless, no earth-directed sidelobes) at the earth's surface as a function of frequency for a number of beam elevation angles.

NOTE: The solid curves are for the geometric-mean galactic temperature, sun noise 10 times the quiet level, the sun in a unity-gain sidelobe, a cool temperate-zone troposphere, 3 K cosmic blackbody radiation, and zero ground noise. The upper dashed curve is for maximum galactic noise (center of galaxy, narrow-beam antenna), sun noise 100 times the quiet level, zero elevation angle, and other factors the same as for the solid curves. The lower dashed curve is for minimum galactic noise, zero sun noise, and a 90° elevation angle. The slight bump in the curves at about 500 MHz is due to the sun noise characteristic. The curves for low elevation angles lie below those for high angles at frequencies below 400 MHz because of the reduction of galactic noise by atmospheric absorption. The maxima at 22.2 GHz and 60 GHz are due to water-vapor and oxygen absorption resonances [1.2].

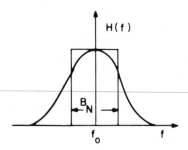

Figure 1.2.3 Equivalent noise bandwidth of a filter.

Noise Density and Power Example

To illustrate the receiver noise calculation, consider an S-band receiver consisting of a low-noise RF amplifier, $F_n = 5$ dB, connected through $L_r = 1$ dB to an antenna pointed at $\theta = 1°$, with a low internal antenna loss $L_a = 0.2$ dB and normal sidelobe characteristics. The receiver is matched to receive 1 μs pulses ($B_n = 1$ MHz). The calculation of noise temperature is as follows:

$$T'_a \approx 65 \text{ K (from Figure 1.2.2)}$$

$T_a = (0.88 \times 65 - 254)/1.047 + 290$		= 102	K
$T_r = 290(1.26 - 1)$		= 75	K
$L_r T_e = 1.26 \times 290(3.16 - 1)$		= 790	K
T_s		= 967	K
$N_0 = 1.38 \times 10^{-23} \times 967$	1.33 $\times 10^{-20}$		W/Hz
B_n		= 10^6	Hz
$N = 1.33 \times 10^{-14}$ W $= -138.8$ dBW		= -108.8	dBm

If used to receive the signal calculated in the previous radar example, with the pulsewidth assumed $\tau = 1$ μs, the received pulse power, energy, and signal-to-noise energy ratio are

$$S = 8 \times 10^{-13} \text{ W} = -121 \text{ dBW} = -91 \text{ dBm}$$

$$E_1 = S\tau = 8 \times 10^{-19} \text{ W-s}$$

$$E_1/N_0 = 60 = +17.8 \text{ dB}$$

Signal-to-Noise Ratio

The optimum filter for detection of a signal in white noise is the *matched filter*, which has a frequency response $H(f)$ equal to the complex conjugate of the signal spectrum $A(f)$, and an impulse response $h(t)$ equal to the time-reversed conjugate of the signal waveform $a(t)$:

$$H(f) = A^*(f)$$
$$h(t) = a^*(-t)$$
(1.2.20)

The theory of the matched filter [1.3] states that the maximum output SNR is equal to the ratio of total received energy to noise spectral density:

$$(S/N)_{mf} = E/N_0$$

If we consider a single received pulse, with the receiving filter matched to the pulse spectrum, this becomes

$$(S/N)_{mf} = E_1/N_0 = S\tau/N_0 = \frac{P_t \tau G_t G_r \lambda^2 \sigma}{(4\pi)^3 R^4 k T_s}$$
(1.2.21)

Other filters will produce lower SNR values

$$S/N = E_1/N_0 L_m$$
(1.2.21a)

where $L_m > 1$ is the matching loss of the filter to the pulse spectrum.

For continuous wave (CW) or coherent pulse radars, which integrate over an observation time t_o in a predetection filter, the received energy ratio and matched filter output SNR is equal to the energy ratio evaluated over t_o:

$$(S/N)_{mf} = E/N_0 = \frac{P_{av} t_o G_t G_r \lambda^2 \sigma}{(4\pi)^3 R^4 k T_s}$$
(1.2.22)

The matching loss for a practical pulsed system of this sort can be expressed as the product of two terms $L_m L_{mf}$, where the first represents the mismatch of the IF filter to the spectrum of each individual pulse, and the second denotes the mismatch loss of the integrating filter to the envelope of the pulse train, extending over t_o.

The advantage of approaching range calculation through the matched-filter SNR is that the effects of modern waveforms and processing, involving pulse compression and integration of multiple-pulse bursts, are readily expressed by the approximately matched filter with small matching loss factors. The energy transmitted in these waveforms, $P_t \tau$ per pulse and $P_{av} t_o$ per target observation, is easily established, as is the noise spectral density $N_0 = kT_s$. All that must be known about the receiving filters and processing steps is how closely they approximate the matched filter. Since matching losses seldom exceed 2 or 3 dB in a practical system, there is little opportunity for serious error or misinterpretation. Alternative procedures, which involve expressions using an input bandwidth B_n and subsequent processing gain factors, are subject to misinterpretation and large errors.

Loss Factors

So far, the radar equation has been derived on the basis of free-space propagation conditions and ideal radar operation. In practice, a number of other factors must be included in arriving at the energy ratio available to the receiver.

(a) *Signal attenuation prior to the receiver.* This comprises: transmission line loss, L_t, between the transmitter tube (at which P_t or P_{av} is measured) and the antenna terminal (at which G_t is measured); antenna losses (usually included in G_t, G_r, or T_s); receiving line and circuit losses (included in T_s through the term L_r); atmospheric attenuation, L_α (to be discussed in Chapter 6); and atmospheric noise (included in T_a' for clear air, but requiring an increase for larger L_α in precipitation).

(b) *Surface reflection and diffraction effects.* In Chapter 6 the pattern-propagation factor F is defined, describing the ratio of one-way field amplitude at range R to that which would have been obtained under free-space conditions in the center of the beam. Hence, the radar equation for a target at a given point in space will contain the factor F^4 in the numerator, accounting for the two-way power ratio.

(c) *Antenna pattern and scanning.* At any given instant, the factor F^4 will give the combined effects of two-way propagation and antenna pattern variation from the peak of the beam. In many cases, however, it is convenient to average the pattern effects over the scan of the antenna, and to include the average loss in received energy as a loss factor L_p in the denominator of the radar equation, either as a separate factor or as an increase in the required beam-center SNR to achieve detection. If L_p is included in this way, for a scan in either coordinate, the pattern-propagation factor F will be calculated for a target on the center of the beam in that coordinate. Details in evaluation of L_p and F will be discussed later.

Applying the appropriate loss factors as described above, the equations for available SNR at the input of the receiver can be written as follows. For the single pulse case,

$$\frac{E_1}{N_0} = \frac{P_t \tau G_t G_r \lambda^2 \sigma F^4}{(4\pi)^3 R^4 k T_s L_t L_\alpha} \tag{1.2.23}$$

For the complete observation over t_o seconds,

$$\frac{E}{N_0} = \frac{P_{av} t_o G_t G_r \lambda^2 \sigma F^4}{(4\pi)^3 R^4 k T_s L_t L_\alpha} \tag{1.2.24}$$

For a pulsed radar, we can see that (1.2.24) gives an energy ratio equal to nE_1/N_0, where $n = t_o f_r$ is the total number of pulses received during the observation. This follows because the product $P_t \tau f_r$ is equal to the average power P_{av}.

Solution for Maximum Range

In Chapter 2 we will derive expressions for the SNR required to achieve target detection with given probabilities of detection and false alarm, and for some additional loss factors which enter into target detection. This will permit us to establish the *detectability factor* D_x, which is the input energy ratio required to achieve detection. For the present, we will assume that this factor is known, and solve for maximum range by setting the energy ratio in (1.2.23) or (1.2.24) equal to the corresponding value $D_x(n)$ or $D_x(1)$. The notation $D_x(n)$ indicates that n samples (or pulses) of signal, each with this energy ratio, will be integrated after envelope detection before being applied to the target detection threshold. Thus, in (1.2.24), where only one coherently integrated sample is available at the envelope detector, the value $D_x(1)$ will be large relative to the value $D_x(n)$, which applies to each one of many received pulses.

In terms of the single-pulse values,

$$R_m^4 = \frac{P_t \tau G_t G_r \lambda^2 \sigma F^4}{(4\pi)^3 k T_s D_x(n) L_t L_\alpha} \tag{1.2.25}$$

while for coherent integration over the total observation,

$$R_m^4 = \frac{P_{av}t_oG_tG_r\lambda^2\sigma F^4}{(4\pi)^3kT_sD_x(1)L_tL_\alpha} \tag{1.2.26}$$

Equation (1.2.25) is essentially the form used by Hall [1.4] in his classic IRE paper, while (1.2.26) is the more general form used in a later work [1.5]. The latter equation may, of course, be applied to radars using post-detection integration, if the appropriate integration loss factors are included in calculation of D_x (1). This will be discussed further in Chapter 2.

In most CW and pulsed doppler radars, the coherent integration time $t_f = 1/B_f$ of the predetection filters will be less than the total observation time t_o of the signal. Hence, during t_o, there will be $n' = t_o/t_f = t_oB_f$ independent outputs from the filter, and these may be integrated after envelope detection. The resulting performance is only slightly below that of the system with predetection filters matched to t_o. If we use, in (1.2.25), the energy $P_{av}t_f$ transmitted in each sample and the detectability factor $D_x(n')$, we can write

$$R_m^4 = \frac{P_{av}t_fG_tG_r\lambda^2\sigma F^4}{(4\pi)^3kT_sD_x(n')L_tL_\alpha} \tag{1.2.27}$$

It is this expression for mixed integration (partially predetection and partially postdetection) which will be applicable in most practical doppler radar problems.

Range Calculation Using the Blake Chart

In the absence of an IEEE or other standard for the radar range calculation procedure, the nearest to an accepted procedure is represented by the range calculation worksheet or Blake chart, Figure 1.2.4. The version shown is a modification of that presented by Blake [1.2], which permits either (1.2.25) or (1.2.27) to be used, and hence is more readily applicable to CW or pulsed doppler radars. The modified chart also directly uses the terms from these equations, in basic units, rather than converting P to kW and λ to c/f with corresponding conversion constants. In the chart, the value D_x (n) is represented by the product $D(n)ML_pL_x$, corresponding to Blake's $V_0C_BL_pL_x$. Calculation and definition of the detectability and matching factors $D(n)$ and M will be discussed in Chapter 2.

The decibel value on line 7 of the chart represents $40 \log R_0$. Table 1, referenced in the chart, simply contains values of $T = T_0(L-1)$, while

PULSE-RADAR RANGE-CALCULATION WORK SHEET

Based on Eqs. (1.2.25) and (1.2.27)

1. **Compute the system input noise temperature** T_s, following the outline in section A below.
2. **Enter range factors** known in other than decibel form in section B below, for reference.
3. **Enter logarithmic and decibel values** in section C below, positive values in the plus column and negative values in the minus column. For display detection, substitute V for D and C_B for M.

| Detection probability, $P_d = 0.$_____ ; False-alarm probability, P_{fa} = ____ ; Case____ target; n = ____ hits |
| Radar antenna height: h_r = m   Target elevation angle: θ = °. |

A. Computation of T_x:	B. Range Factors	C. Decibel Values	Plus (+)	Minus (−)
$T_s = T_a + T_r + L_r T_e$	P_t (or P_{av})	10 log P_t (W)	.	.
(a) Compute T_a.	τ (or t_f)	10 log τ (sec)	.	.
For $T_{ta} = T_{tn}$ = 290 and	G_t	$G_{t(dB)}$	.	.
T_d = 36	G_r	$G_{r(dB)}$	.	.
$T_a = (0.876\,T'_a - 254)/L_a + 290$	σ (sq m)	10 log σ	.	.
$L_{a(dB)}$:_____ L_a:_____	λ	20 log λ (m)	.	.
T_a = _____ K°	T_s (°K)	−10 log T_s	.	.
$\boxed{T_a = \quad °K}$	D	$-D_{(dB)}$	.	.
(b) Compute $T_r = T_{tr}(L_T - 1)$	M	$-M_{(dB)}$	///	.
For T_{tr} = 290 use Table 1.	L_t	$-L_{t(dB)}$	///	.
$L_{r(dB)}$:_____ $\boxed{T_r = \quad °K}$	L_p	$-L_{p(dB)}$	///	.
	L_x	$-L_{x(dB)}$	///	.
(c) Compute $T_e = T_0 (F_n - 1)$	Range-equation constant	75.6	///	///
or using Table 1.	4. Obtain the column totals ——▶		.	.
	5. Enter the smaller total below the larger▶		.	.
$F_{n(dB)}$:_____ T_e:_____ °K	6. Subtract to obtain the net decibels (dB) ▶		+ .	− .

L_r:_____ $\boxed{L_r T_e = \quad °K}$	7. In Table 2 find the range ratio corresponding to this net decibel (dB) value, taking its sign (±) into account. This is R_0 ——▶	☐
Add. $\boxed{T_s = \quad °K}$	8. Multiply R_0 by the pattern-propagation factor $\boxed{F = \quad}$ $F = \sqrt{F_t F_r}$ $R_0 \times F = R'$ ——————▶	☐

9. **On the appropriate curve of Figs. 6.I.2 and 6.I.3**, determine the atmospheric-absorption loss factor, $L_{\alpha(dB)}$, corresponding to R'. This is $L_{\alpha(dB)(1)}$ ——▶ ☐

10. **Find the range factor** δ_1 corresponding to $-L_{\alpha(dB)(1)}$ from the formula δ = antilog $(-L_{\alpha(dB)}/40)$ or by using Table 2. ——▶ ☐

11. **Multiply R' by δ_1.** This is a first approximation of the range R_1. ——▶ ☐

12. **If R_1 differs appreciably from R'**, on the appropriate curve of Figs I-2 and I-3, find the new value of $L_{\alpha(dB)}$ corresponding to R_1. This is $L_{\alpha(dB)(2)}$ ——▶ ☐

13. **Find the range-increase factor (Table 2)** corresponding to the difference between $L_{\alpha(dB)(1)}$ and $L_{\alpha(dB)(2)}$. This is δ_2. ——▶ ☐

14. **Multiply R_1 by δ_2.** This is the radar range in kilometers. R_m ——▶ ☐

Note: If the difference between $L_{\alpha(dB)(1)}$ and $L_{\alpha(dB)(2)}$ is less than 0.1 dB, R_1 may be taken as the final range value, and steps 12 through 14 may be omitted. If $L_{\alpha(dB)(1)}$ is less than 0.1 dB, R' may be taken as the final range value, and steps 9 through 14 may be omitted. (For radar frequencies up to 10,000 megahertz, correction of the atmospheric attenuation beyond the $L_{\alpha(dB)(2)}$ value would amount to less than 0.1 dB.)

Figure 1.2.4 Modified range calculation worksheet.

Table 2 converts $40 \log R_0$ to R_0. The conversion constant used in the chart is

$$-30 \log 4\pi - 10 \log k - 120 = 75.6 \text{ dB}$$

where the 120 dB gives the range R_0 in km rather than in m. For other units of range, the constant is

195.6 dB for R_0 in m

64.9 dB for R_0 in nmi

When the chart is used to solve (1.2.27) for pulsed doppler or CW radar, the peak power P_t is replaced by average power P_{av}, pulsewidth τ by the averaging time $t_f = 1/B_f$ of the predetection filter, and number of pulses $n = f_r t_o$ by the number of filter output samples $n' = t_o/t_f$ remaining to be integrated after envelope detection. The matching factor M must now include combined losses comparable to the product $L_m L_{mf}$ discussed under (1.2.22). In the limit, with pure predetection integration (matched filtering), $t_f = t_o$ and $n' = 1$, for which (1.2.26) applies.

The major value of the standardized chart is that it makes clear what loss and signal integration assumptions have been used in performing the range calculation, as well as the values of major radar parameters. The major shortcoming is that the loss L_x must include a large number of components, not all of them well specified in most radar programs.

Example of Blake Chart Calculation

Using the previous radar example, the Blake chart may be filled in as shown in Figure 1.2.5. We will take the required detection probability $P_d = 0.9$, and false-alarm probability $P_{fa} = 10^{-6}$, and assume a Case 1 target with $n = 24$ hits per scan. Chapter 2 will show that the detectability factor $D_1(n) = 11.0$ dB for this case (with simple video integration). Allowances for practical receiver matching to the pulse spectrum, $M = 0.8$ dB, for transmission line loss, beamshape loss, $L_p = 1.3$ dB, and miscellaneous processing loss, $L_x = 3$ dB, will also be assumed. This results in a requirement $D_x = +16.1$ dB. The transmission loss is 1.0 dB, and the sum of all gains and losses gives an initial range estimate of

$$40 \log R_0 = +78.6 \text{ dB}$$
$$R_0 = 92.3 \text{ km}$$

| Detection probability, P_d = 0.9 ; False-alarm probability, P_{fa} = 10^{-6}; Case 1 target; n = 24 hits |||||
| Radar antenna height: h = 10 m. Target elevation angle: θ = 1°. |||||

A. Computation of T_s:	B. Range Factors	C. Decibel Values	Plus (+)	Minus (−)
T_s $T_a \cdot T_r \cdot L_r T_c$	P_t (or P_{av}) $\;10^5$ W	10 log P_t (W)	50.0	.
	(or t_f) $\;10^{-6}$ s	10 log τ (sec)	.	60.0
(a) Compute T_a.	G_t $\;10^4$	$G_{t(dB)}$	40.0	.
For T_{tg} T_{ta} = 290 and	G_r $\;10^4$	$G_{r(dB)}$	40.0	.
T_g = 36	$\sigma_{(sq\,m)}$ $\;1.0$	10 log σ	0.0	
$T_a = (0.876\,T'_a - 254)/L_a + 290$	λ $\;0.1$	20 log λ (m)	.	20.0
$L_{a(dB)}$: 0.2 L_a: 1.047	T_s (°K) $\;967$	−10 log T_s	.	29.9
T'_a = 65 K°	D $\;12.6$	−D (dB)	.	11.0
$\boxed{T_a = 102\,°K}$	M $\;1.2$	−M (dB)	//////	0.8
	L_t $\;1.26$	$-L_{t(dB)}$	//////	1.0
(b) Compute T_r = $T_{tr}(L_r - 1)$	L_p $\;1.35$	$-L_{p(dB)}$	//////	1.3
For T_{tr} = 290 use Table 1.	L_x $\;2.0$	$-L_{x(dB)}$	//////	3.0
$L_{r(dB)}$: 1.0 $\boxed{T_r = 75\,°K}$	Range-equation constant		75.6	//////
	4. Obtain the column totals ————→		205.6	127.0
(c) Compute T_c = T_0 (F_n −1)	5. Enter the smaller total below the larger→		127.0	.
or using Table 1.	6. Subtract to obtain the net decibels (dB) —→		+78.6	− .
$F_{n(dB)}$: 5.0 T_c: 625 °K	7. In Table 2 find the range ratio corresponding to this net decibel (dB) value, taking its sign (±) into account. This is R_0 ——→			$\boxed{92.3}$
L_r: 1.26 $\boxed{L_r T_c = 790\,°K}$			
	8. Multiply R_0 by the pattern-propagation factor			
Add. $\boxed{T_s = 967\,°K}$	$\boxed{F = 1.0}$ $F = \sqrt{F_t F_r}$		
	$R_0 \cdot F$ R' ————→			$\boxed{92.3}$

9. On the appropriate curve of Figs. I-2 and I-3, determine the atmospheric-absorption loss factor, $L_{\alpha(dB)}$, corresponding to R'. This is $L_{\alpha(dB)(1)}$. ————→ $\boxed{1.3}$

10. Find the range factor δ_1 corresponding to $-L_{\alpha(dB)}$ from the formula δ antilog $(-L_{\alpha(dB)} 40)$ or by using Table 2. ————→ $\boxed{0.9279}$

11. Multiply R' by δ_1. This is a first approximation of the range R_1. ————→ $\boxed{85.6}$

12. If R_1 differs appreciably from R', on the appropriate curve of Figs I-2 and I-3, find the new value of $L_{\alpha(dB)}$ corresponding to R_1. This is $L_{\alpha(dB)(2)}$. ————→ $\boxed{1.2}$

13. Find the range-increase factor (Table 2) corresponding to the difference between $L_{\alpha(dB)(1)}$ and $L_{\alpha(dB)(2)}$. This is δ_2. ————→ $\boxed{1.0058}$

14. Multiply R_1 by δ_2. This is the radar range in kilometers. R_m. ————→ $\boxed{86.1}$

Figure 1.2.5 Example of range calculation using work sheet.

Assuming, for the present, that a reflection-free surface is present under the path, with the target at the center of the elevation beam ($\theta_t = \theta_b = 1°$), we have $F = 1$ and $R' = R_0 = 92.3$ km. Atmospheric attenuation data may be found in Chapter 6, indicating a loss at this range of $L_\alpha = 1.3$ dB. This decreases the radar detection range by a factor of 0.9279, to 85.6 km. The chart allows for one iteration to account for the slight reduction in L_α at this range, leading to a final range $R_m = 86.1$ km.

This calculation will be repeated for a coherent integration process in Chapter 5. Further discussion of the propagation effects is given in Chapter 6.

Computer Program RGCALC

The calculation of range according to Blake's procedure has been programmed for personal computers [1.7]. This program, titled RGCALC, uses as inputs the entries of Figure 1.2.5, except that transmitter power is in kilowatts, pulsewidth is in microseconds, and frequency in megahertz is used instead of wavelength. The detectability factor is calculated for the given number of pulses and target case (limited to Cases 0–4, without diversity), and the value is printed out along with the radar range. The antenna temperature T_a and atmospheric loss L_α are calculated by the computer, rather than being read from graphs by the user. The printed inputs and outputs for the example of Figure 1.2.5 are shown in Figure 1.2.6.

1.3   THE SEARCH RADAR EQUATION

Derivation for Uniform Search

The potential performance of a search radar, distributing its energy uniformly over an assigned solid angle, can be determined from its average power, receiving aperture area, and system temperature, without regard to frequency and waveform. The steps in deriving the equation for optimum search performance, starting from (1.2.26), are as follows.

(a) Assume uniform search, without overlap, of an assigned solid angle ψ_s in a frame time t_s, using a rectangular beam that has a solid angle of

$$\psi_b = \theta_a\theta_e = \frac{4\pi}{G_tL_n} \ll \psi_s = A_m\,(\sin\theta_m - \sin\theta_0) \qquad (1.3.1)$$

where θ_a and θ_e are the 3-dB beamwidths, A_m is the azimuth sector searched, θ_m and θ_0 are the upper and lower elevation search limits, and L_n is a pattern constant (1.2 to 1.6) accounting for power radiated outside the idealized main lobe.

(b) Express the observation time for a target as

$$t_o = \frac{t_s\psi_b}{\psi_s} = \frac{4\pi t_s}{G_t\psi_sL_n} \qquad (1.3.2)$$

and assume that all signal energy reaching A_r during t_o is integrated for one detection decision. Note that the definition of two-coordinate beam-shape loss, included below in search loss L_s, is consistent with (1.3.2).

```
Radar and Target Parameters (inputs) --

Peak Pulse Power (kilowatts) ..................    100.0
Pulse Duration (usec) .........................    1.0000
Transmit Antenna Gain (dB) ....................    40.0
Receive Antenna Gain (dB) .....................    40.0
Frequency (MHz) ...............................    3000.0
Receiver Noise Factor (dB) ....................    5.0
Bandwidth Correction Factor (dB) ..............    .8
Antenna Ohmic Loss (dB) .......................    .2
Transmit Transmission Line Loss (dB) ..........    1.0
Receive  Transmission Line Loss (dB) ..........    1.0
Scanning-Antenna Pattern Loss (dB) ............    1.3
Miscellaneous Loss (dB) .......................    3.0
Number of Pulses Integrated ...................    24
Probability of Detection ......................    .900
False-Alarm Probability (Negative Power of Ten)    6.0
Target Cross Section (Square Meters) ..........    1.0000
Target Elevation Angle (Degrees) ..............    1.00
Average Solar and Galactic Noise Assumed
Pattern-Propagation Factors Assumed = 1

       ***********************************

Calculated Quantities (Outputs) --

Noise Temperatures, Degrees Kelvin --
       Antenna (TA) ...........................    111.3
       Receiving Transmission Line (TR) .......    75.1
       Receiver (TE) ..........................    627.1
       TE X Line-Loss Factor = TEI ............    789.4
       System (TA + TR + TEI) .................    975.8
Two-Way Attenuation Through Entire Troposphere (dB)   2.8


Swerling     Signal-     Tropospheric   Range,     Range,
Fluctuation  to-Noise    Attenuation,   Nautical   Kilometers
Case         Ratio, dB   Decibels       Miles
---------    ---------   -----------    --------   ----------

    1          10.98        1.24          46.5        86.1
```

Figure 1.2.6 Example of range calculation using RGCALC.

(c) Substitute (1.2.6), (1.3.1), and (1.3.2) into (1.2.26), and assume $F = 1$ (free-space propagation), to obtain the search radar equation:

$$R_m^4 = \frac{P_{av} A_r t_s \sigma}{4\pi\psi_s k T_s D_0(1) L_s} \tag{1.3.3}$$

where the total search loss L_s includes L_n and all the loss factors from the radar equation and the equation relating the detectability factor $D_x(1)$ to the basic detectability factor $D_0(1)$ for the steady target.

Neither wavelength nor waveform appears directly in (1.3.3), although the combination of wavelength and aperture size must permit (1.3.1) to be satisfied for concentration of search energy within ψ_s, and the loss terms will vary with wavelength, waveform, and scan procedure.

Effective Angle for Cosecant-Squared Coverage

In air surveillance applications, the required coverage often follows the type of contour shown in Figure 1.3.1, with a maximum range R_m, a maximum altitude h_m, and a maximum elevation angle θ_2. Full-range coverage is required only from the horizon to $\theta_1 = \sin^{-1}(h_m/R_m)$. This coverage can be produced by an antenna pattern having a cosecant-squared gain function:

$$G(\theta) = G(\theta_1)(\csc^2\theta)/\csc^2\theta_1, \quad \theta_1 < \theta < \theta_2 \tag{1.3.4}$$

If such a pattern is used for transmitting and receiving, the effective solid angle is calculated from

$$\theta_m = \theta_1[2 - \theta_1 \cot\theta_2] \tag{1.3.5}$$

and the effective receiving aperture is reduced by the ratio θ_m/θ_1. This effective sector height approaches $2\theta_1$, for $\theta_2/\theta_1 \gg 1$.

There are two other approaches to $\csc^2$ search coverage. In a stacked-beam radar, elevation beams using the full aperture A_r (and hence with constant G_r) can be used from the horizon to θ_2, and the broad transmitter pattern can be shaped as $\csc^4\theta$. The effective search sector is then calculated from

$$\sin\theta_m = \frac{4}{3}\sin\theta_1 \left[1 - \frac{1}{4}\left(\frac{\sin\theta_1}{\sin\theta_2}\right)^3\right]$$

(1.3.6)

This effective elevation sector height approaches $(4/3)\theta_1$ for $\theta_2/\theta_1 \gg 1$. The same equation applies when a scanning pencil beam is used with constant $G_t G_r$ and with $P_t \tau$ adjusted according to $csc^4\theta$.

In calculating the extra time required to perform this search, when a scanning beam is used with reduced power and with interpulse period matched to the coverage contour, the integration leads to a different value:

$$\sin\theta_m' = \sin\theta_1 \left[1 + \ln\frac{\sin\theta_2}{\sin\theta_1}\right]$$

(1.3.7)

The solid angle for calculation of power-aperture product or detection range, using (1.3.3), will be evaluated from (1.3.1) with $\sin\theta_m$ given by (1.3.6).

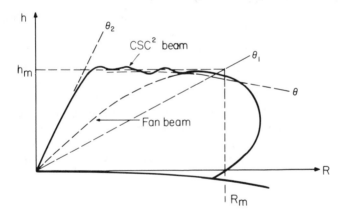

Figure 1.3.1 Cosecant-squared coverage pattern.

Significance of the Search Radar Equation

The search radar equation states, in essence, that the performance of the radar (as measured by the SNR which can be developed on each target during the frame time) is dependent on the product of average power and receiving aperture, divided by the search rate in steradians per second.

The radar's performance is also dependent on such factors as target cross section, system temperature, and quality of the detection performance (as measured by the probabilities of detection and false alarm, which determine detectability factor and some loss components). Apart from the brute force factors of power and aperture, the radar characteristics enter the equation primarily through the several components of the total search loss factor L_s.

The number of transmitting beam positions into which the search sector is divided does not enter the equation because the higher gain resulting from a smaller transmitting beam is canceled by the reduced time-on-target. In order to use a given receiving aperture at shorter wavelengths, it may be necessary to cover the transmitting beam angle with several smaller receiving beams, but it makes no difference whether these receiving beam positions are all illuminated simultaneously or sequentially by the transmitting beam, as long as there is efficient integration of received energy. In principle, the transmitting beam can illuminate the entire search sector, with multiple receiving beams staring at each point within this sector and integrating the received echoes over the entire frame time t_s. In practice, integration over time periods in excess of a few tens of milliseconds is quite difficult and likely to be accompanied by large losses.

Since the search radar equation leaves open the choice of wavelength and waveform for the radar, in achieving the required SNR under idealized conditions, we must look to other requirements to establish the constraints that dominate the system design process. These other requirements are imposed primarily by the clutter and multiple-target environments in which most radars must operate. Once the general scale of the radar has been determined from the search radar equation, the details of design will be determined by these environmental factors, not by the need for SNR.

Search Loss Budgets

Before proceeding to the environments of clutter and jamming, it is appropriate to define the many components of the search loss factor L_s, and also those loss components that normally appear as contributors to other terms in the search radar equation, some of which have already been discussed. These losses are shown in Figure 1.3.2, and are defined as follows. The first five components represent actual reductions in the signal power or energy ratio available to the receiver.

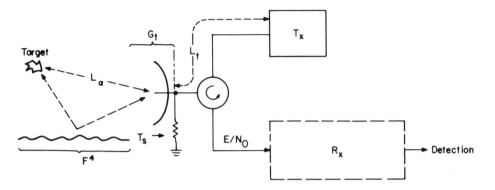

Figure 1.3.2 RF and other loss factors in radar system.

(a) *Transmission line loss, L_t.* This was included in (1.2.24) as a component which reduces the actual signal power received from the target.

(b) *Antenna losses, L_a and L_n.* The first was considered in connection with (1.2.15) and (1.2.24). In the receiving path, L_a is normally included as a component of T_s. In the transmitting path, L_a may be included as a component of L_n in (1.3.2), reducing the gain without expanding the solid angel covered by the transmitting beam, or L_a may be included in L_t. The minimum value of L_n for a constrained-feed array, with no spillover and with ohmic losses included in L_t, is 0.48 dB. This corresponds to an expression for antenna gain (4.2.11):

$$G_t = 11.3/\theta_a\theta_e \text{ or, for } \theta \text{ in degrees, } G_t = 37,100/\theta_a\theta_e \qquad (1.3.8)$$

For a horn-fed reflector, $G_t = 7.6/\theta_a\theta_e$, or for θ in degrees, $G_t = 25,000/\theta_a\theta_e$, corresponding to $L_n = 1.6$, or 2 dB.

It is important to note the difference between the antenna losses used in the search radar equation and those in the Blake chart or (1.2.26). In the Blake chart or (1.2.26) antenna gains and observation time are used to calculate received signal energy, and antenna loss appears in the gain and noise temperature factors. In the search radar equation, the transmitting gain and observation time are assumed to be related by (1.3.2), and the factor L_n must be included in the total search loss factor L_s.

(c) *Atmospheric attenuation, L_α.* This is the two-way loss in passage of the radar wave through the atmosphere to the target, discussed in Chapter 6. Atmospheric loss is included as a separate factor in the Blake chart and in (1.2.26), but as a component of total search loss L_s in the search radar equation.

(d) *Receiving line loss, L_r.* This loss is included as a contributor to the system temperature, T_s, according to (1.2.14) to (1.2.16), rather than as a component of L_s.

(e) *Beamshape loss, L_p.* This loss is a component of L_s, representing the reduction in signal energy received from the average target during the search scan, as compared with what would have been achieved if the beam were rectangular in shape or were to point directly at the target for a time interval t_o. In the search radar equation, beamshape loss is normally assigned a value of 2.5 dB to represent the fact that the target may be off the beam axis in both of the two angle coordinates. In the Blake chart, beamshape loss is stated separately from other losses, while in (1.2.26), beamshape loss is included in the detectability factor $D_x(1)$.

The remaining components represent inefficiencies in processing the received energy, and may depend strongly on the objective of the radar (e.g., the required detection and false-alarm probabilities). Hence, these losses are sometimes characterized as statistical losses to distinguish them from the pure attenuation factors of the first group.

(f) *Filter matching factor, M.* This factor represents the amount by which the input signal energy must be increased to compensate for the fact that the ideal (matched) filter is not used in the receiver. When the actual filter bandwidth is narrower than that of the matched filter, the factor M takes into account the resulting reduction in output false-alarm rate, and hence is somewhat less than the loss L_m, which describes only the SNR at the filter output. When the filter is wider than optimum, the factor M takes into account the usual postdetection bandwidth reduction, which increases the number of signal-plus-noise samples available for integration during each input pulse, again resulting in a lower loss than would be found by considering only the SNR at the filter output. The matching factor is stated as a separate loss in the Blake chart, and included as a component of $D_x(1)$ in (1.2.26). In the search radar equation, it appears as a component of total search loss L_s.

(g) *Integration loss, L_i.* This is the loss resulting from failure to integrate all signal pulses coherently in the receiving filter. The integration loss acutally is caused by the passage of the signal through the envelope detector at a lower level than would have been attained through coherent (predetection) integration, and the cross-modulation of noise and signal in the envelope detector prevents the subsequent integrator from regaining the SNR corresponding to input energy ratio of the pulse group. In the Blake chart, the integration loss appears as a component of the detectability factor $D(n)$, while in (1.2.26) the loss is assumed to equal unity for coherent integration. In the search radar equation, all received energy is assumed to be integrated for one detection decision with detectability factor $D_0(1)$, and the integration loss must be included as a component of total search loss L_s.

(h) *Collapsing loss, L_c.* This loss results from use of any video integration process in which extra noise samples are combined with the desired signal-plus-noise samples. Examples are integration in range gates or display elements wider than the signal pulse, collapsing of data from separate receivers or filters into a single video integration channel, or extension of the integration period beyond the duration of the signal pulse train. In the Blake chart or (1.2.26), collapsing loss is included as a component of L_x or in $D_x(1)$, while in the search radar equation this loss is a component of L_s.

(i) *Fluctuation loss, L_f.* All practical radar targets fluctuate in amplitude as a function of time and radar frequency. Fluctuation loss is the radar engineering term for the fade margin necessary to achieve adequate detection performance when the target drops below its average amplitude. In the Blake chart or (1.2.26), the fluctuation loss is included as a component of the detectability factor $D(n)$ or $D_x(1)$, while in the search radar equation it becomes a component of the total search loss L_s.

(j) *Processing losses, L_x.* Many other steps in the practical signal processing chain involve compromises in approaching the optimum process. These compromises result in requirements for greater input energy ratio in order to achieve the desired detection performance level, and the total increase is described as processing loss. It may be divided into components resulting from straddling of the signal by range gates, doppler filters, or integration windows, from action of the constant-false-alarm rate (CFAR) circuits, from MTI or doppler filter stopbands, from coarse quantization in digital systems, and from sundry other departures from the ideal processing steps. Processing losses are stated separately in the Blake chart and as a component of $D_x(1)$ in (1.2.26), but in the search radar equation these losses become components of total search loss L_s.

(k) *Scan distribution loss, L_d.* This loss is unique to the use of the search radar equation as a performance reference. Scan distribution loss describes the relative inefficiency of scanning the search volume more than once, when the detections from successive scans are combined in the simple cumulative detection process, rather than by integration before the detection threshold.

(l) *Other losses.* In particular systems, there may be several other sources of loss, relative to the idealized performance represented by the search radar equation. These will be discussed in subsequent chapters.

Example of Search Radar Calculation

As an example of a search radar range calculation, consider the S-band radar for which the Blake chart calculation was performed. The reflector antenna gain of 40 dB corresponds to a beam solid angle:

$$\psi_b = \frac{4\pi}{G_t L_n} = \frac{4\pi}{10^4 \times 1.6} = 0.00079 \text{ sr} = 2.6 \text{ degrees-squared}$$

for example, $\theta_a = 1.3°$, $\theta_e = 2.0°$. The effective receiving aperture is

$$A_r = \frac{G_r \lambda^2}{4\pi} = 10^4 \times 10^{-2}/4\pi = 8 \text{ m}^2$$

The search solid angle, for a single elevation beam, is

$$\psi_s = 2\pi \sin 2° = 0.22 \text{ sr}$$

If the system is to achieve the $n = 24$ hits per scan assumed in calculation of the detectability factor for the Blake chart, at a rotation rate $\omega_a = 2\pi/t_s = 1.05 \text{ r/s} = 60°/\text{s}$, for $t_s = 6 \text{ s}$, we have

$$t_o = t_s \theta_a/360° = 6 \times 1.3/360 = 0.0217 \text{ s}$$

$$f_r = n/t_o = 24/0.0217 = 1108 \text{ Hz}$$

$$D_u = f_r \tau = 0.0011$$

$$P_{av} = P_t D_u = 110 \text{ W}$$

The search loss budget, for targets at the peak of the elevation beam, is as follows:

L_t	= 1.0 dB	M	= 0.8 dB
L_n	= 2.0 dB	L_i	= 3.2 dB
L_α	= 1.2 dB	L_f	= 8.4 dB
L_p	= 1.3 dB	L_x	= 3.0 dB

Total search loss, $L_s = 20.9$ dB. The range calculation then gives

$$R_m^4 = \frac{110 \times 8 \times 6 \times 1}{4\pi \times 0.22 \times 1.38 \times 10^{-23} \times 967 \times 20.9 \times 123}$$

$$= 5.56 \times 10^{19} \text{ m}^4$$

$$R_m = 86.4 \text{ km}$$

For targets distributed uniformly between the horizon and two-degree elevation, the beamshape loss for the elevation coordinate adds 1.3 dB to the loss budget, reducing the average range to $R_m = 80.2$ km.

In a more realistic search radar example, the elevation coverage would be extended to about 10°, requiring the use of five vertically stacked beams. The transmitter power could be split equally among these beams, with resulting reduction in G_t by a factor of five. If the 80.2 km average detection range were to be maintained, average power would be increased by this same factor of five (maintaining the ratio P_{av}/ψ_s in (1.3.3), or the $P_t G_t \tau$ product in the Blake chart and (1.2.25)). If this average power were obtained by increasing the peak power or pulsewidth, this would also maintain the number of hits per scan at the azimuth scan rate of 60°/s. Numerous other options are available, such as using a pulse segmented into five subpulses at different frequencies, to generate the five beams with a frequency-scanning antenna. The peak power of the transmitter would remain at 100 kW, and the pulsewidth would be 5 μs. A wideband receiver front end would then feed five 1.0-MHz receiver channels tuned to the frequencies of the subpulses, effectively creating five radars operating in parallel, each transmitting 100 W average power and covering a two-degree elevation sector to 81.2 km in range.

1.4 RADAR RANGE WITH JAMMING

Equivalent Temperature of a Noise Jammer

One of the most effective forms of ECM is the continuous noise jammer, which radiates random noise with bandwidth B_j that is wider than the radar receiver bandwidth B_n. The effect is to raise the total spectral density of the background noise in the receiver from N_0 to $N_0 + J_0$, where the received jamming spectral density is

$$J_0 = \frac{P_j G_j G_r \lambda^2 F_j^2}{(4\pi)^2 B_j R_j^2 L_{\alpha j}} \tag{1.4.1}$$

Here, P_jG_j represents the ERP of the jammer, P_jG_j/B_j is the effective radiated power density at the frequency to which the radar is tuned, F_j^2 is the one-way pattern-propagation factor from the jammer into the radar antenna, R_j is the range to the jammer, and $L_{\alpha j}$ is the one-way atmospheric attenuation.

For the case of a stand-off jammer, which maintains a constant range from the radar, it is convenient to consider the jammer as one more component of noise temperature to be added to T_s at the receiver input:

$$T_j = \frac{P_jG_jG_r\lambda^2F_j^2}{(4\pi)^2kB_jR_j^2L_{\alpha j}} \tag{1.4.2}$$

If (1.2.25) to (1.2.27) or the Blake range chart are then modified to replace T_s with $T_s + T_j$, the target detection range of the radar can be calculated directly. In many cases, when the main lobe of the radar antenna looks at the jammer ($F_j^2 \rightarrow 1$), the resulting T_j is so great that the calculated range is negligible. However, in search sectors where only sidelobes see the jammer ($F_j^2 \ll 1$), T_j may be comparable to T_s and useful detection range is still obtained. For two or more jammers, individual values T_{ji} may be calculated, and the total input temperature will be the sum $T_s + T_{j1} + T_{j2} + \ldots$ All receiver processing considerations remain the same for this jamming as for receiver noise, so long as B_j covers the bandwidth B_n of the radar waveform.

Jamming Effectiveness

Two ways in which jamming may depart from thermal noise are in spectral distribution and amplitude distribution. If $B_j < B_n$, the jamming will cover only a portion of the signal spectrum, but with greater spectral density. The result is generally less effective in obscuring targets than is white noise of equal power. Similarly, if the jamming transmitter is operated in the saturated mode, the transmitter's output will not follow the Gaussian distribution that characterizes random noise. Although a higher average power may be available from the transmitter in this mode, the resulting jamming may not produce as great an effect in the receiver as the same random noise power. This variation in jamming effectiveness is discussed in Chapter 3. In using the equations for effective temperature of a jammer, the ERP of the jamming transmitter should be reduced to account for the effectiveness of the transmitter actually used.

Another factor which may reduce the effective ERP of the jammer is the need for polarization diversity. If more than one radar is to be jammed, it may be necessary to radiate jamming on both orthogonal polarizations (e.g., horizontal and vertical, or right- and left-hand circular). The power radiated in the polarization orthogonal to that of the radar receiving antenna is then rejected by the radar, reducing the effective jammer ERP.

Noise Jammer Example

Consider the case of a stand-off barrage noise jammer, directed at the S-band surveillance radar used in the example range calculation of Section 1.2. We will assume the following jammer parameters:

$$\left.\begin{array}{l} P_j G_j = 100 \text{ kW} \\ B_j = 300 \text{ MHz} \end{array}\right\} \frac{P_j G_j}{B_j} = 333 \text{ W/MHz} = 3.33 \times 10^{-4} \text{ W/Hz}$$

$$R_j = 200 \text{ km}$$
$$L_{\alpha j} = 1.0 \text{ dB}$$

With this jammer in the main lobe of the radar, $F_j = 1$, we find, from (1.4.2), the effective jamming temperature:

$$T_j = \frac{10^5 \times 10^4 \times 10^{-2} \times 1}{(4\pi)^2 \times 1.38 \times 10^{-23} \times 3 \times 10^8 \times (2 \times 10^5)^2 \times 1.25}$$
$$= 3 \times 10^8 \text{ K}$$

The resulting receiver noise level is 3.2×10^5 or $+55$ dB above receiver noise, reducing range by a factor of 23.7 to only 3.9 km.

However, if the jammer is in a region where average receiving sidelobe levels are 50 dB below the main lobe gain ($F_j^2 = 10^{-5}$), the jammer temperature will be $T_j = 3 \times 10^3$ K, leading to $T_s + T_j = 3967$ K. This is only 6.1 dB above the normal receiver noise level. The detection range will be reduced by a factor of 1.42 to 61.2 km, providing significant surveillance performance in all sectors where the low sidelobe levels are directed toward the jammer.

Deception Jamming

In some cases, rather than emitting continuous noise to force an increase in the radar detection threshold, the jammer emits signals which simulate the radar transmission, producing false echoes in the radar receiver. These false echoes may be generated at range delays and with doppler shifts other than those of the real target. If the jamming is sufficiently strong to penetrate the sidelobes of the radar receiving antennas, the false echoes may appear when the antenna is looking at angles far removed from that of the jamming source. The received energy for each pulse of this type of jamming is given by (1.4.1), with $B_j = 1/\tau_j$, where τ_j is the radar pulsewidth replicated by the jammer. After processing in the receiver, the false echo output will have a J/N power ratio equal to $J_0/N_0 L_m$, and will be otherwise indistinguishable from a target echo, except for imperfections in the replicated waveform and possible angle-of-arrival information implicit in the pattern-propagation factor F_j^2. Methods of recognizing and rejecting such jamming pulses will be discussed in Chapter 10.

Self-Screening and Escort Jamming

The self-screening jammer is located on the target, such that $R_j = R$ and $F_j = F$. In calculating the range at which a target echo becomes detectable above the jamming (e.g., for ranging purposes), the radar must obtain an energy ratio:

$$\frac{E_1}{J_0} = \frac{P_t \tau G_t \sigma F^2 B_j}{4\pi P_j G_j R_{bt}^2 L_t L_{aj}} = D_x(n) = 40.7 = +16.1 \text{ dB} \tag{1.4.3}$$

Solving for burnthrough range R_{bt}, we obtain

$$R_{bt}^2 = \frac{P_t \tau G_t \sigma F^2 B_j}{4\pi P_j G_j L_t L_{aj} D_x(n)} \tag{1.4.4}$$

This computation rarely yields a useful burnthrough range. In our previous radar example, if the self-screening jammer has an ERP of 10 kW and $B_j = 300$ MHz, we find

$$R_{bt}^2 = \frac{10^5 \times 10^{-6} \times 10^4 \times 1 \times 1 \times 3 \times 10^8}{4\pi \times 10^4 \times 1.26 \times 1.26 \times 40.7} = 3.39 \times 10^4 \text{ m}^2$$

$R_{bt} = 192$ m

The escort jammer closes in range with the target, but may lie outside the main lobe in which the target is to be detected. If the sidelobes directed toward the escort jammer are at a level F_j^2 relative to the main lobe, (1.4.4) will have F^4 instead of F^2 in the numerator, and F_j^2 in the denominator, greatly increasing the burnthrough range on the target. In our example, if $F_j^2 = 10^{-4}$ or -40 dB (for an escort jammer outside the first sidelobes), we find

$$R_{bt}^2 = 3.39 \times 10^8 \text{ m}^2$$
$$R_{bt} = 18.4 \text{ km}$$

If the escort jammer is in the skirt of the main lobe or in one of the principal sidelobes, even this range may not be available.

In the case of penetrating jammers, the radar system response is not to await burnthrough, but to identify the angles of the jammers and to attempt triangulation with data from adjacent radar sites. The jammers are then tracked passively in angle, and missile guidance is accomplished by using track-on-jam or a home-on-jam seeker, when the target is judged to be within missile range.

1.5 RADAR RANGE WITH CLUTTER

The calculation of target detection range in a background of clutter is one of the most difficult calculations to perform with accuracy. This is so for several reasons. The first reason is that the reflectivity of clutter sources, and the corresponding propagation factors, are only approximately known. The second is that the variation of clutter power with range is extremely uncertain, especially in the case of land clutter. The third is that the amplitude statistics of clutter may depart significantly from the Rayleigh distribution that characterizes random noise, and for which $D_x(n)$ has been determined. The fourth is that attenuation of clutter in MTI or doppler processors is difficult to predict with good accuracy. As a result of all these factors, range equations that purport to predict a range at which targets can be detected must be used with caution. Nonetheless, we will present here some equations which are useful when the output of the receiver-processor is dominated by clutter, with suggestions as to when these equations can be used. Equations that predict the signal-to-clutter ratio will also be used in the next section, when we consider multiple simultaneous sources of interference.

Volume Clutter: Rain and Chaff

These clutter sources are considered first because they are more homogeneous in spatial distribution and more nearly noise-like (being essentially Rayleigh-distributed in amplitude) than other clutter types. The radar resolution cell, at a given clutter range R_c, is characterized by its volume (Figure 1.5.1a):

$$V_c = \frac{R_c\theta_a}{L_p} \frac{R_c\theta_e}{L_p} \frac{\tau_n c}{2} \tag{1.5.1}$$

where θ_a and θ_e are the azimuth and elevation beamwidths in radians, $L_p = 1.33$ is the beamshape loss, τ_n is the width of the processed pulse (e.g., after pulse compression, if that technique is used), and c is the velocity of light. The radar cross section σ_c of clutter filling this volume is

$$\sigma_c = V_c\eta_v \tag{1.5.2}$$

where η_v is the volume clutter reflectivity (in units of m^2/m^3). If η_v and radar parameters are known, the received clutter power from this cell may be calculated:

$$C = \frac{P_t\tau G_t G_r \lambda^2 \sigma_c F_c^4}{(4\pi)^3 R_c^4 L_t L_{\alpha c}} = \frac{P_t\tau G_t G_r \lambda^2 \theta_a \theta_e (\tau_n c/2)\eta_v F_c^4}{(4\pi)^3 R_c^2 L_t L_{\alpha c} L_p^2} \tag{1.5.3}$$

where F_c^4 and $L_{\alpha c}$ are the pattern-propagation factor and atmospheric attenuation for the radar-to-clutter path. Both of these factors should be averaged over the resolution cell, in cases where they would change significantly with elevation angle. For an upward-looking narrow beam filled with clutter, the factor $F_c^4 \approx 1$, since the average effect of antenna pattern shape has been included as L_p^2. However, for horizontal beams which illuminate an underlying reflecting surface, the factor may take the form:

$$F_c(\theta) = 2 \sin(2\pi h_r \theta/\lambda)$$

and the average of F_c^4 is then equal to six. Intermediate values between one and six will apply to beams which are tilted slightly above the horizontal, or for operation over surfaces which are not totally reflecting.

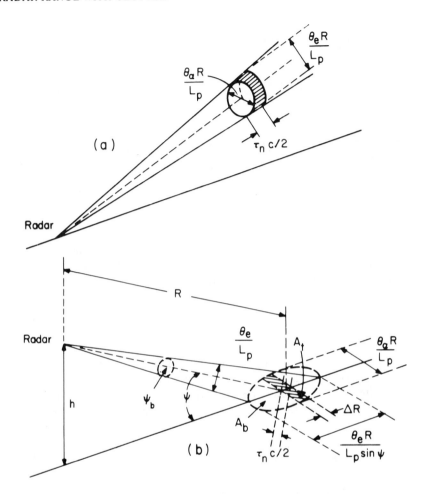

Figure 1.5.1 Geometry of clutter elements: (a) volume clutter; (b) surface clutter.

If the clutter competes with a target at range R, the input C/S ratio will be

$$C/S = \frac{R^4\theta_a\theta_e(\tau_n c/2)L_\alpha\eta_v F_c^4}{\sigma F^4 L_p^2 R_c^2 L_{\alpha c}} \tag{1.5.4}$$

When the radar has no range ambiguities in which clutter is present, $R_c = R$, $L_{\alpha c} = L_\alpha$, and we obtain

$$C/S = \frac{R^2\theta_a\theta_e(\tau_n c/2)\eta_v F_c^4}{\sigma F^4 L_p^2} \tag{1.5.5}$$

When clutter is present in multiple range ambiguities,

$$C/S = \frac{R^4\theta_a\theta_e(\tau_nc/2)L_\alpha}{\sigma F^4 L_p^2} \sum_{i=1}^{j} \frac{\eta_{vi}F_{ci}^4}{R_{ci}^2 L_{\alpha ci}} \tag{1.5.6}$$

where each of the factors in the summation can have a separate value in each of the j ambiguities. In many cases, the $1/R_{ci}^2$ variation of terms in the summation permits us to omit all but the first term, and (1.5.4) can be used with the values for $i = 1$.

Detection Range in Volume Clutter

If we assume that the clutter, after possible reduction by MTI or doppler processing, remains the dominant source of background interference at the radar output, we may write

$$(S/C)_{\text{out}} = \frac{I_m}{C/S} = D_{xc}(n), \quad \text{or } C/S = \frac{I_m}{D_{xc}(n)} \tag{1.5.7}$$

where I_m is the MTI or doppler improvement factor and $D_{xc}(n)$ is the clutter detectability factor, defined as the value of $(S/C)_{\text{out}}$ required to achieve the desired detection performance (means of calculating this factor are given in Chapter 2). For example, with no range ambiguities,

$$R^2 = \frac{I_m\sigma F^4 L_p^2}{D_{xc}(n)\theta_a\theta_e(\tau_nc/2)\eta_v F_c^4} \tag{1.5.8}$$

When the first range ambiguity is the dominant clutter source,

$$R^4 = \frac{\sigma F^4 L_p^2}{D_{xc}(n)\theta_a\theta_e(\tau_nc/2)L_\alpha} \left[\frac{R_{c_1}^2 L_{\alpha_1} I_{m_1}}{\eta_{v_1} F_{c_1}^4} \right] \tag{1.5.9}$$

When subsequent ambiguities are significant sources, the term in brackets is replaced by

$$\left[\sum \frac{\eta_{vi}F_{ci}^4}{R_{ci}^2 L_{\alpha ci} I_{mi}} \right]^{-1} \tag{1.5.10}$$

Note that the improvement factor may be dependent on the range or spectral spread of the clutter in each ambiguity, and hence must be included in the summation.

Example of Detection Range in Rain

Assume a low-PRF radar (with no visible clutter beyond the unambiguous range $R_u = c/2f_r$), with rain clutter ($\eta_v = 2 \times 10^{-8}$ m^2/m^3) filling the beam within R_u, the radar characteristics being those used previously in Figure 1.2.5. No effective MTI improvement will be assumed ($I_m = 1$) because of the wind velocity of the rain cloud. However, the clutter spectrum will be assumed narrow enough to reduce the integration gain to that for a system integrating only 4.5 pulses: $D_{xc}(24) = D_x(4.5) = +21.1$ dB. From previous calculations, $\theta_a = 1.3° = 0.023$ r, $\theta_e = 2.0°$ deg $= 0.035$ r, $\tau_n = \tau = 1$ μs. An upward tilted beam will be assumed, over a poorly reflecting surface, to give $F_c^4 = 1$, and the target will be assumed at the beam elevation, $F = 1$. Thus,

$$R^2 = \frac{1 \times 1 \times 1 \times 1.33^2}{129 \times 0.023 \times 0.035 \times 150 \times 2 \times 10^{-8} \times 1}$$

$$= 5.7 \times 10^6 \text{ m}^2$$

$$R = 2380 \text{ m}$$

In order to obtain any significant target detection range, the radar would have to use a circularly polarized antenna, achieving a net improvement in S/C of about 18 dB. Then,

$$R^2 = 5.7 \times 10^6 \times 63 = 3.6 \times 10^8 \text{ m}^2$$

$$R = 19 \text{ km}$$

Clearly, since clutter reduced the radar range by a factor greater than four, some type of doppler signal processing would be essential in this radar to provide useful surveillance range in rain.

Surface Clutter: Land and Sea

We may approach range calculation with a simple, homogeneous clutter model, in which the clutter fills the azimuth beamwidth out to a horizon range given by

$$R_h = \sqrt{2kah_r} \qquad (1.5.11)$$

where ka is the effective earth radius (8.5×10^6 m) and h_r is the height of the radar antenna above the clutter surface. A more refined model, which takes into account the propagation factor for land and sea clutter, is described in Chapter 3.

The area of surface within the radar resolution cell (Figure 1.5.1b) is

$$A_c = \frac{R\theta_a}{L_p} \frac{\tau_n c}{2} \sec\psi$$

(1.5.12)

where for surface-based radar $\sec\psi \approx 1$. The clutter cross section is

$$\sigma_c = A_c\sigma^0$$

(1.5.13)

where σ^0 is the surface clutter reflectivity (a dimensionless quantity). The best simple model for reflectivity, at a grazing angle ψ, is

$$\sigma^0 = \gamma \sin\psi = \gamma h_r/R_c$$

(1.5.14)

where $\gamma \approx 0.1$ for land (values from 0.03 to 0.15 characterize different terrain types). For the sea, γ is dependent on wind conditions and radar wavelength (see Chapter 3). Using this model, σ_c is constant with range:

$$\sigma_c = \frac{h_r\theta_a}{L_p} \frac{\tau_n c}{2} \gamma$$

(1.5.15)

The clutter-to-signal ratio is

$$C/S = \frac{R^4 h_r \theta_a (\tau_n c/2) L_\alpha \gamma F_c^4}{\sigma F^4 L_p R_c^4 L_{\alpha c}}$$

(1.5.16)

Then, for a system with no range ambiguities containing clutter, the C/S ratio becomes

$$C/S = \frac{h_r \theta_a (\tau_n c/2) \gamma F_c^4}{\sigma L_p F^4}$$

(1.5.17)

In the region well within the horizon, where $F = F_c = 1$, the output S/C ratio will be

$$(S/C)_{\text{out}} = I_m S/C = \frac{\sigma L_p I_m}{h_r \theta_a (\tau_n c/2) \gamma}$$

(1.5.18)

and if this exceeds the required $D_{xc}(n)$, the target will be detectable throughout this region. If the target is not detectable in the region where $F_c = 1$, the target may become detectable at longer range, where the F_c is reduced more rapidly than F. Models that predict this effect are given in Chapter 3.

When range ambiguities are occupied by clutter, (1.5.16) becomes

$$C/S = \frac{R^4 h_r \theta_a (\tau_n c/2) L_\alpha \gamma}{\sigma F^4 L_p} \sum_{i=1}^{j} \frac{F_{ci}^4}{R_{ci}^4 L_{\alpha ci}} \tag{1.5.19}$$

As with volume clutter, it is usually the first ambiguity, $i = 1$, which dominates the result, and (1.5.16) can then be used with clutter values for that ambiguity.

Example of Detection Range over Land

Consider first the case of unambiguous range. For example, with $I_m = 30$ dB, $h_r = 10$ m, $\gamma = 0.1$, we have

$$(S/C)_{\text{out}} = \frac{1 \times 1.33 \times 10^3}{10 \times 0.023 \times 150 \times 0.1} = 386, \quad \text{or } +25.9 \text{ dB}$$

Targets of 1 m^2 would be detectable above clutter at all ranges in this case. With $\sigma = 0.3$ m^2 or $I_m = 25$ dB, and $D_{xc}(n) = 21.1$ dB, the detection requirement is almost met at all ranges within the clutter horizon, and would be exceeded beyond this horizon. With yet lower σ or I_m, the target is below the detection threshold at short ranges, but may become detectable when F^4/F_c^4 increases, near and beyond the horizon.

Now consider a high-PRF land-based radar, operating at $f_r = 100$ kHz, for which $R_u = 1500$ m. For a long-range target appearing in the center range gate (e.g., $R = 750$ m $+ iR_u$), and assuming $F_c = L_{ac} = 1$ at 750 m, (1.5.16) becomes

$$C/S = \frac{R^4 h_r \theta_a (\tau_n c/2) L_\alpha \gamma}{\sigma F^4 L_p (750)^4} = I_m/D_{xc}(n)$$

$$R^4 = \frac{I_m \sigma F^4 L_p (750)^4}{D_{xc}(n) h_r \theta_a (\tau_n c/2) L_\alpha \gamma}$$

Assuming $I_m = 80$ dB for a high-PRF doppler processor, and $F = L_\alpha = 1$, with $D_{xc}(n) = D_x(n) = +16.1$ dB, we have

$$R^4 = \frac{10^8 \times 1 \times 1.33 \times 750^4}{40.7 \times 10 \times 0.023 \times 150 \times 0.1} = 3.0 \times 10^{17}$$

$$R = 23.4 \text{ km}$$

This shows the extreme requirements for doppler improvement factor in a range-ambiguous system.

Land Clutter Detectability Factor

As we will show in Chapter 3, most land clutter exhibits a broad amplitude distribution, requiring that the threshold be set higher relative to the average clutter output than for random noise, if false alarms are to be controlled. In addition, the pulse-to-pulse correlation of land clutter, if not destroyed by effective MTI or doppler processing, can lead to less effective integration. Both of these effects lead to an increase in the clutter detectability factor: $D_{xc}(n) > D_x(n)$. This must be taken into account in calculating target detection performance over land.

1.6 RADAR RANGE WITH COMBINED INTERFERENCE SOURCES

We have seen that different types of interference (noise, jamming, volume clutter, and surface clutter) vary as different functions of range, especially when the pattern-propagation factors F_c depart from unity. In addition, the detectability factors for different types of clutter may have values greater than D_x for noise, and MTI improvement factors may vary with clutter range. As a result, we cannot expect to write an equation for the ratio of signal to combined interference, set this equal to D_x, and solve in closed form for range. There is, however, a graphical procedure which not only indicates the expected detection range, but also the sensitivity of this range to parameters of the radar, the clutter models, and the propagation models.

Basic Signal Level Plot

The graphical procedure starts with the chart of Figure 1.6.1, in which the levels of noise, signal, and several interference sources are plotted as

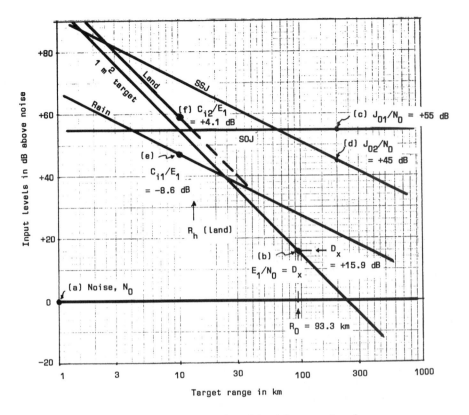

Figure 1.6.1 Free-space input signal and interference levels.

functions of target range, with propagation factors and atmospheric attenuations set to unity. It is convenient to use signal energy and interference
spectral density levels, referred to the receiver input, and calculated as
follows.

(a) The noise level N_0 (W/Hz) is used as the zero-dB reference level,
near the bottom of the plot.

(b) A Blake chart calculation of R_0 is done, leading to the point (b)
on the plot at R_0, with a level $E_1 = D_x(n)$ above N_0. Recall that $D_x(n) =
D(n)ML_pL_x$, expressed in dB as the sum of these factors. A straight line
for E_1 *versus* range is constructed with a slope of -40 dB/decade through
point (b). The illustrations here will use an MTI radar with noncoherent
integration. For coherent integration, a signal energy E_f based on $P_{av}t_f$ of
(1.2.27) would replace E_1, and $D_x(n')$ would replace $D_x(n)$.

(c) The level J_{01} of a stand-off jammer in the main lobe ($F_j = 1$) is calculated from (1.4.1) or (1.4.2), giving point (c) at R_j. A horizontal line through this point indicates that the receiver sees the same jamming level at all target ranges. The example used in Section 1.4 is used in Figure 1.6.1.

(d) The level J_{02} of a self-screening jammer in the main lobe is also calculated from (1.4.1) or (1.4.2), using the presumed lower ERP of that jammer and leading to point (d) at R_j. A line through that point with a slope of -20 dB/decade indicates the increasing J_{02} as that target approaches the radar.

(e) The input level of volume clutter C_{i_1} is calculated as follows: input C/S is computed from (1.5.5), at any convenient range R_c (10 km will be used here); input $C_{i_1} = (C/S)E_1$ is then plotted as point (e) at range $R_c = 10$ km. A line through this point at -20 dB/decade indicates the range variation of volume clutter filling the beam at target range. In Figure 1.6.1, a rainfall rate of 10 mm/h and use of a linearly polarized antenna are assumed.

(f) The level of surface clutter C_{i_2} is calculated in the same way from (1.5.17), leading to point (f) at $R_c = 10$ km. For the constant-γ clutter model, a line at -40 dB/decade indicates the constant S/C ratio as a function of range for this type of clutter. However, the line should not extend beyond R_h as given by (1.5.11), and we would see that it is to be further modified by the pattern-propagation factor F_c.

(g) Additional components of jamming or clutter can be added as required, using the procedures given above.

Pattern-Propagation and Processing Factors

The second chart, Figure 1.6.2, is prepared on the basis of Figure 1.6.1, but it includes the antenna sidelobe levels and the range-dependent pattern-propagation and signal processing factors.

(a) The noise level is unchanged.

(b) Signal level E_1 is calculated for several ranges at which F may vary because of position of the target within the beam. At this step, it is preferable to omit the variation of F with reflection lobing near the horizon, and to concentrate on the free-space antenna pattern component of F. A horizon range for the target may be indicated, given by

$$R_{ht} = \sqrt{2ka} \left(\sqrt{h_t} + \sqrt{h_r} \right) \quad \text{m} \tag{1.6.1}$$

In Figure 1.6.2, the radar antenna height is 10 m and the target is assumed to be at 200 m altitude, giving $R_{ht} = 71$ km. The two-degree elevation

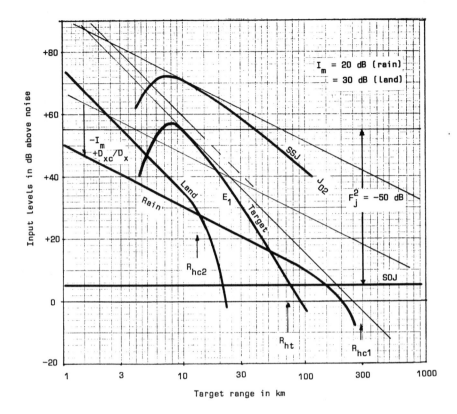

Figure 1.6.2 Levels adjusted for pattern-propagation and signal processing factors.

beam is centered at 1°, and the target passes this elevation at $R = 10$ km.

(c) For a stand-off jammer in the sidelobe region ($F_j^2 \ll 1$), the level J_{01} is drawn at the level $20 \log F_j$ relative to the level of point (c). In Figure 1.6.2, $F_j^2 = -50$ dB is assumed for a stand-off jammer in the far sidelobes of a low-sidelobe antenna.

(d) The self-screening jammer line is drawn below its original level by $20 \log F$, indicating the departure of the one-way path to the radar from its beam-center value.

(e) The volume clutter pattern-propagation factor F_c^4 is calculated as a function of range, considering the location of the clutter cloud relative to the center of the beam, the extent of filling of the beam, and possible enhancement by reflection lobing. Two other adjustments are then needed to find the output clutter spectral density. First, the clutter power must be multiplied by the ratio D_{xc}/D_x, to account for the higher detection threshold needed with clutter residue as the output interference (see Equation (2.6.1)). Then, the output clutter is divided by the improvement factor resulting from MTI or doppler processing, which in general will vary with clutter range, to find $C_{01} = (C_{i_1}/I_m)(D_{xc}/D_x)$. This average of clutter level relative to the free-space beam center value is converted to decibels and is used to adjust the original curve for C_{i_1}. In Figure 1.6.2, rain is assumed to extend to $h = 5$ km, giving $R_{hc_1} = 290$ km. To illustrate the method, it will be assumed that a constant $I_m = 20$ dB is available from MTI against rain clutter, and that the output residue is partially correlated to give six independent samples for integration ($D_{xc} = +20.2$ dB).

(f) The surface clutter factor F_c^4 is calculated in a similar way, including reflection-interference effects at low grazing angles and diffraction effects near and beyond the horizon (see Chapter 3). The original curve for C_{i_2} is adjusted accordingly to give C_{02}. In Figure 1.6.2, a constant $I_m = 30$ dB with six independent output samples for integration is assumed for land clutter.

Atmospheric Attenuation

The third chart, Figure 1.6.3, follows the second, but includes atmospheric attenuations. These are readily found as a function of range and elevation angle by using the curves of Chapter 6. For radars at VHF and UHF, and for short-range microwave systems in clear weather, this step may not be necessary. Under rain conditions, however, this method can provide valuable insight for all microwave radars, and it is essential for systems at X-band or above. Two or more charts at different rainfall rates may be needed. The solid curves in Figure 1.6.3 are for the S-band radar used as an example, with 10 mm/h rain. Dashed curves show how the greater X-band attenuation would reduce range in that band. The rain clutter level is left unchanged to represent the result of using circular polarization to overcome the larger X-band reflectivity.

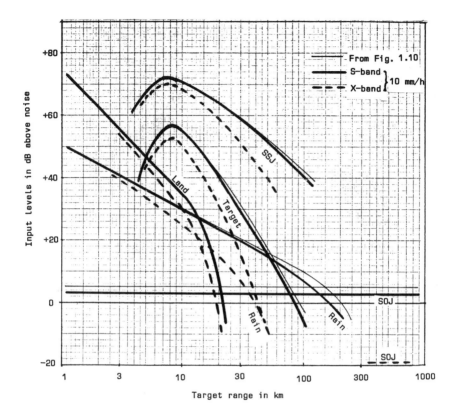

Figure 1.6.3 Levels adjusted for atmospheric attenuation.

Combined Interference

It is now possible to draw, on the plots from the previous step, a curve representing the combined interference level:

$$I_0 = N_0 + J_{0_1} + J_{0_2} + C_{0_1} + C_{0_2}$$

The adding of components is performed, of course, with energy levels, not their decibel values. In its simplified form, this combined interference curve may follow the highest of the five individual components of Figure 1.6.3. A smooth curve drawn through these points and asymptotic to the highest of the individual curves will then be an accurate indication of combined interference (Figure 1.6.4).

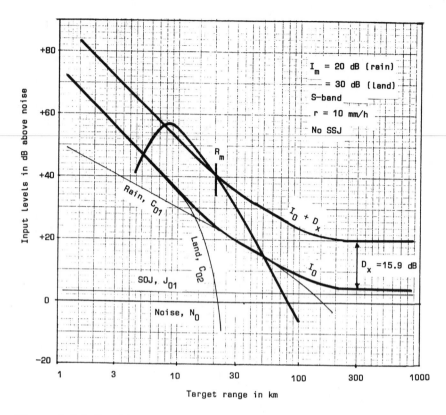

Figure 1.6.4 Combined interference and resulting threshold.

Detection Threshold Plot

   The final curve in Figure 1.6.4 is drawn at a level $D_x(n)$ above I_0,
indicating the level at which E_1 will provide the required detection performance. If desired, three D_x levels may be drawn, one for the original
P_d used in the Blake chart and two for other significant levels (e.g., $P_d =$
0.90, 0.50, and 0.10). This method will indicate ranges at which detection
may be achieved, even if the probability is not as high as desired. In this
example, a maximum detection range $R_m = 20$ km is found, limited by a
combination of rain and land clutter.

Lobing Plot

Where significant lobing effects on the target are expected (e.g., over water or smooth, barren land surfaces), it is now appropriate to compute F for elevation angles of concern. The smooth-sphere diffraction contribution to F may also be included to arrive at a more accurate range estimate near the horizon. In some cases, knife-edge diffraction calculations should also be made (see Chapter 6), to determine whether there is significant detection probability extending into the shadow zone below a mask.

Ambiguous-Range Plot

For medium-PRF doppler radars, in which short-range surface and volume clutter may affect long-range target detection, the C_{01} and C_{02} plots from Figure 1.6.3 may be used as a basis for plotting the clutter interference in successive range ambiguities, as in Figure 1.6.5. The value of the detection threshold set by short-range clutter $C_0(R_c) + D_x$, for $0 < R_c < R_u$, is repeated at ranges $R_c + iR_u$, for $i = 1, 2, \ldots$ It immediately becomes apparent whether there is a significant contribution from clutter originating beyond R_u, or whether the clutter from the first range interval is dominant in successive intervals. Having plotted the ambiguous response to the short-range clutter, a curve $D_x(n)$ above this level will define the ranges at which E_1 reaches the given detection threshold. In Figure 1.6.5, the value of doppler improvement factor has been increased by 40 dB compared with Figures 1.6.1 to 1.6.8, and jamming has been omitted. Even with the high doppler improvement, the target is only barely detectable in the second range ambiguity.

Multiple-Time-Around Clutter Plot

In some cases (e.g., when a magnetron transmitter or staggered PRF is used with MTI), clutter from beyond the first range ambiguity remains uncanceled and may appear at short range above the level of short-range clutter (Figure 1.6.6). Here, the effective input clutter level $C_i(R_c + iR_u)D_{xc}(n)/D_x(n)$ is used to reflect the absence of cancelation, and this multiple-time-around input is then shifted in range to become $C_0(R_c)$ in

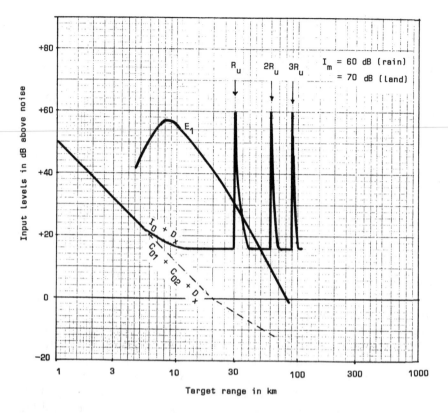

Figure 1.6.5 Medium-PRF signal and interference levels.

the first range interval. For the case shown, this uncanceled second-time-around clutter is the dominant interference component for short-range as well as long-range targets. In Figure 1.6.6, an MTI improvement factor of 40 dB had been assumed, which would have given excellent detection performance were it not for the uncanceled ambiguous clutter that limits the detection range to 20 km. This calculation is especially important if ground clutter appears from beyond the normal horizon, as a result of ducted propagation (see Chapter 6).

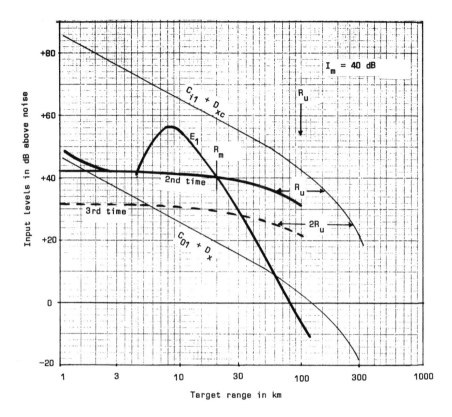

Figure 1.6.6 Uncanceled clutter in distant ambiguities.

Sensitivity Time Control Plot

If a plot is prepared showing target curves for successively smaller targets (Figure 1.6.7), it is apparent that moving objects of very small size (passing the MTI filter) may become detectable at short and medium ranges. In Figure 1.6.7, the power of the radar has been increased by 20 dB relative to Figure 1.6.4, and the improvement factors likewise by 40 dB, to make possible detection of small targets. If the system cannot tolerate so many detections of birds and land vehicles, the use of sensitivity time control (STC) is required. Receiver gain is reduced at short range, and circuit noise from the later amplifier stages is allowed to establish the system noise and threshold levels. This STC also protects the receiver from

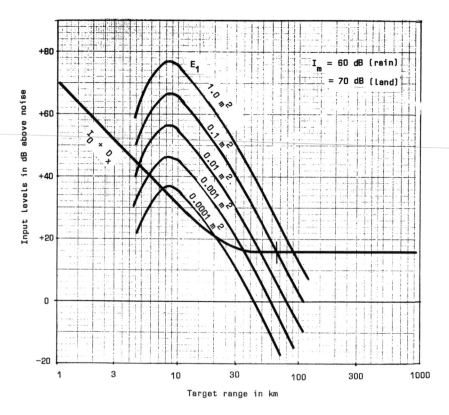

Figure 1.6.7 Detection of small targets.

overload on short-range clutter (see upper dashed curve in Figure 1.6.8). The resulting detection threshold, as a function of range, is shown in Figure 1.6.8. Targets of the intended size (e.g., $\sigma > 0.1$ m^2) remain detectable to 56 km, almost the range that may have been achieved with full receiver gain (65 km), and smaller objects are largely rejected. This is the necessary mode of operation for most low-PRF radars in which the basic blind speed is too low to permit rejection of birds and land vehicles on the basis of their doppler velocities. If used in medium- or high-PRF radar, STC extends the eclipsed range regions beyond the length of the transmitted pulse, and hence reduces the clear regions in which targets may be detected. Hence, such systems are normally designed with sufficient dynamic range and clutter attenuation so that STC is unnecessary.

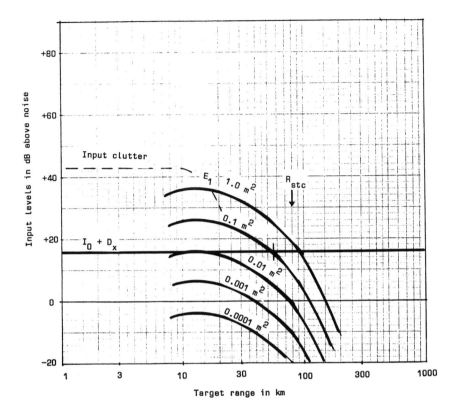

Figure 1.6.8 Use of STC to reject small targets.

Sensitivity to Parameters and Models

The sensitivity to variation in radar parameters and environmental models may be visualized by simply shifting the levels of the target and the individual interference components. In particular, when the target energy curve runs almost parallel to the combined interference curve over an extended range region, it indicates that no accurate closed-form expression for detection range is possible.

REFERENCES

[1.1]   IEEE Standard Letter Designations for Radar-Frequency Bands, *IEEE Std. 521-1984*, 1984.

[1.2]   L.V. Blake, *Radar Range-Performance Analysis*, Artech House, 1986.

[1.3]   D.O. North, An analysis of the factors which determine signal/ noise discrimination in pulsed carrier systems, *RCA Laboratories Tech. Rep. PTR-6C*, June 25, 1943; reprinted in *Proc. IEEE* **51**, No. 7, July 1963, pp. 1016–1027.

[1.4]   W.M. Hall, Prediction of pulse radar performance, *Proc. IRE* **44**, No. 2, February 1956, pp. 224–231; reprinted in D.K. Barton (ed.), *Radars*, Vol. 2, *The Radar Equation*, Artech House, 1974.

[1.5]   W.M. Hall, General radar equation, *Space/Aeronautics R and D Handbook*, 1962–63; reprinted in D.K. Barton (ed.), *Radars*, Vol. 2, *The Radar Equation*, Artech House, 1974.

[1.6]   D.K. Barton and H.R. Ward, *Handbook of Radar Measurement*, Prentice-Hall, 1969; Artech House, 1984.

[1.7]   J.E. Fielding and G.D. Reynolds, *RGCALC: Radar Range Detection Software and User's Manual*, Artech House, 1987.

Chapter 2
The Theory of Target Detection

The radar signal can be represented by a sinusoidal RF carrier with narrowband modulation of its amplitude and phase superimposed by the transmitter, the antenna, and the target. Narrowband, in this context, means that the modulation bandwidth is a small fraction of the carrier frequency. Detection of a target requires that the receiver output, consisting of signals, noise, and interference, be processed in such a way that the desired signal causes an output, or alarm, with high probability, while noise and interference produce random false alarms with low probability. The probabilities of detection and false alarm are determined by the amplitude distribution (probability density function, or pdf) of the noise and interference, and that of the signal plus these unwanted components. In this chapter, we will consider, primarily, the pdf values which result from thermal (Gaussian) noise, added to the signal sinusoid. The theory for this case will then be generalized for application to interference with other distributions.

Procedures will be developed by which the detection probability, P_d, can be calculated when the available S/N and acceptable false-alarm probability P_{fa} are known. In addition, procedures will be given to find the required SNR, or *detectability factor*, D, when the required values of P_d and P_{fa} are given. Both S/N and D will refer to the signal-to-noise power ratio at the input to the envelope detector, the difference between them being that S/N is the result of applying the radar equation (e.g., Equation (1.2.21a)), while D is a required value calculated from P_d and P_{fa}. The basic definition of detectability factor is, from [2.1]:

> In pulsed radar, the ratio of single-pulse signal energy to noise power per unit bandwidth that provides stated probabilities of detection and false alarm, measured in the intermediate-frequency amplifier and using an intermediate-frequency filter matched to the single pulse, followed by optimum video integration.

Thus, the energy ratio actually required per pulse, or for the central pulse of a group modulated by the antenna pattern, will be increased to the value D_x used in (1.2.25) and (1.2.26) to account for mismatch of the filter, beamshape loss, and other nonoptimum conditions in the processing.

2.1 NOISE AND FALSE ALARMS

Noise Statistics

In Section 1.2, the thermal noise was described by an equivalent input spectral density N_0, resulting from a receiving system input termination at a temperature T_s (see Figure 2.1.1). Thermal noise is described as white, Gaussian noise, the term *white* referring to its uniform spectral density over the receiver passband, and *Gaussian* to its pdf. For simplicity in system analysis, the receiving system was modeled as a cascade of ideal components, converting the input signal of energy E, along with noise N_0, to intermediate frequency (IF) in a filter noise bandwidth B_n with unity gain and without adding any further noise. As a result, the noise at the output of the IF system has a power $N = N_0 B_n$. This noise is described as band-limited Gaussian noise. In an actual receiver, of course, there would be a large gain for both signal and noise components, but it is convenient to set this gain to unity so that we may use the symbols S and N in expressing the components applied to the phase or envelope detector.

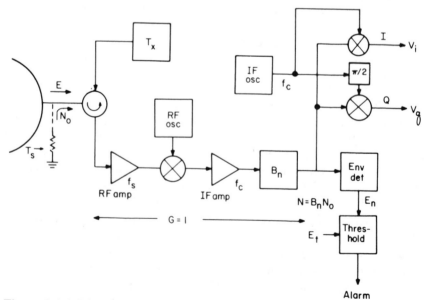

Figure 2.1.1 Idealized receiver block diagram.

The noise statistics can be measured by connecting phase-sensitive detectors to the IF output, as shown in Figure 2.1.1. The in-phase (I) and quadrature (Q) detectors will each produce a bipolar noise output v characterized by a Gaussian pdf:

$$dP_{vi} = dP_{vq} = dP_v = \frac{1}{\sqrt{2\pi N}} \exp(-v^2/2N) \, dv \qquad (2.1.1)$$

where dP is the probability that the instantaneous value of the noise lies between v and $v + dv$, and N is the mean noise power.

The corresponding envelope of the IF noise, which appears at the output of the envelope detector in Figure 2.1.1, has a Rayleigh pdf:

$$dP_e = \frac{E_n}{N} \exp(-E_n^2/2N) \, dE_n, \quad E_n > 0 \qquad (2.1.2)$$

A square-law envelope detector would produce an output proportional to the IF power $x = E_n^2$, having an exponential pdf:

$$dP_x = (1/N) \exp(-x/N) \, dx \qquad (2.1.3)$$

These three pdf distributions are illustrated in Figure 2.1.2.

False-Alarm Probability

A simple detection process, shown in Figure 2.1.1, consists of the envelope detector and an amplitude threshold, set at a voltage level E_t. Such a detector can be used in certain modern radars which transmit only one pulse per beam position, or as the first step in the binary integration process to be described later. In single-pulse detection, the threshold level E_t is set to produce the required low probability of a false alarm caused by noise:

$$P_{fa} = \int_{E_t}^{\infty} (E_n/N) \exp(-E_n^2/2N) \, dE_n = \exp(-E_t^2/2N) \qquad (2.1.4)$$

From this, the threshold voltage can be expressed as

$$E_t = \sqrt{2N\ln(1/P_{fa})} \qquad (2.1.5)$$

If the receiver output is applied continuously to the threshold, independent samples of noise at a rate B_n will give an average *false-alarm rate* $B_n P_{fa}$, with a corresponding *false-alarm time* given by

$$t_{fa} = 1/B_n P_{fa} \tag{2.1.6}$$

It is actually this false-alarm time that is of primary interest to the user of the radar, and which establishes the requirements on the detection threshold.

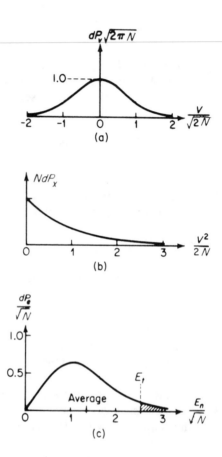

Figure 2.1.2 Probability density functions of noise: (a) Gaussian distribution of IF noise components; (b) Rayleigh distribution of detected envelope; (c) exponential distribution of IF noise power.

2.2 DETECTION OF ONE SAMPLE OF SIGNAL WITH NOISE

The Rician Distribution

Let us now introduce at the receiver input a sinusoidal signal E_s $\cos(2\pi f_s t)$, the duration of which is such that its peak amplitude at the IF output reaches E_s. The output envelope of signal plus noise, at the time of peak signal output, will have a pdf which is described as *Rician*, after S. O. Rice, whose early work [2.2] provided the basis for modern signal detection theory:

$$dP_s = \frac{E_n}{N} \exp[-(E_n^2 + E_s^2)/2N]\, I_0(E_n E_s/N)\, dE_n \qquad (2.2.1)$$

where I_0 is the Bessel function with imaginary argument. The probability of detection P_d, for a sample taken at the time of peak signal output, is the area under this pdf and above the threshold E_t, as illustrated in Figure 2.2.1 for different signal-to-noise power ratios $S/N = E_s^2/2N$.

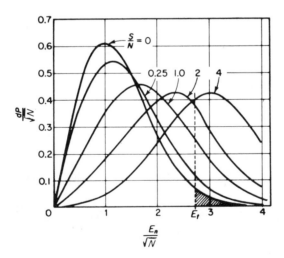

Figure 2.2.1 Probability density functions of envelope of signal plus noise.

Using (2.1.5) to express E_t as a function of P_{fa}, we can use Rice's results to plot P_d as a function of S/N for different values of P_{fa}, as in Figure 2.2.2. This family of curves will serve as the basis for calculation of detection probabilities and detectability factors for all types of radar signals and detection procedures. This plot is used directly for the single-pulse detection process.

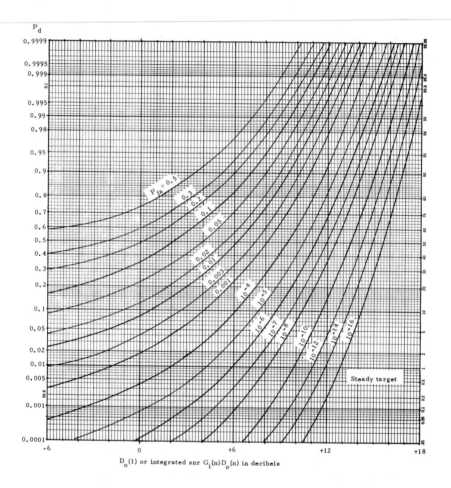

Figure 2.2.2 Detectability factor for a steady target.

For example, assume that $P_d = 0.9$ is required, with $P_{fa} = 10^{-6}$, a typical radar requirement. For a single pulse of steady, sinusoidal signal, the required SNR can be described as the detectability factor $D_0(1)$, where the subscript denotes the Case 0 (steady) target and the argument (1) denotes the single-pulse process. From Figure 2.2.2, we find

$$D_0(1) = +13.2 \text{ dB}$$

Detector Loss Relative to Ideal System

The envelope detector is used in radar when the phase of the received signal is unknown. Ideally, if the signal phase were known *a priori*, the IF oscillator of Figure 2.1.1 would be operated in phase with the signal, and the I-channel phase detector would produce a positive output reaching a peak equal to $\sqrt{2} \, E_s$. The in-phase noise component v_i would add or subtract from this level, producing the pdf distributions shown in Figure 2.2.3. Calculations of P_{fa} and P_d for this ideal system can be made by using tabulated values for the integral of the normal distribution (or of the error function), and its inverse, as shown in Table 2.1. Table 2.2 defines and relates the probability integral $Q(E)$, the error function $\text{erf}(V)$, the complementary error function $\text{erfc}(V)$, and their inverse functions.

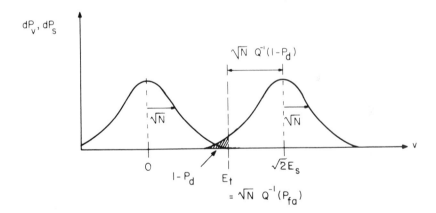

Figure 2.2.3 Probability density functions for coherent detection.

The ideal detectability factor $D_c(1)$ for coherent or synchronous detection is also given in terms of these inverse functions. All of these relationships are exact for Gaussian noise. Accurate analytic approximations are available for these functions [2.3].

When the ideal $D_c(1)$ is compared to $D_0(1)$ for envelope detection, it is found that the lack of *a priori* phase data has caused a detector loss

$C_x(1)$, which indicates the loss in information for target detection purposes. This loss, which describes the slightly higher input SNR required for the envelope detector case, can be defined and expressed as follows [2.4]:

$$C_x(1) \equiv D_0(1)/D_c(1) \approx \frac{(S/N) + 2.3}{(S/N)} \qquad (2.2.2a)$$

$$C_x(1) \equiv D_0(1)/D_c(1) \approx \frac{D_0(1) + 2.3}{D_0(1)} \qquad (2.2.2b)$$

For this single-pulse case, $C_x(1)$ is quite small, amounting only to 0.8 dB for $P_d = 0.5$, $P_{fa} = 10^{-4}$, and to 0.4 dB for $P_d = 0.9$, $P_{fa} = 10^{-6}$. The effect is sometimes described as *small-signal suppression*, and it becomes increasingly important when we consider postdetection integration of pulses with small SNR. The detector loss is plotted in Figure 2.2.4, as a function of SNR at the envelope detector.

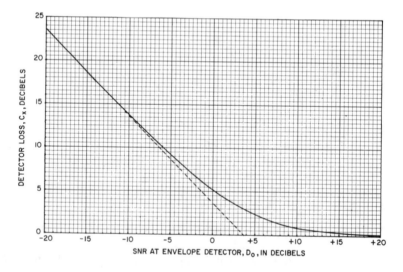

Figure 2.2.4 Envelope detector loss *versus* SNR.

The coherent detector has zero response to the quadrature noise component. The envelope detector, for reasonably high SNR, also rejects the quadrature component of noise, since the resultant of $E_s + v_i + v_q$, shown in Figure 2.2.5, has an amplitude determined essentially by

Table 2.1 Equations for Coherent Detection

pdf of IF noise (I or Q component)	$dP_v = \dfrac{1}{\sqrt{2\pi N}}\exp(-v^2/2N)\,dv$	(2.1.1)
False-alarm probability in I channel	$P_{fa} = \displaystyle\int_{E_t}^{\infty} dP_v = \dfrac{1}{\sqrt{2\pi N}}\int_{E_t}^{\infty}\exp(-v^2/2N)\,dv$	(2.2.3)
	$\quad = Q(E_t/\sqrt{N}) = (1/2)\,\mathrm{erfc}(E_t/\sqrt{2N})$	
Threshold voltage	$E_t = \sqrt{N}\,Q^{-1}(P_{fa}) = \sqrt{2N}\,\mathrm{erfc}^{-1}(2P_{fa})$	(2.2.4)
pdf of signal + noise in I channel	$dP_s = \dfrac{1}{\sqrt{2\pi N}}\exp[-(v+E_s)^2/2N]\,dv$	(2.2.5)
Detection probability in I channel	$P_d = \displaystyle\int_{E_t}^{\infty} dP_s = \dfrac{1}{\sqrt{2\pi N}}\int_{E_t}^{\infty}\exp[-(v+E_s)^2/2N]\,dv$	
	$\quad = Q[(E_t - \sqrt{2S})/\sqrt{N}] = (1/2)\,\mathrm{erfc}[(E_t - \sqrt{2S})/\sqrt{2N}]$	
	$\quad = Q[Q^{-1}(P_{fa}) - \sqrt{2S/N}]$	
	$\quad = (1/2)\,\mathrm{erfc}[\mathrm{erfc}^{-1}(2P_{fa}) - \sqrt{S/N}]$	(2.2.6)
Required signal voltage	$E_s = E_t + \sqrt{N}\,Q^{-1}(1 - P_d)$	
	$\quad = \sqrt{N}\,[Q^{-1}(P_{fa}) + Q^{-1}(1 - P_d)]$	(2.2.7)
Detectability factor for coherent process	$D_c(1) = (1/2)\,[Q^{-1}(P_{fa}) - Q^{-1}(P_d)]^2$	
	$\quad = [\mathrm{erfc}^{-1}(2P_{fa}) - \mathrm{erfc}^{-1}(2P_d)]^2$	(2.2.8)

NOTE: *See* Table 2.2 for definitions of functions.

Table 2.2 Definitions of Probability Functions and their Relationships

$$Q(E) = (1/\sqrt{2\pi}) \int_E^\infty \exp(-v^2/2)\, dv = P \quad Q^{-1}(P) = E$$

$$\mathrm{erf}(V) = (2/\sqrt{\pi}) \int_0^V \exp(-v^2)\, dv = P \qquad \mathrm{erf}^{-1}(P) = V$$

$\mathrm{erfc}(V) = 1 - \mathrm{erf}(V) = 1 - P$	$\mathrm{erfc}^{-1}(1 - P) = V$
$\mathrm{erf}(-V) = -\mathrm{erf}(V)$	$\mathrm{erf}^{-1}(-P) = \mathrm{erf}^{-1}(P)$
$\mathrm{erfc}^{-1}(-V) = 1 + \mathrm{erf}(V)$	$\mathrm{erfc}^{-1}(1 + P) = -\mathrm{erf}^{-1}(P)$
$Q(E) = (1/2)\,\mathrm{erfc}(E/\sqrt{2}) = P$	$Q^{-1}(P) = \sqrt{2}\,\mathrm{erfc}^{-1}(2P)$
$\mathrm{erfc}(V) = 2Q(\sqrt{2V}) = P$	$\mathrm{erfc}^{-1}(P) = (1/\sqrt{2})\,Q^{-1}(P/2)$
$Q(-E) = 1 - P$	$Q^{-1}(1 - P) = -E$

$$Q(0) = 0.5$$
$$Q(E > 0) < 0.5$$
$$Q(E < 0) > 0.5$$

$$Q^{-1}(0.5) = 0$$
$$Q^{-1}(P < 0.5) > 0$$
$$Q^{-1}(P > 0.5) < 0$$

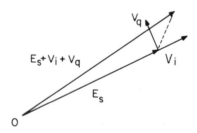

Figure 2.2.5 Envelope of signal plus noise.

$E_s + v_i$. The quadrature component of noise causes primarily a phase modulation, which is ignored by the envelope detector. Thus, there is little reason to attempt coherent detection unless the SNR is less than about 10 dB (for which $C_x \approx 1$ dB).

Approximations for Detection Probability

The exact value of detection probability for the envelope detector with one signal sample is given by integrating (2.2.1), and can be expressed in terms of the *incomplete Toronto function*, T:

$$P_d = \int_{E_t}^{\infty} dP_s = 1 - T_{\sqrt{\ln(1/P_{fa})}} (1, 0, \sqrt{S/N}) \qquad (2.2.9)$$

Various approximations have been proposed for easy calculation of this function, and two of these, with their inverse expressions, are given in Table 2.3. The North approximation is very accurate, and is simple enough for use in a subroutine for a pocket calculator [2.5] to find $D_0(1)$.

Matched Filters and Matching Loss

The equations for detection probability involve the voltage ratio $E_s/\sqrt{2N}$, or the equivalent power ratio S/N at the input to the envelope detector. As discussed in Section 1.2, this ratio can achieve its maximum value when the IF filter is matched to the signal:

$(S/N)_{mf} = E_1/N_0$ for a single pulse

$(S/N)_{mf} = E/N_0$ for a filter matched to the entire observed signal

For other filters, there will be a mismatch loss L_m with respect to the pulse spectrum, and also possibly a loss L_{mf} with respect to the spectrum of the envelope of the pulse train. The IF output SNR will then be

$S/N = E_1/N_0L_m$ for a single pulse

$S/N = E/N_0L_mL_{mf}$ for the entire observed signal

The effect of these mismatch losses is to increase the required value of energy ratio, D_x, at the receiver input, in order to meet the SNR requirements at the envelope detector. However, in many cases a low-pass filter will be placed between the envelope detector and the threshold, reducing the noise variance at the threshold. This reduces the system loss by introducing integration over the duration of the envelope-detected signal. The resulting integration gain will partially compensate for the mismatch loss, as described in later sections of this chapter.

Table 2.3 Equations and Approximations for Single-Pulse Noncoherent Detection

pdf of IF noise envelope (Rayleigh)	$dP_e = (E_n/N) \exp(-E_n^2/2N) \, dE_n, \ E_n > 0$	(2.1.2)
False-alarm probability	$P_{fa} = \int_{E_t}^{\infty} (E_n/N) \exp(-E_n^2/2N) \, dE_n = \exp(-E_t^2/2N)$	(2.1.4)
Threshold voltage	$E_t = \sqrt{2N \ln (1/P_{fa})}$	(2.1.5)
pdf of IF signal-plus-noise envelope (Rician)	$dP_s = (E_n/N) \exp[-(E_n^2 + E_s^2)/2N] I_0(E_n E_s/N) \, dE_n$	(2.2.1)
Detection probability	$P_d = \int_{E_t}^{\infty} dP_s = 1 - T_{\sqrt{\ln(1/P_{fa})}}(1, 0, \sqrt{S/N})$	(2.2.9)
North's approximation [2.6]	$P_d \approx Q[\sqrt{2 \ln(1/P_{fa})} - \sqrt{2(S/N) + 1}]$ $= (1/2) \, \mathrm{erfc}[\sqrt{\ln (1/P_{fa})} - \sqrt{(S/N) + 1/2}]$	(2.2.10)
	$D_0(1) \approx [\sqrt{\ln(1/P_{fa})} - (1/\sqrt{2})Q^{-1}(P_d)]^2 - 1/2$ $= [\sqrt{\ln(1/P_{fa})} - \mathrm{erfc}^{-1}(2P_d)]^2 - 1/2$	(2.2.11)
DiFranco and Rubin [2.7, p. 316]	$P_d \approx Q[\sqrt{2 \ln (1/P_{fa})} - \sqrt{2S/N}]$	(2.2.12)
	$D_o(1) \approx [\sqrt{\ln(1/P_{fa})} - \mathrm{erfc}^{-1}(2P_d)]^2$ $= [\sqrt{\ln(1/P_{fa})} - (1/\sqrt{2})Q^{-1}(P_d)]^2$	(2.2.13)

2.3   INTEGRATION OF PULSE TRAINS

When a train of pulses is received from the target over the observation time t_o, there will be $n = f_r t_o$ pulses available for processing. Similarly, if a continuous signal is received for a time t_o in a receiver whose bandwidth $B_n > 1/t_o$, there will be $n = B_n t_o$ samples of signal and independent noise available. There are four distinct ways in which the information from these n pulses or samples may be processed to improve detection performance:

(a) *Coherent integration,* in which the pulses are added prior to envelope detection;

(b) *Noncoherent* (or *video*) *integration,* in which each pulse is envelope detected, and the resulting video pulses are added together prior to application of thresholding;

(c) *Binary integration,* in which each pulse is applied to a threshold, and the number m of threshold crossings is used as the criterion for an output alarm;

(d) *Cumulative detection,* in which $m = 1$ is the alarm criterion.

The methods are listed in order of declining efficiency of integration, and also declining complexity of implementation.

Coherent Integration

If the IF filter is matched to the entire signal observed over t_o, coherent integration is obtained in that filter. This requires that the signal have a predictable phase relationship (coherence) over this period, and that the phase response of the filter be such as to bring all signal components into the same phase while they are added. If the signal is a pulse train, then the filter must be matched to the pulse-to-pulse phase change (doppler frequency) of the target. The signal spectrum of a coherent train of rectangular pulses at repetition rate f_r is a series of narrow lines, as shown in Figure 2.3.1. The matched filter for a fixed target has response bands of width $1/t_o$ at $f = f_0 \pm if_r$, extending over the infinite spectral envelope $\mathrm{sinc}(\pi f \tau)$. For targets with doppler shift $|f_d| > 1/2t_o$, similar filters are required with response bands offset by f_d. Implementation of the coherent transmitter, receiver, and matched filter will be discussed in Chapter 5.

For an ideal coherent system, $S/N = E/N_0$, and Equations (2.2.6) and (2.2.8), give the resulting P_d and $D_c(1)$. The requirement for each pulse or sample of the coherent train is simply $D_c(n) = D_c(1)/n$. Thus, the coherent integration gain, which is the ratio of the effective SNR for detection purposes to that of the single pulse, is exactly n. For a system with a filter spectral envelope which fails to match the pulse spectrum

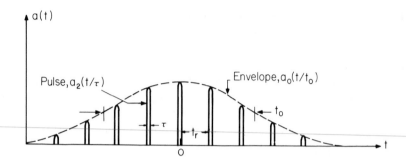

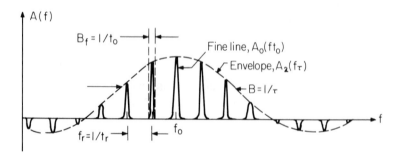

Figure 2.3.1 (a) Waveform and (b) spectrum of coherent pulse train.

$A_2(f\tau)$, there will be a loss L_m, and for one having passbands that fail to match the line shape $A_0(ft_o)$, there will be a further loss L_{mf}. It can be assumed that the receiving filter operates at the same repetition frequency $f_r = 1/t_r$ as the transmission. Then, ignoring the effects of possible post-detection low-pass filtering to recover a portion of the matching loss, the required total energy ratio will be

$$E/N_0 = D_c(1)L_mL_{mf} \qquad (2.3.1)$$

where $D_c(1)$ is given by (2.2.8). The energy ratio for each of n equal pulses will be

$$E_1/N_0 = D_c(n)L_mL_{mf}/n \qquad (2.3.2)$$

If the pulse-train envelope is established by a scanning antenna, with t_o taken between the half-power points of the one-way beam, there will be a further increase in the required energy ratio of the central pulse by the beamshape loss L_p, as described in Section 1.2.

Video Integration

If n signal samples are to be added without controlled phase relationships, the samples must first be passed through the envelope detector to remove the random phase. The effective (integrated) SNR is increased in the video integrator by the factor n, but not before the pulses experience an increased detector loss, given by (2.2.2a) with the reduced value $S/N = D_0(n)$:

$$C_x(n) = \frac{D_0(n) + 2.3}{D_0(n)} \qquad (2.3.3)$$

where $D_0(n)$ denotes the detectability factor (required SNR) for each pulse of the n-pulse train. As the signal input to the envelope detector decreases, the small-signal suppression loss increases, and subsequent video integration cannot restore the SNR represented by total signal energy ratio. As a result, there is a requirement for more energy per pulse, which has been termed *integration loss*. It is important to recognize, however, that the loss is not the result of the integration, but of the location of the integrator after the envelope detector, and the resulting increased small-signal suppression in the nonlinear process of envelope detection. This integration loss may be expressed as

$$L_i(n) \equiv nD_0(n)/D_0(1) = C_x(n)/C_x(1) \qquad (2.3.4)$$

This loss is thus the ratio of total signal energy required of the n-pulse train to that which would have been required if a single pulse had been transmitted and processed, or if a matched filter for the n-pulse train had been used.

The *integration gain* for the noncoherent process is

$$G_i(n) \equiv D_0(1)/D_0(n) = n/L_i(n) \qquad (2.3.5a)$$

and the effective SNR of the train, to be used in entering Figure 2.2.2 to find detection performance, is $D_0(1)_{\text{eff}} = D_0(n)G_i(n)$. The n-pulse detectability factor is thus

$$D_0(n) = \frac{D_0(1)L_i(n)}{n} \qquad (2.3.5b)$$

In calculating $C_x(n)$ and integration loss or gain, (2.3.3) requires that $D_0(n)$ be known. The family of curves in Figure 2.3.2 represents the solution to this problem when the required P_d and P_{fa}, and hence $D_0(1)$, are given along with n. For example, assume that $P_d = 0.9$, $P_{fa} = 10^{-6}$, and $n = 24$ are given. The total signal energy ratio and the ratio per pulse, for a system using a filter matched to the spectrum of each pulse, are found from the graphs as follows:

$$\left.\begin{array}{l} P_d = 0.9 \\ P_{fa} = 10^{-6} \end{array}\right\} \qquad \text{Fig. 2.2.2} \rightarrow D_0(1) = +13.2 \text{ dB}$$

$$n = 24, \qquad\qquad \text{Fig. 2.3.2} \rightarrow +L_i(n) = \underline{+\ \ 3.2 \text{ dB}}$$

$$\text{Total energy ratio,} \qquad\qquad\qquad nD_0(n) = +16.4 \text{ dB}$$
$$-10 \log n = \underline{-13.8 \text{ dB}}$$

$$\text{Single-pulse energy ratio,} \qquad\qquad\qquad D_0(n) = +\ \ 2.6 \text{ dB}$$

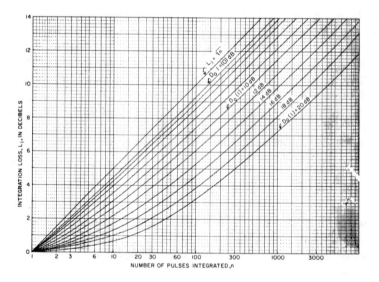

Figure 2.3.2 Integration loss *versus* number of pulses integrated after envelope detection.

The calculation may also be made easily without graphs, using procedures described in [2.4]. We note that the effective coherent energy ratio is

$$D_c(1) = nD_0(n)/C_x(n) = nD_0^2(n)/[D_0(n) + 2.3] \qquad (2.3.6)$$

The value of $D_c(1)$ can be found from (2.2.8). Solving for $D_0(n)$ in terms of $D_c(1)$, we find

$$D_0(n) = [D_c(1)/2n][1 + \sqrt{1 + 9.2n/D_c(1)}] \qquad (2.3.7)$$

The corresponding integration loss is

$$L_i(n) = \frac{1 + \sqrt{1 + 9.2n/D_c(1)}}{1 + \sqrt{1 + 9.2/D_c(1)}} \qquad (2.3.8)$$

The problem must sometimes be worked in the other direction, where SNR is known and detection probability is to be obtained. The concept of detector loss and its simple (although approximate) relationship to IF SNR provides a direct method for this calculation, given S/N and n:

Eq. (2.2.2a), $C_x(n) = \dfrac{(S/N) + 2.3}{(S/N)}$

Effective $D_c(n) = (S/N)/C_x(n)$

Effective $D_c(1) = n(S/N)/C_x(n) = n(S/N)^2/[(S/N) + 2.3]$

From Eq. (2.2.6), $P_d = Q\{Q^{-1}(P_{fa}) - \sqrt{2n(S/N)^2/[(S/N) + 2.3]}\}$

$$(2.3.9)$$

These procedures are accurate to about 0.3 dB in practical cases. Analytic approximations are also given in [2.7, pp. 366–367]. Greater accuracy is possible by using the work of J.I. Marcum [2.8], which has been programmed for calculators and personal computers by Skillman [2.9, 2.10].

Binary Integration

The video integrator is often implemented in digital form, by placing an analog-to-digital (A/D) converter after the envelope detector. With fine enough granularity in the converter, the results of the previous section will be obtained. A simplification in equipment results when the number of converter bits is reduced to one, corresponding to a threshold detector operating directly on the output of the envelope detector, followed by an accumulator which counts up to m threshold crossings before generating an output alarm. The resulting performance is within about 1.5 dB of ideal video integration, comparing favorably with most practical video integration circuits.

Analysis of performance for the binary integrator operating with n equal pulses of signal is based on the binomial distribution:

$$P(j) = \frac{n!}{j!(n - j)!} p^j (1 - p)^{n-j} \qquad (2.3.10)$$

where $P(j)$ is the probability that exactly j crossings of the threshold will occur in n trials, each with threshold crossing probability p. The probability of an output alarm will be

$$P(j \geq m) = P(m) + P(m + 1) + \ldots + P(n) \qquad (2.3.11)$$

The output false-alarm probability is found by setting p equal to the probability that noise alone will exceed the threshold on one trial:

$$p_{fa} = \exp(-E_t^2/2N) \qquad (2.3.12)$$

$$P_{fa} = \sum_{j=m}^{n} \frac{n!}{j!(n-j)!} p_{fa}^j (1 - p_{fa})^{n-j} \qquad (2.3.13)$$

$$\approx \frac{n!}{m!(n-m)!} p_{fa}^m, \quad p_{fa} \ll 1 \qquad (2.3.14)$$

Then, the single-pulse p_d is found from Figure 2.2.1 or Table 2.3, and output P_d is found from (2.3.13) with p_d replacing p_{fa}.

The range of optimum m is quite broad, lying near the value determined by Schwartz [2.11]:

$$m_{opt} = 1.5 \sqrt{n} \qquad (2.3.15)$$

A comparison of required SNR for optimum binary integration, $D_b(n)$, and $D_0(n)$ is shown in Figure 2.3.3, along with a comparison of integration losses.

Cumulative Detection

The least efficient way of combining information from n signal samples is to make independent detection decisions on each, and to rely on the cumulative probability of detection to reach the required output value. This is equivalent to operating a binary integrator with $m = 1$. The detectability factor and loss from this process are also plotted in Figure 2.3.3. The single-pulse false-alarm probability in this case must be set at P_{fa}/n, while single-pulse p_d is given by

$$p_d = 1 - (1 - P_d)^{1/n} \qquad (2.3.16)$$

When the n pulses are distributed over separate antenna scans, the extra loss in the cumulative process, compared to video integration, can be

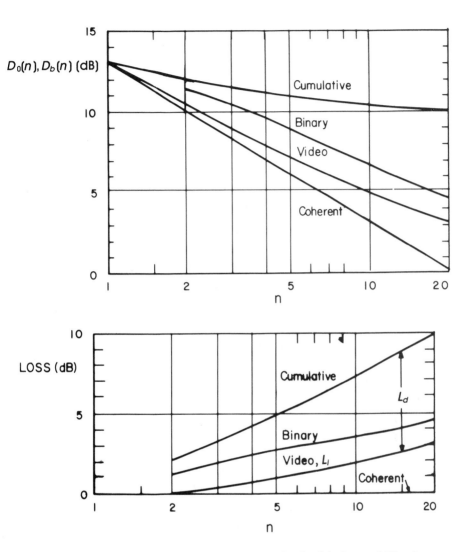

Figure 2.3.3 Comparison of integration methods: (a) detectability factors
and (b) integration loss.

described as *scan distribution loss*. It is apparent that the gain from cumulative detection of n signals on a steady target is small, but when fluctuating targets are considered, there may be significant gain.

Integrator Weighting Loss

When video integration is used, the optimum weighting of the n pulses is matched to their individual SNRs. The calculations of integration gain and loss were based on pulses of equal SNR, with uniform weighting over the train. If the pulse train envelope is the result of antenna scanning, it will have an approximately Gaussian shape:

$$a_0(t/t_o) = \exp(-2.77t^2/t_o^2) \qquad (2.3.17)$$

for which the amplitude is 0.5 at $t = \pm t_o/2$. The energy in this train is nE_1/L_p, where E_1 is the energy in the central pulse and $L_p = 1.33$ (or 1.24 dB) is the beamshape loss. The matched video integrator will apply this same function as a weight (or a_0^2 for a square-law detector) to the signals in a moving-window process.

Practical integrators are more likely to use uniform weighting of the most recent k pulses, where $k < n$, or recursive processes in which the weighting function is exponential or the convolution of two exponential functions (the two-pole integrator). Such weighting will introduce an additional loss, which will be small if the time constant is properly selected [2.12]. The often used value [2.13, 2.14] of beamshape loss $L_p = 1.6$ dB is actually the combined beamshape and weighting loss for uniform weighting of $0.84n$.

False-Alarm Time

When integration is used, the number of independent detection decisions is reduced. In each interval t_o, a decision is made in each range-doppler resolution cell. If there are n_t range gates or cells, and n_f doppler filters, the decisions are made at a rate:

$$\nu = n_t n_f/t_o \qquad (2.3.18)$$

The average false-alarm rate is νP_{fa}, and the average time between false alarms is

$$t_{fa} = 1/\nu P_{fa} = t_o/n_t n_f P_{fa} \qquad (2.3.19)$$

For a noncoherent system, in which $n_f = 1$ and all range cells are used for detection, we have

$$n_t = t_r/\tau = 1/f_r\tau$$

$$t_o = nt_r = n/f_r$$

$$t_{fa} = n\tau/P_{fa} \approx n/B_nP_{fa} \tag{2.3.20}$$

where the approximation applies to a matched system. For a coherent system, using all doppler cells, we have

$$n_f = n$$

$$t_{fa} = \tau/P_{fa} \approx 1/B_nP_{fa} \tag{2.3.21}$$

The numbers of cells will be reduced, and false-alarm time increased, if only certain ranges and dopplers are used for detection.

Collapsing Loss

Practical radars seldom preserve their full RF signal resolution through the integration and thresholding processes, for which reduced video bandwidth and broadened range gates may prove economical. The term *collapsing loss* is applied to describe the increase in required input SNR when noise from adjacent resolution cells or channels is combined with the signal and noise of the channel actually containing the target. The n video samples of signal plus noise will then be integrated along with m extra samples of noise alone, giving a *collapsing ratio* ρ defined by

$$\rho \equiv (n+m)/n \tag{2.3.22}$$

The effect, when using a square-law detector, is the same as if the signal energy had been redistributed over $\rho n = n + m$ pulses, leading to a larger value of integration loss $L_i(\rho n)$. The collapsing loss is then defined as

$$L_c(\rho,n) = L_i(\rho n)/L_i(n) \tag{2.3.23}$$

Equations for collapsing ratio are given in Table 2.4 for different processes. In cases where the number of threshold decisions is reduced by the factor ρ, with false-alarm time t_{fa} held constant ($P'_{fa} = P_{fa}/\rho = $ constant), the loss is reduced slightly, such that

$$L_c(\rho,n) = \frac{L_i(\rho n)}{L_i(n)} \times \frac{D_0(\rho n) \text{ for } \rho P_{fa}}{D_0(\rho n) \text{ for } P_{fa}} \qquad (2.3.24)$$

All these relationships apply to the square-law detector. When a linear detector is used, the collapsing loss is increased [2.15].

Calculations of collapsing loss may be made, for the square-law detector, by increasing the detector loss in (2.3.3) to

$$C_x(\rho n) = \frac{D_0(\rho n) + 2.3}{D_0(\rho n)} \qquad (2.3.25)$$

$$L_c(\rho n) = \frac{(S/N) + 2.3\rho}{(S/N) + 2.3} \qquad (2.3.26)$$

Equations (2.3.9) and (2.3.7) then become

$$P_d = Q\{Q^{-1}(P_{fa}) - \sqrt{2\rho n(S/N)^2/[(S/N) + 2.3]}\} \qquad (2.3.27)$$

$$D_0(n) = [D_c(1)/2n] [1 + \sqrt{1 + 9.2\rho n/D_c(1)}\,] \qquad (2.3.28)$$

The collapsing loss may be calculated directly as

$$L_c(\rho n) = \frac{1 + \sqrt{1 + 9.2\rho n/D_c(1)}}{1 + \sqrt{1 + 9.2n/D_c(1)}} \qquad (2.3.29)$$

The collapsing loss should be used to calculate the actual increase in required energy ratio, denoted by the *matching factor M*, in cases where the IF bandwidth is greater than its matched value, if the video bandwidth is matched to the pulse. From Table 2.4, we have

$$\rho = B\tau + 1$$

$$M = L_c(\rho n) = \frac{1 + \sqrt{1 + 9.2n(B\tau + 1)/D_c(1)}}{1 + \sqrt{1 + 9.2n/D_c(1)}} \qquad (2.3.30)$$

For this case, the false-alarm probability is related to false-alarm time by (2.3.20), using $\tau \approx 1/2B_v$ rather than the smaller $1/B_n$.

Table 2.4 Equations for Collapsing Ratio*

Cases for which P_{fa}/ρ remains constant:

(a) Restricted CRT sweep speed s, where d is spot diameter and τ is pulsewidth.

$$\rho = \frac{d + s\tau}{s\tau}$$

(b) Restricted video bandwidth B_v, where $B = 1/\tau$ is IF signal bandwidth.

$$\rho = \frac{2B_v + 1/\tau}{2B_v} = \frac{2B_v + B}{2B_v}$$

(c) Collapsing of coordinates onto the display, where $2\Delta_r/c$ is the time-delay interval displayed per display cell, $\omega_e t_v$ and $\omega_a t_v$ are elevation and azimuth scans during the integration time t_v, and θ_e and θ_a are the beamwidths.

$$\rho = \frac{2\Delta_r}{c\tau} \quad \text{or}$$

$$\rho = \frac{\omega_e t_v}{\theta_e} \quad \text{or}$$

$$\rho = \frac{\omega_a t_v}{\theta_a}$$

Cases for which P_{fa} remains constant:

(d) Excessive IF bandwidth $B > 1/\tau$ followed by matched video.

$$\rho = \frac{B\tau + 1}{B}$$

(use L_c in place of L_m)

(e) Receiver outputs mixed at video, where m is the number of receivers.

$$\rho = m$$

(f) IF filter followed by gate of width τ_g and by video integration.

$$\rho = \frac{1}{B\tau} + \frac{\tau_g}{\tau}$$

---

*Collapsing ratio $\rho \equiv \dfrac{m + n}{n}$

2.4 DETECTION OF FLUCTUATING TARGETS

Single-Sample Detection

Having established procedures for calculating detection performance on a steady target signal, we must now extend the result to the more likely fluctuating target model. Essentially all radar target objects produce echo signals having amplitudes that are Rayleigh distributed (exponential distribution of power, or cross section). In his classic work on fluctuating targets [2.16], Swerling designated such targets as Case 1 (when the fluctuation was slow) or Case 2 (when it was fast). The Rayleigh distribution is a special case of the chi-square distribution, with two degrees of freedom (DOF). When n independent samples of the rapidly fluctuating (Case 2) target are considered, it becomes a chi-square distribution with $2n$ DOF. Swerling also considered a Case 3 target (4 DOF) and Case 4 ($4n$ DOF), but the only practical application of these models lies in the use of Case 3 to represent a Case 1 target observed by dual-diversity radar.

The pdf and detection equations for the Case 1 target are shown in Table 2.5, for the single-sample case. The signal-plus-noise pdf is the same as that for noise alone, except that the mean power is now $S + N$ rather than N. The expressions for P_d and $D_1(1)$ are exact, as well as being simple. In fact, many analyses (see, for example, [2.17]) start with this expression, extend it to n-pulse integration through integrals of the chi-square distribution, and approach the steady target case for DOF $\rightarrow \infty$. However, as we shall see, there are advantages to starting with the steady target detectability factor $D_0(1)$, from Figure 2.2.2, and proceeding through the n-pulse integration calculation before incorporating target fluctuation effects.

Table 2.5 Equations for Fluctuating Target

Noise envelope and P_{fa} as in Table 2.3.

pdf of IF signal-plus-noise envelope (Rayleigh)	$dP_s = [E_n/(S + N)]$ $\times \exp[-E_n^2/2(S + N)]\, dE_n$ (2.4.1)
Detection probability	$P_d = \displaystyle\int_{E_t}^{\infty} dP_s$ $= \exp[-E_t^2/2(S + N)]$ $= \exp[\ln P_{fa}/(1 + S/N)]$ (2.4.2)
Detectability factor	$D_1(1) = \dfrac{\ln P_{fa}}{\ln P_d} - 1$ (2.4.3)

Fluctuation Loss

Curves for the Case 1, single-pulse detectability factor are shown in Figure 2.4.1. When these are compared to Figure 2.2.2, we will see that a larger average signal is required for the fluctuating target than for the steady target, if high P_d is to be obtained. The extra signal requirement is similar to the fade margin required in communications channels, but in radar this requirement is referred to as a *fluctuation loss*. It is defined [2.1] as

The apparent loss in radar detectability or measurement accuracy for a target of given average echo power return due to target fluctuation. It may be measured as the increase in required average echo return power of a fluctuating target as compared to a target of constant echo return, to achieve the same detectability or measurement accuracy.

Thus, in our notation, we have

$$L_f(1) = D_1(1)/D_0(1) \tag{2.4.4}$$

This loss, plotted in Figure 2.4.2 for the single-pulse case, is primarily a function of P_d, but also depends weakly on P_{fa}.

Integration of Case 1 Signals

The result of integrating n pulses or samples of fluctuating signal plus noise will depend on whether there is only one independent sample of the signal, or more than one. If the target amplitude remains constant for the n-pulse train, either video or coherent integration can be used as with the steady target. The integration gain will be the same as for the steady-target case, namely $n/L_i(n)$, where $L_i = 1$ for coherent integration, or as given in (2.3.4) for video integration. It is important to note that the integration loss depends not on the average SNR, $D_1(n)$, but rather on the effective steady-target value $D_0(n) = D_1(n)/L_f(1)$. This is because the detector loss increases as the target fades. To achieve a high P_d, the signal reduced by fading must be adequate, not just its average value. Thus, we can write the simplified equation for the n-pulse detectability factor of the Case 1 target as

$$D_1(n) = D_0(n)L_f(1) = \frac{D_0(n)L_f(1)}{n} = \frac{D_0(1)L_i(n)L_f(1)}{n} \tag{2.4.5}$$

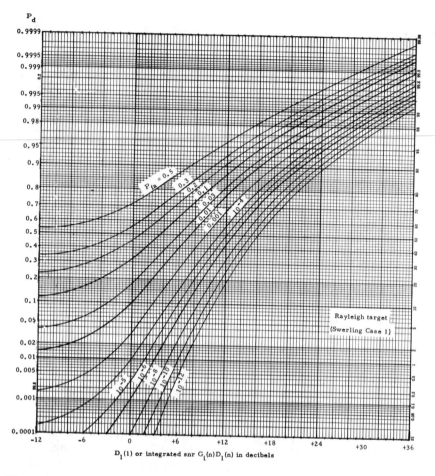

Figure 2.4.1 Detectability factor for single-pulse Case 1 target.

For example, in calculating the Case 1 detectability factor for $P_d = 0.9$, $P_{fa} = 10^{-6}$, $n = 24$, we follow the procedure of Section 2.3, but with the addition of a fluctuation loss:

$$\left.\begin{array}{l} P_d = 0.9 \\ P_{fa} = 10^{-6} \end{array}\right\} \qquad \text{Fig. 2.2.2} \rightarrow D_0(1) = +13.2 \text{ dB}$$

$$n = 24, \qquad \text{Fig. 2.3.2} \rightarrow +L_i(n) = \underline{\quad +3.2 \text{ dB}}$$

Total energy ratio, $\qquad\qquad\qquad\qquad nD_0(n) = +16.4 \text{ dB}$

$$-10\log n = -13.8 \text{ dB}$$

Single-pulse energy ratio, $\qquad\qquad D_0(n) = +2.6 \text{ dB}$

Fluctuation loss, $\qquad\qquad\qquad +L_f(1) = \underline{\quad +8.4 \text{ dB}}$

Average single-pulse, $\qquad\qquad\quad D_1(n) = +11.0 \text{ dB}$

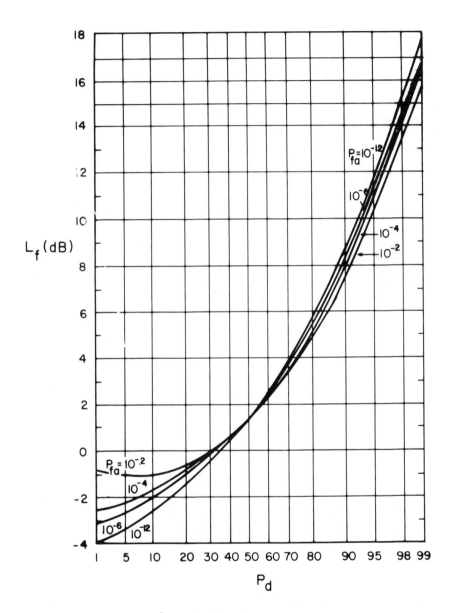

Figure 2.4.2 Fluctuation loss for Case 1 target.

Note that the total average energy ratio is increased to 24.8 dB for the Case 1 target.

If the average input SNR is given, and calculation of P_d is required, the problem becomes more difficult. Both fluctuation and integration losses depend on the final value of P_d, which is unknown. In [2.5], an iterative procedure was used. In hand computations, a convenient approach is to calculate $D_1(n)$ for $P_d = 0.5$, compare this with the available S/N, and make one or two further calculations of $D_1(n)$ for P_d either higher or lower than 0.5. When these are plotted on probability paper, interpolation will quickly yield the correct P_d for the given S/N. Alternatively, the analytical approximations from [2.7 pp. 406–207] or the many curves of [2.18] may be used.

Integration with Independent Signal Samples

In many cases, the target amplitude does not remain constant over the observation time t_o, but varies between the beginning and end of the interval. Although Swerling solved this problem for the case of very rapid fluctuation, such that n independent signal samples were received (Case 2), there was no solution for the general case where $n_e < n$ independent signal samples were integrated along with n noise samples. An empirical expression was developed [2.4] to make possible calculations for such cases. By analyzing the Swerling Case 2 results, along with limited data on dual-diversity reception systems, an equation for fluctuation loss with n_e independent signal samples was obtained:

$$L_f(n_e) = [L_f(1)]^{1/n_e}$$ (2.4.6)

or, with losses expressed in dB,

$$L_f(n_e)_{dB} = (1/n_e)L_f(1)_{dB}$$ (2.4.7)

This agrees closely with the Case 2 results, when $n_e = n$, and also permits the intermediate cases, not covered in other literature, to be evaluated. The general expression for detectability factor with n pulses, within which there are n_e independent signal samples, becomes

$$D_e(n, n_e) = D_0(n)L_f(n_e) = \frac{D_0(n)L_f(n_e)}{n}$$

$$= \frac{D_0(1)L_i(n)L_f(n_e)}{n}$$ (2.4.8)

In dB format, we can write

$$D_e(n, n_e)_{dB} = D_0(1)_{dB} + L_i(n)_{dB} + (1/n_e)L_f(1)_{dB} - 10 \log n$$ (2.4.9)

In the example calculated above, if there were $n_e = 2$ independent signal samples during the n pulses, the fluctuation loss in dB would be cut in half, and the required energy ratio would be reduced by 4.2 dB. For Case 2 signals, with n_e = 24, the loss would be only 0.4 dB and $D_2(24)$ = +3.0 dB. This last figure may be compared with the precise Case 2 value of +3.12 dB. Further work by Kanter [2.19] confirms the accuracy of the procedure represented by (2.4.6) to (2.4.9).

Chi-Square Target Models

Using the empirical expression for fluctuation loss, we are now in a position to solve for detectability factor of the generalized chi-square distributed target. If the number of degrees of freedom is $2K$, then n_e is increased by the factor K, and the fluctuation loss becomes

$$L_f(Kn_e)_{dB} = (1/Kn_e)L_f(1)_{dB} \tag{2.4.10}$$

The factor K represents half the number of independent Gaussian components added together to form a target signal, while n_e represents the number of independent signals integrated during n pulses.

The steady target and all Swerling models are special cases of the generalized model:

Case 0, Steady Target:	$n_e \to \infty,$	$K \to \infty$
Case 1 Target:	$n_e = 1,$	$K = 1$
Case 2 Target:	$n_e = n,$	$K = 1$
Case 3 Target:	$n_e = 2,$	$K = 2$
Case 4 Target:	$n_e = 2n,$	$K = 2$

Diversity Gain

The theory of fluctuation loss provides a quantitative approach to determining the beneficial effects of diversity on target detection. The reduction in fluctuation loss may be considered as a *diversity gain* for a system taking samples over intervals in time or frequency. The diversity gain can be defined as

$$G_d(n_e) \equiv L_f(1)/L_f(n_e) = [L_f(1)]^{1 - 1/n_e} \tag{2.4.11}$$

or, in dB notation,

$$G_d(n_e)_{dB} = (1 - 1/n_e)L_f(1)_{dB} \tag{2.4.12}$$

Four diversity cases may be distinguished:

(a) *Time diversity*, in which independent samples are received at intervals equal to the correlation time of the target;

(b) *Frequency agility*, in which independent samples are obtained rapidly, by changing transmitter frequency;

(c) *Frequency diversity*, in which parallel channels are used to obtain independent samples;

(d) *Combined diversity*, in which both time and frequency effects are used to increase the number of samples.

Time diversity requires that the integration interval t_o exceed the correlation time of the target, t_c. The number of samples available is

$$n_e = 1 + t_o/t_c \qquad (2.4.13)$$

For rigid targets, the correlation time may be estimated as

$$t_c = \lambda/2\omega_a L_x \qquad (2.4.14)$$

where λ is radar wavelength, ω_a is the rate of rotation of the target about the line of sight, and L_x is the target dimension normal to the line of site and the axis of rotation. In a typical aircraft case, ω_a may be a few hundredths of one radian per second, and L_x may be a few tens of meters. Even at microwave frequencies (e.g., $\lambda = 0.03$ m), the resulting correlation time is tens of milliseconds, and there is not much time diversity effect unless long dwells are used for detection ($t_o \gg t_c$). The Case 2 target will require $t_c < t_r = 1/f_r$, a condition that is hardly ever encountered. If integration can be carried out over several scans of the antenna, then time diversity becomes more practical. However, as targets move between scans, it becomes difficult to perform the integration in narrow range cells, and the cumulative detection process is preferred.

Frequency agility radar can approach the Case 2 condition, if the agile bandwidth of the transmitter is great enough and if pulse-to-pulse frequency change is introduced. The number of independent target samples available in a bandwidth Δf is

$$n_e = 1 + \Delta f/f_c \qquad (2.4.15)$$

where the correlation frequency of the target is

$$f_c = c/2L_r \qquad (2.4.16)$$

The velocity of light is c, and L_r is the radial length of the target. For a typical aircraft, L_r is 15 to 30 m, and the resulting $f_c = 10$ to 5 MHz. To obtain a Case 2 target in our previous example, with $n = 24$, the agile bandwidth would have to be 115 to 230 MHz, with the pulses equally distributed over that band. This is quite possible in microwave radar, provided there is no requirement for MTI or doppler processing to reject clutter. If MTI is required, the agility may be on a burst-to-burst basis, with reduction in the number of independent samples ($n_e = n/m$, where m-pulse bursts are used).

Frequency diversity radar offers a solution to the problem of large fluctuation loss, without compromising MTI or doppler performance. A number of channels (often two) are operated in parallel, spaced by at least f_c within the radar band. Pulses are received in each of n_e channels, so the integration gain is increased as fluctuation loss is decreased.

Combined diversity obtains samples in both time and frequency, so that the total number of samples becomes

$$n_e = (1 + t_o/t_c)(1 + \Delta f/f_c) \qquad (2.4.17)$$

It is necessary to ensure that the transmissions are uniformly distributed over the time-frequency space, to avoid correlation between samples which would reduce n_e.

It should be noted that diversity gain is only possible if the nondiverse system has a fluctuation loss. From Figure 2.4.2, if the value of P_d is less than 0.33, diversity is not advantageous. On the other hand, if high probabilities are required, some form of diversity is almost essential.

Binary Integration on Fluctuating Targets

In (2.3.10) to (2.3.14), the occurrences of first threshold crossings were assumed to be independent. This is certainly true for a steady target in noise, but not for fluctuating targets. The question then arises as to how the detectability factor $D_b(n)$ or detection probability P_d for the fluctuating target should be determined. A straightforward approach to determine $D_b(n)$ is to perform the calculation for the steady target, and then increase the SNR by the fluctuation loss for the required value of P_d. However, as in the case of video integration, there is no direct approach to determining P_d from a known average SNR, since the fluctuation and integration losses both depend on the final value of P_d. A solution is to perform the calculation for video integration, and to assume that the additional integration loss from the binary process is the same as that shown in Figure 2.3.3.

Cumulative Detection on Fluctuating Targets

In Section 2.3, we noted that the cumulative detection process was relatively ineffective, when compared to video or binary integration. In the case of fluctuating targets, however, the scan-to-scan build-up of cumulative detection probability provides a significant benefit in reduced fluctuation loss, without the necessity of integration within a moving range cell. For early warning purposes, single-scan probabilities of detection as low as 0.33 can lead to cumulative probability $P_c > 0.9$ within six scans, and even lower values of P_d can be useful. Fluctuation loss is then eliminated or actually converted into a gain factor. Optimum scan and integration procedures for search radar will be discussed further in Chapter 7.

2.5 CONSTANT-FALSE-ALARM-RATE DETECTION

The definition of constant-false-alarm rate (CFAR) in radar is [2.1]:

A property of threshold or gain control devices specially designed to suppress false alarms caused by noise, clutter, or ECM of varying levels.

The use of CFAR is essential if the output data is fed directly to communications channels or automatic data processors. If a human operator is used for detection and track initiation, the person's mental processes will often provide the equivalent of a CFAR function, although even here some circuits to avoid saturation of the display will prove valuable.

The basic CFAR process is to form an estimate of the noise and interference level in the cell where target detection is being carried out, and to set the detection threshold based on that estimate, rather than at some constant level determined in advance of operation. There are two basic approaches to performing this estimate:

(a) The level may be estimated by averaging over adjacent reference cells in range, doppler, angle, or some combination of radar coordinates;
(b) The level may be estimated by averaging the output of the detection cell itself over several scans.

The estimate may involve only the mean level, or higher moments of the pdf of the interference. In some cases, the detection procedure may be designed to be independent of the form of the pdf (nonparametric CFAR).

Cell-Averaging CFAR

Averaging over reference cells is often done in range, because that coordinate has better resolution than angles, in most radars. A typical cell-averaging CFAR, with analog implementation, is shown in Figure 2.5.1. The envelope-detected outputs of m adjacent range cells are available simultaneously in a tapped delay line, and the center tap represents the detection cell. The m reference taps are averaged to form an estimate w of the local noise and interference within the radar beam, and the ratio of the detection cell amplitude x^S to this average is used as the video output. In effect, the threshold level E_t is scaled to the estimate of local noise, rather than to an *a priori* value as in previous discussions.

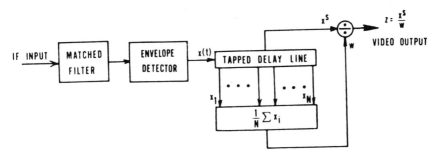

Figure 2.5.1 Range cell averaging CFAR [2.20].

NOTE: The effective number of reference samples, m_{eff}, is calculated as follows, where m is the number of taps in Figure 2.5.1.

For m-cell averaging, Figure 2.5.1:

$m_{eff} = m$, for square-law detector,
$= (m + k)/(1 + k)$, where $k = 0.09$ for envelope detector,
$k = 0.65$ for log detector.

When using greatest-of selection, Figure 2.5.3:

$k = 0.37$ for square-law detector,
$= 0.5$ for envelope detector,
$= 1.26$ for log detector.

For CFAR using hard limiting, add 1 dB limiting loss, and

$m_{eff} = (B_w/B_n) - 1$, for Dicke-fix receiver,
$= B_\tau - 1$, for dispersive or pulse compression CFAR.

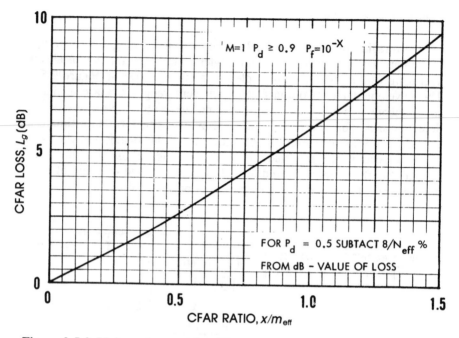

Figure 2.5.2 Universal curve for CFAR loss in single-hit detection, for steady or Rayleigh target [2.20].

If the number of reference cells is large enough, and the actual noise level is constant, the estimate will be accurate enough that detection performance matches the results presented above. In practice, however, it is necessary to restrict the number of cells so that the threshold follows a varying noise level, and the estimate will in turn vary around its proper value. To avoid excessive false alarms when the estimate deviates on the low side, the threshold multiplier $E_t/\sqrt{N}$ must be increased, resulting in a decrease in detection probability. The increase in required SNR, to restore the desired P_d, is called the *CFAR loss*, and this is a function of m, P_f, and to a lesser extent of other parameters. Gregers Hansen [2.20] has presented a universal CFAR loss curve, Figure 2.5.2, which gives the loss, L_g, as a function of a CFAR ratio x/m_{eff}, where $x = -\log P_{fa}$. Rules for determining m_{eff} are given in the note under the figure. In most cases, it is impractical to attempt control of P_{fa} below about 10^{-4}, in which case $m = 8$ results in $L_g < 3$ dB.

When n-pulse integration is used, the number of reference samples for noise and most types of jamming will be increased to nm. As a result, fewer reference cells may be used without encountering large losses. However, if the interference background is from clutter, it may remain correlated over the n-pulse repetition intervals, and this benefit disappears. Hence, the usual practice is to use $m > 10$ and to accept the slower response to varying interference levels. In a region containing only noise, the estimate will be improved and L_g will be small.

A special problem arises when there is an abrupt change in interference level. This may happen when a jammer is turned on and off, or when clutter is received in a naval radar from a shore line. The cell-averaging CFAR of Figure 2.5.1 requires that the delay line be filled with the new interference before the threshold is correct. While the second half of the line is being filled, the high level is present in the detection cell, and a burst of false alarms occurs. To avoid this, the delay line can be divided into an early and a late portion, and the larger of the two separate averages can be used to control the threshold (Figure 2.5.3). Only a slight increase in false-alarm rate is then encountered when only half the reference cells are filled. The cost in terms of loss is also slight, as the note under Figure 2.5.2 indicates that m_{eff} is not reduced by the factor of two.

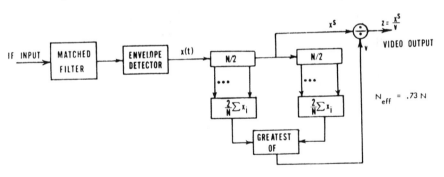

Figure 2.5.3 Greatest-of selection CFAR [2.20].

One of the hazards of using cell-averaging CFAR is that a formation of targets may act to suppress all detections. If the formation occupies enough reference cells, the resulting threshold will be so high that no target is detected. To avoid this, and other suppression effects from random strong signals, the outputs of the reference cells may be edited to remove the largest outputs, and the remaining cells averaged to control the threshold. This, of course, reduces the value of m_{eff} and increases the false-alarm rate.

In doppler radar, the cell averaging may be performed in doppler filters rather than in range gates, or both range and doppler may be used to increase the value of m without extending too far in any direction. The well known Dicke-fix receiver, which protects against certain types of jamming, is another special case of cell-averaging CFAR, in which reference samples are taken from frequencies over a band B_w surrounding the signal bandwidth B_n. Another form of CFAR uses hard limiting of a dispersed pulse of width τ, followed by pulse compression to an output pulsewidth $1/B$. This is equivalent to a range-cell-averaging CFAR.

Time-Averaging CFAR

With the advent of digital processing and large memory capacities, it has become possible to store large amounts of data in clutter maps, where the interference in each resolution cell can be averaged for several scans of the antenna. Instead of averaging over adjacent cells, to estimate the interference level in the detection cell, that cell itself can be observed over several scans, and an appropriate threshold set without regard to other cells [2.21]. In ground clutter, this is an especially advantageous procedure, because the spatial correlation of clutter amplitudes from one cell to the next may be small. Clearly, the technique is not adequate for ECM, which may change within the period of a single scan. In addition, the time-averaging technique may lead to suppression of slowly moving targets, unless high spatial resolution or long scan periods are used.

Two-Parameter CFAR

In the presence of clutter which is not Rayleigh distributed, an evaluation of average level may not lead to a high enough threshold to prevent large numbers of false alarms. False-alarm control in such cases requires that the second moment of the interference pdf be measured and used to adjust the threshold multiplier [2.22]. Unfortunately, when such a procedure is followed, the detection performance is seriously degraded. The correct threshold setting for constant P_{fa} on clutter with log-normal or Weibull distributions (see Chapter 3) leads to a *clutter distribution loss*, shown in Figure 2.5.4. Land clutter, viewed from a surface-based radar, may have a log-normal deviation $\sigma_y = 15$ dB, or, equivalently, a Weibull parameter $a = 4.5$ (for which the mean-to-median ratio is about 28 dB).

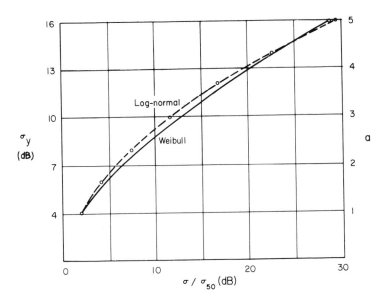

Figure 2.5.4(a) Ratio of mean to median for log-normal and Weibull distributed clutter.

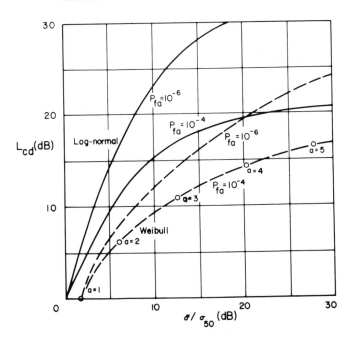

Figure 2.5.4(b) Clutter distribution loss for log-normal and Weibull clutter distributions.

The corresponding L_{cd} is in the neighborhood of 20 dB for $P_{fa} = 10^{-4}$, and this penalty is intolerable for most radars. As a result, the two-parameter CFAR does not provide radar performance much better than a simple blanker, operating in regions containing significant land clutter.

Nonparametric CFAR

Another approach to CFAR in the presence of non-Rayleigh interference is the nonparametric CFAR [2.21]. This technique also has the problem of serious losses, but it has found application in the simple form of a median detector [2.23]. The median detector is simply a binary integrator with the second threshold $m = n/2$. Since this is not far from the optimum value for detection in thermal noise, detection performance is not seriously degraded. The problem is to obtain an adequate number of independent clutter samples, without extending the reference cells too far from the detection cell. Other implementations of nonparametric CFAR have been used effectively, but the resulting performance is not greatly different from that of more conventional processes. In general, the more robust the CFAR action in varied environments, the greater is the loss in target detection.

2.6 LOSS FACTORS IN DETECTION

In Chapter 1, several loss factors in the radar equation were discussed. First, in Section 1.2, factors which attenuated or otherwise reduced the received target signal were listed. To these factors were added, in determining the detectability factor D_x, loss components introduced in signal processing: the filter matching factor, M; integration loss, L_i; collapsing loss, L_c; fluctuation loss, L_f; and other processing losses lumped together as L_x. In Section 1.3, a scan distribution loss, L_d, was added, describing the inefficiency in the cumulative detection process. These losses have been described in more detail in the preceding sections of this chapter, with procedures for their quantitative evaluation. In addition, the loss components of L_x resulting from binary integration, integrator weighting, and CFAR detection have been discussed. These will not be the only processing losses, but their effects on detection performance will be quite significant, and measures to minimize them will be as important as steps to achieve higher power and greater antenna gain in modern radar systems.

In listing a loss budget, it is important to separate factors which attenuate the RF signal and clutter, affecting the S/N and S/J ratios, but not the S/C ratio. These losses are placed in the denominator of the radar equation as loss factors L, or are included in determining T_s and the antenna gain terms G_t and G_r. Another type of loss acts to degrade the detection efficiency in the later stages of the receiver and signal processor, and this loss is included in determining the detectability factors D_x and D_{xc}. In addition, there are certain losses that are present only in clutter, which increase D_{xc} relative to D_x. For example, if two-parameter CFAR is used, the clutter distribution loss L_{cd} (Figure 2.5.4(b)) will apply as a function of the actual clutter distribution presented to the CFAR circuit. When clutter is correlated over the n samples to be integrated, less integration gain will be obtained, and the gain $G_i(n)$ for noise will be replaced by $G_i(n_c)$. The effective number of clutter samples n_c is given by (2.4.17), in which t_c is now the correlation time of the clutter as it is presented to the integrator, and f_c is the correlation frequency of the clutter.

The increase in D_{xc} caused by these factors may be partially offset by absence of a matching factor M, if that factor resulted from excess noise bandwidth, $B_n\tau > 1$. As a result, we may write

$$D_{xc}(n) = D_x(n) \frac{L_{cd}G_i(n)}{MG_i(n_c)} \tag{2.6.1}$$

The only situation for which the two detectability factors are equal is one in which (a) frequency agility or doppler processing (MTI or pulsed doppler) has uncorrelated clutter residue at its output, (b) the detection threshold is set according to noise statistics, and (c) the clutter does not suppress signals as a result of receiver nonlinearity.

Approaches to loss minimization will be discussed in the chapters on search radar and tracking radar (Chapters 7 and 8).

REFERENCES

[2.1]   IEEE Standard Dictionary of Electrical and Electronic Terms, *ANSI/IEEE Std. 100–1984*, 1984.

[2.2]   S.O. Rice, Mathematical analysis of random noise, *Bell System Technical Journal* **23**, No. 3, July 1944, pp. 282–332, and **24**, No. 1, January 1945, pp. 46–156.

[2.3]   M. Abramowitz and I.A. Stegun (eds.), *Handbook of Mathematical Functions*, National Bureau of Standards, June 1964, pp. 932–933.

[2.4]   D.K. Barton, Simple procedures for radar detection calculation, *IEEE Trans.* **AES-5**, No. 5, September 1969, pp. 837–846; reprinted in D.K. Barton (ed.), *Radars*, Vol. 2, *The Radar Equation*,

Artech House, 1974.

[2.5]   D.K. Barton and P. Peterson, Radar detection hinges on probabilities, *Microwaves,* September 1979, pp. 70–75.

[2.6]   D.O. North, An analysis of the factors which determine signal/noise discrimination in pulsed carrier systems, *RCA Labs. Tech. Rep. PTR-6C,* June 23, 1943; reprinted in *Proc. IEEE* **51**, No. 7, July 1963, pp. 1015–1027.

[2.7]   J.V. DiFranco and W.L. Rubin, *Radar Detection,* Prentice-Hall, 1968; Artech House, 1980.

[2.8]   J.I. Marcum, A statistical theory of target detection by pulsed radar, *Rand Research Memo. RM-754,* December 1947, with Appendix, *RM-753,* July 1948; reprinted in *IRE Trans.* **IT-6**, No. 2, April 1960.

[2.9]   W.A. Skillman, *Radar Calculations Using the TI-59 Programmable Calculator,* Artech House, 1983.

[2.10] W.A. Skillman, *Radar Calculations Using Personal Computers,* Artech House, 1984.

[2.11] M. Schwartz, A coincidence procedure for signal detection, *IRE Trans.* **IT-2**, No. 4, December 1956, pp. 135–139.

[2.12] B.H. Cantrell and G.V. Trunk, Angular accuracy of a scanning radar employing a two-pole filter, *IEEE Trans.* **AES-9**, No. 5, September 1973, pp. 649–653.

[2.13] L.V. Blake, The effective number of pulses per beamwidth for a scanning radar, *Proc. IRE* **41**, No. 6, June 1953, pp. 770–774.

[2.14] W.M. Hall, Antenna beam-shape factor in scanning radars, *IEEE Trans.* **AES-4**, No. 3, May 1968, pp. 402–409.

[2.15] G.V. Trunk, Comparison of the collapsing losses in linear and square-law detectors, *Proc. IEEE* **60**, No. 6, June 1972, pp. 743–744.

[2.16] P. Swerling, Probability of detection for fluctuating targets, *Rand Research Memo. RM-1217,* March 17, 1954; reprinted in *IRE Trans.* **IT-6**, No. 2, April 1960.

[2.17] P.A. Bakut, *et al., Problems in the Statistical Theory of Radar,* Vol. 1, Soviet Radio Publishing House, 1963 (in Russian; translation available as AD608462).

[2.18] D.P. Meyer and H.A. Mayer, *Radar Target Detection,* Academic Press, 1973.

[2.19] I. Kanter, Exact detection probability for partially correlated Rayleigh targets, *IEEE Trans.* **AES-22**, No. 2, March 1986, pp. 184–196.

[2.20] V. Gregers Hansen, Constant false alarm rate processing in search radars, *Radar—Present and Future,* IEE Conf. Pub. No. 105, October 1973, pp. 325–332.

[2.21] A. Farina and F.A. Studer, A review of CFAR detection tech-
        niques in radar systems, *Microwave J.*, September 1986, pp.
        115–128.
[2.22] M. Sekine, *el al.*, Suppression of Weibull-distributed clutter using
        a cell-averaging log/CFAR receiver, *IEEE Trans.* **AES-14**, No. 5,
        September 1978, pp. 823–826.
[2.23] G.V. Trunk and S.F. George, Detection of targets in non-Gaus-
        sian sea clutter, *IEEE Trans.* **AES-6**, No. 5, September 1970, pp.
        620–628.

Chapter 3
Targets and Interference

The performance of a radar system in detection and measurement can be analyzed and predicted with confidence only when the target characteristics agree with their predicted values. Since a wide variety of targets may be encountered in use of radar systems, we must devise target descriptions of sufficient latitude to accommodate wide variation in characteristics of individual targets at specific times. The importance of the signal-to-noise power ratio and energy ratio has been shown in the previous two chapters. Knowledge of the radar cross section, or backscattering coefficient, of the target is essential in any such calculation. Although there are a few cases in which this coefficient is a constant, it will generally be found to vary considerably for each target as the aspect angle changes, as internal motions of the target change its shape, and as radar frequency is varied. These changes force us to use statistical methods to describe the radar target. Similarly, the interfering signals backscattered from clutter (objects other than the desired target) will be described statistically, and as functions of radar-clutter path geometry, radar wavelength, and type of clutter.

In measurement of target position and velocity, there are factors other than echo amplitude which are also important. The echo signal may consist of many components of energy scattered from points distributed over the surface of the target. The amplitude and phase of each component will vary as a function of time, aspect angle, and radar frequency, and the interaction of these components will affect the radar measurement process. This chapter will summarize these target characteristics and describe them in ways which will permit accurate analysis of radar system performance.

3.1   DEFINITION OF RADAR CROSS SECTION

The first important characteristic of any radar target is a measure of its ability to reflect energy to the radar receiving antenna. The parameter used to describe this ability is the *radar cross section* or RCS of the target, also termed the *effective echoing area* or the *backscattering* (or forward-scattering, bistatic-scattering) *coefficient*. This is defined [3.1, p. 716] as

> 4π times the ratio of the power per unit solid angle scattered in a specified direction to the power per unit area in a plane wave incident on the scatterer from a specified direction . . . Three cases are distinguished: (1) monostatic or backscattering RCS when the incident or pertinent scattering directions are coincident but opposite in sense, (2) forward-scattering RCS when the two directions and senses are the same, and (3) bistatic RCS when the two directions are different.

Thus, if the target were to scatter power uniformly over all angles, its RCS would be equal to the area from which power was extracted from the incident wave. Since the sphere has this ability to scatter isotropically, it is convenient to interpret the RCS in terms of an equivalent sphere.

Equivalent Sphere

A sphere with radius a that is large compared to the radar wavelength will intercept power contained in an area πa^2 of the incident wave, and will scatter this power uniformly over 4π steradians of solid angle. The RCS of the sphere, using the preceding definition, is therefore equal to its projected area πa^2 (if we neglect the forward-scattering RCS, to be discussed later). That portion of the surface which actually returns power to a monostatic radar is located close to the point where the wave first strikes the sphere, where the surface lies parallel to the wavefront. A physical interpretation of the RCS, for sufficiently large objects, is then: the RCS of a given target is the projected area of a conducting sphere which would produce the same signal as the target, if placed in the same position relative to the radar. For small targets, the equivalent sphere has a radius so small that a resonance phenomenon sets in, and the physical interpretation becomes inadequate.

The RCS of a sphere is exactly one-fourth the total surface area, and it can be shown that the average cross section of any large object which consists of continuous, curved surfaces will be one-fourth of its total surface area. This relationship can be used in estimating the RCS of elongated or irregular bodies, provided that the irregularities are not in the form of sharp edges or structures which are resonant at the radar wavelength. It should be kept in mind, however, that the average RCS may represent largely the contribution of one or more narrow lobes of large peak amplitude, leaving very low values over most aspect angles.

Equivalent Antenna

Another physical interpretation of RCS, which can be derived from the equivalent sphere, relates it to the flat-plate area A_p, which would provide an equal return signal if located at the target position and oriented normal to the radar beam, for monostatic radar (or along the bisector of the two paths in bistatic radar). Since no portion of the sphere is actually flat, we shall take that portion which lies within a distance $\lambda/4\pi$ of the wavefront as it arrives at the sphere (Figure 3.1.1). This is a circular disk of radius $a_1 = \sqrt{a\lambda/2\pi}$, the area of which is $A_p = \pi a_1^2 = a\lambda/2$. If the nearly flat surface is considered as an antenna, the power intercepted will be reradiated toward the radar with gain $G = 4\pi A_p/\lambda^2$. The resulting RCS is given by the product of gain times interception area:

$$\sigma = A_p G = (a\lambda/2)(4\pi a\lambda/2\lambda^2) = \pi a^2 \qquad (3.1.1)$$

It can be shown that the area lying within the distance $\lambda/4\pi$ of the wavefront when it first makes contact with any smooth target will give the approximate flat-plate area A_p for that object, from which RCS can be calculated by using (3.1.1).

The analysis based on equivalent flat-plate area demonstrates the use of antenna theory to describe the reradiation properties of targets. Whether the power which illuminates the surfaces of the target originates in the radar or at some other point makes no difference in the resulting radiation pattern, so long as the proper phase relationships are established. Thus, a rectangular flat plate oriented normal to the radar beam can be equated to a uniformly illuminated aperture with constant phase, producing a radiation lobe directed at the radar. As the aspect angle changes by θ, the main RCS lobe will move in space by 2θ, so this and the sidelobes will be half as wide in the RCS pattern as in the antenna pattern for the same size aperture. The RCS for this normal incidence is

$$\sigma(0) = AG = 4\pi A^2/\lambda^2 \qquad (3.1.2)$$

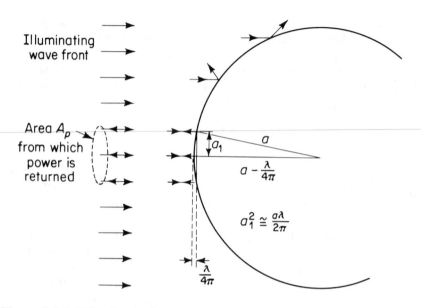

Figure 3.1.1 Effective flat-plate area of sphere.

and the RCS pattern will follow the $(\sin^2 x)/x^2$ pattern of a rectangular antenna

$$\sigma(\theta) = \sigma(0) \left[\frac{\sin(kL \sin\theta)}{kL \sin\theta} \cos\theta \right]^2, \quad (k = 2\pi/\lambda) \qquad (3.1.3)$$

This pattern is shown in Figure 3.1.2, with a main-lobe width $\Delta = 0.44\lambda/L$ between its half-power points, $2\Delta_0 = \lambda/L$ between nulls, and sidelobe widths $\Delta_0 = \lambda/2L$ near broadside. For other shaped plates, the equivalent antenna patterns can be used, with the angle scale doubled for reflection.

3.2   RADAR CROSS SECTION   OF SIMPLE OBJECTS

Equations describing the peak RCS, lobe widths, and total numbers of lobes presented by the target during a complete rotation are given in Table 3.1. Except for the dipole, these equations are subject to the limitation that target dimensions or radii of curvature must be larger than the wavelength. In some cases, more complex objects may be represented by combinations of these shapes, approximating the actual surface, at least to get an approximation of RCS characteristics. A cylinder, for example, may be represented by the lobe structure of Figure 3.1.2, uniformly distributed around the longitudinal axis and dependent on the length L, plus

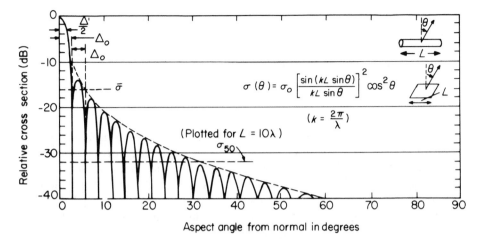

Figure 3.1.2 RCS lobe pattern of flat plate or cylinder.

Table 3.1 RCS of Simple Bodies

Object	$\sigma_{\max}$	$\sigma_{\min}$	Number of lobes	Major lobe width
Sphere	πa^2	πa^2	1	2π
Ellipsoid ($k = a/b$)	πa^2	$\dfrac{\pi b^2}{k^2}$	2	$\approx \dfrac{b}{a}$
Cylinder	$\dfrac{2\pi a L^2}{\lambda}$	null	$\dfrac{8L}{\lambda}$	$\dfrac{\lambda}{L}$
Flat plate	$\dfrac{4\pi A^2}{\lambda^2}$	null	$\dfrac{8L}{\lambda}$	$\dfrac{\lambda}{L}$
Dipole	$0.88\lambda^2$	null	2	$\dfrac{\pi}{2}$
Infinite cone (half-angle α)	$\dfrac{\lambda^2 \tan^4 \alpha}{16\pi}$	null		
Convex surface	$\pi a_1 a_2$			
Square corner reflector	$\dfrac{12\pi a^4}{\lambda^2}$		4	$\dfrac{\pi}{4}$
Triangular corner reflector	$\dfrac{4\pi a^4}{3\lambda^2}$		4	$\dfrac{\pi}{4}$

an end-lobe structure caused by the flat ends and dependent on the diameter D. In the region between side and end aspect, these two patterns will interfere with each other, producing irregular lobes which depend on the ratio of length to diameter.

Wavelength Dependence of RCS

The peak RCS values shown in Table 3.1 depend in different ways on the radar wavelength λ. The RCS of flat plates and corner reflectors (which appear as flat plates oriented normal to any radar aspect angle) vary as λ^{-2}. The RCS of a cylinder varies as λ^{-1}, and those of large doubly curved objects are independent of wavelength. Dipoles and pointed objects have RCS varying as λ^2. It has also been shown [3.2, pp. 178–179] that curved edges have RCS varying as λ, while discontinuities in radii of curvature have values varying as higher powers of λ.

The corner reflector, listed in the table, provides an interesting illustration of how RCS can be much larger than the physical area of a target. The action of the three orthogonal surfaces in the reflector is such as to reflect incident energy back toward the sources, regardless of aspect angle (Figure 3.2.1). In effect, the corner reflector is a flat plate oriented always normal to the incident ray, but with varying size, depending on how much of the ray experiences the triple bounce. In certain aspects, double or single bounce may replace triple bounce, giving the sharp lobes of the figure, but over most angles there is a broad, triple-bounce lobe. At the center of this lobe, the RCS (for a triangular corner) will approximate $4a^4/\lambda^2$. Such reflectors, mounted on navigation buoys, typically have $a = 0.5$ m, and at X-band ($\lambda = 0.03$ m) will give RCS = 280 m^2. Yet, the target fits within a sphere having a diameter of only one meter.

At the other extreme is the typical missile nose cone, the diameter of which may also be 1 m, length 2 m, and cone angle about 30°. The tip RCS, corresponding to the infinite cone in Table 3.1, is about 10^{-6} m^2 at X-band. With a rounded base, having discontinuities in curvature as it merges with the cone, RCS near $0.1a\lambda$ may appear (0.0015 m^2 at X-band), while with a flat base, much higher values will be seen, even near nose aspect. Only at aspects near normal to the conical surface, and to the rear, will there be RCS near or above the physical area of the target (Figure 3.2.2).

Resonant phenomena also introduce wavelength dependence, even in the sphere which we have previously considered as having constant RCS. Figure 3.2.3 shows the RCS dependence of the conducting sphere on the ratio of circumference to wavelength. Three regions can be identified:

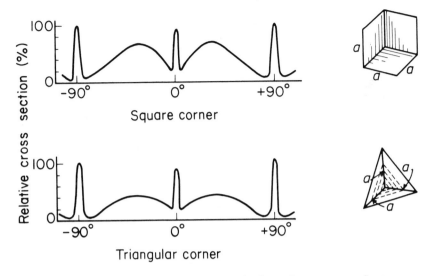

Figure 3.2.1 RCS patterns of square and triangular corner reflectors.

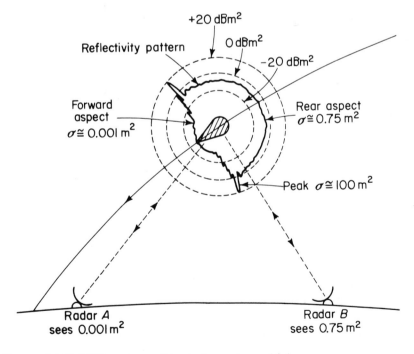

Figure 3.2.2 RCS pattern of typical reentry vehicle.

(1) Optical region (large spheres, $2\pi a/\lambda > 10$), where the RCS closely approaches the projected area;
(2) Resonant region ($0.5 < 2\pi a/\lambda < 10$), where large oscillations in RCS occur, reaching a peak of $3.7\pi a^2$ when $2\pi a = \lambda$;
(3) The Rayleigh region (small spheres, $2\pi a/\lambda < 0.5$), where the RCS drops rapidly.

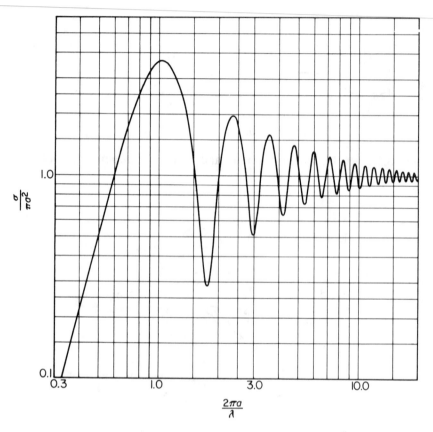

Figure 3.2.3 Normalized RCS of conducting sphere.

   Physically, the resonant region can be explained by the interference between the incident wave and the creeping wave, which circles the sphere and either adds to or subtracts from the total field at the leading surface. The Rayleigh region corresponds to wavelengths so long that only a fraction of the field gradient can excite currents on the surface.

Metallic spheres are often used in radar calibration. A typical calibration sphere, designed to be carried aloft on a small weather balloon, has a radius of 7.6 cm and an optical RCS of 0.018 m², or −17.4 dBm². At C-band, λ = 5.3 cm, the ratio $2\pi a/\lambda$ = 9, and the RCS will be within about 15% (or 0.7 dB) of its optical value. For accurate calibration, the exact wavelength would have to be known, and Figure 3.2.3 used to correct the optical RCS at C-band or longer wavelengths, while at X-band and shorter wavelengths no correction would be required.

The small sphere, in the Rayleigh region, exhibits RCS varying as the sixth power of its radius, and inversely as the fourth power of wavelength:

$$\sigma/\pi a^2 = 14{,}000\ (a/\lambda)^4 = 9(2\pi a/\lambda)^4, \quad (\lambda > 10a) \tag{3.2.1}$$

This dependence explains the steep variation of rain RCS with wavelength in the microwave bands. The small circular disk has a similar dependence, but a smaller RCS, due to its reduced volume of interaction with the incident wave:

$$\sigma/\pi a^2 = 1125(a/\lambda)^4 = 0.72(2\pi a/\lambda)^4, \quad (\lambda > 10a) \tag{3.2.2}$$

As an example of RCS of a small sphere, we can consider a test target sometimes used with sensitive CW radars. An air rifle pellet (a = 2.2 mm) fired near the radar will have enough velocity to pass the clutter rejection filter. At X-band, λ = 0.03m, its RCS, from (3.2.1), will be 6×10^{-6} m². An individual raindrop will have an even smaller RCS value, in most cases, but huge numbers of them will lie within the radar resolution cell.

Polarization Dependence of RCS

Long cylinders also show a resonant phenomenon that depends on the circumference and the incident polarization, as illustrated in Figure 3.2.4. When the E-vector is perpendicular to the axis, there is a pronounced resonance at $2\pi a/\lambda \approx 0.8$. For parallel polarization, there is a steady increase in the ratio of RCS to its normal cylinder value $2\pi a L^2/\lambda$, although the absolute RCS decreases. Short, thin cylinders have resonances that depend on their length, as shown in Figure 3.2.5. The RCS here is normalized to the square of the resonant wavelength, so increasing values of $2\pi L/\lambda$ correspond to reduced wavelengths, higher frequencies. The main resonant peak corresponds to the value $0.88\lambda^2$ given in Table 3.1, and resonances at higher frequencies are reduced by factors of 7, 11, and more.

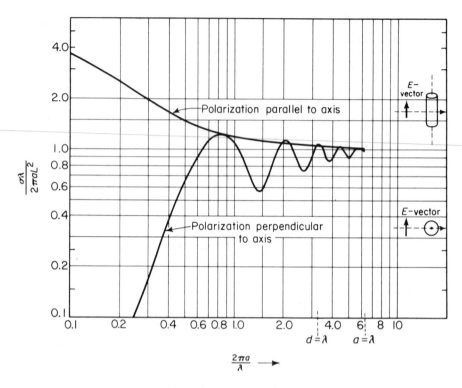

Figure 3.2.4 Wavelength dependence of RCS of large cylinder.

Thin dipoles will be discussed further when we turn to chaff clutter, later in this chapter.

Target RCS must be defined and evaluated as a function of transmitted and received polarizations, as well as aspect angle and wavelength. The thin cylinder will scatter only with polarization along its axis. A sharp edge on an extended surface will scatter with similar linear polarization. A sphere or smooth surface will reflect waves with the same polarization as received, but in the case of circular polarization the sense will be reversed. Hence, the received linear polarization will be the same as transmitted, giving the full RCS value shown in Table 3.1, while the component received with the circular polarization as transmitted will be near zero (except for any edge effects). The echo received in circular polarization of opposite sense will be that given in the table, for smooth surfaces and corner reflectors.

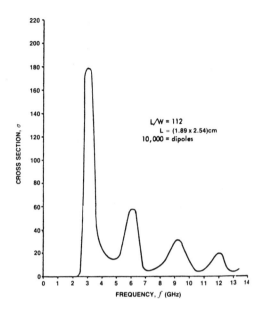

Figure 3.2.5 Wavelength dependence of RCS of dipole.

3.3 RADAR CROSS SECTION OF COMPLEX TARGETS

In the case of simple objects, we have seen that RCS has no single relationship to the physical area of the target. Most targets of interest present an even more complicated relationship, so that attempts to write equations based on physical dimensions are seldom successful. Prediction techniques based on elaborate computer programs can lead to useful results when theories of physical optics and the geometrical and physical theories of diffraction are applied [3.2, Chapter 5]. In most cases, however, full-scale or small models of the actual targets are measured [3.3] to determine RCS, or at least to validate the results of prediction programs.

A small aircraft target ($L = 15$ m), viewed by X-band radar ($\lambda = 0.03$ m), will have $8L/\lambda = 4000$ lobes in a great circle cut passing through the nose and tail. Over all solid angles, there will be roughly $8(L/\lambda)^2 = 2 \times 10^6$ lobes, and a measurement program would have to take at least four data points per lobe for an accurate description of the pattern. Such measurements would have to be made at many frequencies, and with polarizations of interest, leading to a huge data base, for only a single radar band. Even if measured and stored on tape, such a data base would be of limited value to most of those involved in radar performance analysis, unless it could be summarized in statistical form. This has been done on many targets of interest, leading to simple characterizations: mean RCS and probability density function (pdf), over different sectors of aspect angle.

Swerling Target Models

In his classic paper on target modeling [3.4], Peter Swerling established four statistical models.

Cases 1 and 2: Exponential pdf of RCS, Rayleigh pdf of signal amplitude.

$$\mathrm{d}P_\sigma = (1/\overline{\sigma}) \exp(-\sigma/\overline{\sigma}) \, \mathrm{d}\sigma \tag{3.3.1}$$

$$\mathrm{d}P_e = (E_s/S) \exp(-E_s^2/2S) \, \mathrm{d}E_s \tag{3.3.2}$$

where $S = \overline{E_s^2}$ is the average signal power.

Cases 3 and 4: Chi-square, four-degree-of-freedom pdf of signal amplitude.

$$\mathrm{d}P_\sigma = (4\sigma/\overline{\sigma^2}) \exp(-2\sigma/\overline{\sigma}) \, \mathrm{d}\sigma \tag{3.3.3}$$

Cases 1 and 3 refer to slow fluctuation, in which the signal sample remains constant over integration of n pulses during observation time t_o, while Cases 2 and 4 have independent signal samples on each of n pulses. The detection performance corresponding to these four cases was discussed in Chapter 2.

Physically, the Case 1 or 2 (Rayleigh) model corresponds to a target in which many scattering points are added with random phases. The amplitude pdf has the same form as that of the random noise envelope (Figure 2.1.2(b)). When plotted in terms of the decibel value of RCS, the pdf of Figure 3.3.1 results. Values more than 10 dB above the mean are seldom seen, while fades of 10 dB or more below the mean occur with about 10% probability. Virtually all targets in the real world are very close to this model, since many scattering sources are involved.

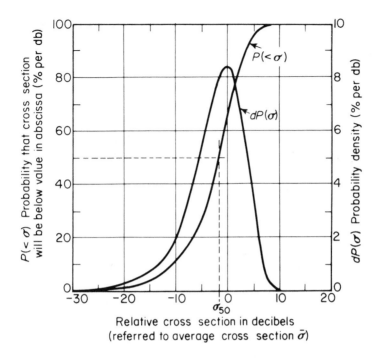

Figure 3.3.1 Log plot of Rayleigh distribution.

The Swerling Case 3 target has a pdf which is less broadly spread than the Case 1 about its mean value. This would lead to reduce fading and fluctuation loss, but such a pdf does not correspond to real targets that have been measured. In Figure 3.3.2, we show measured data [3.5] on two types of aircraft at L, S, and X-band. Within each 10° azimuth sector, over several flight segments, data were fitted with a chi-square pdf having $2K$ degrees of freedom, with K and mean RCS plotted as the output in the figure. For both aircraft, $K \approx 1.0$, corresponding to the Rayleigh target, at all aspect angles and frequencies. The variation in mean with aspect angle is expected, with high values near broadside and approximately constant, low values at other aspects. It should be noted that the large increase in mean for sectors near broadside makes it difficult to model the target over all aspect angles with a single pdf, unless distributions with wider spreads than the Rayleigh are used. However, in most cases it is unnecessary (and undesirable) to include the broadside peaks in the statistical description, since to do so would distort the probabilities of detection over the other angles that are of greatest interest. Numerous targets have been measured, with results similar to those shown in Figure 3.3.2, leading to the conclusion that the Case 3 model cannot be applied to real targets except as a result of dual frequency diversity operation.

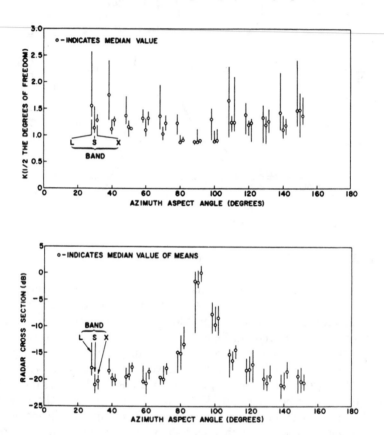

Figure 3.3.2(a) Mean RCS and K for jet fighter in three radar bands [3.5].

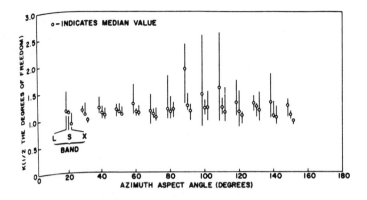

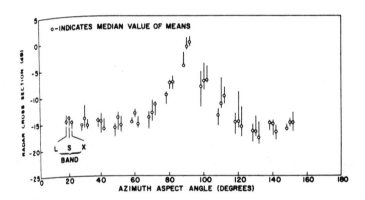

Figure 3.3.2(b) Mean RCS and K for propeller aircraft in three radar bands [3.5].

Generalized Target Models

In special analyses, it may be desirable to use more general target models, based on the chi-square or log-normal families of distributions. The pdf of the first is given, for $2K$ degrees of freedom, by

$$dP_\sigma \left[(K/\bar{\sigma})/(K-1)!\right](K\sigma/\bar{\sigma})^{K-1} \exp(-K\sigma/\bar{\sigma}) \, d\sigma \qquad (3.3.4)$$

As K increases, the pdf becomes narrower about the mean, and as $K \to \infty$ the steady target (Case 0) is obtained. Going in the other direction, Weinstock [3.6] has shown that targets with even greater fading than the Rayleigh target can be modeled with $K < 1$.

The log-normal model gives a normal pdf of the decibel value of RCS, with a standard deviation σ_y in dB. An approximation for this value can be obtained from measured values of the median RCS, σ_{50}, and the mean:

$$10 \log \overline{\sigma} = 10 \log \sigma_{50} + 0.115 \, \sigma_y^2 \qquad (3.3.5)$$

Since most targets remain close to the Rayleigh pdf, there is little application of these generalized distributions, except in describing radar clutter.

Target Spectrum and Correlation Time

The rate at which RCS changes is important in both detection and measurement applications. This rate, and the corresponding target correlation time t_c, will determine radar performance. When signals are integrated over an observation time t_o, the resulting output pdf will depend on whether only one sample ($t_c \gg t_o$) or several independent samples ($t_c < t_o$) have been observed, and fading of the output will be most pronounced with only a single sample (Case 1). When a cumulative probability of detection is to be determined, the usual assumption is that the successive trials at intervals t_s are independent ($t_c < t_s$). In sequentially sampled measurement processes, the best performance is obtained if the target remains constant during the measurement ($t_c \gg t_o$).

The fluctuation rate of the target is the rate at which RCS lobes pass across the line of sight to the radar. For a target having scatterers that are uniformly distributed over the span L_x across the line of sight, the spectrum of the received signal will be rectangular with a width:

$$f_{\max} = 2\omega_a L_x / \lambda \qquad (3.3.6)$$

where ω_a is the rate of change of aspect angle. The correlation function will be

$$\rho(t) = \frac{\sin(\pi f_{\max} t)}{\pi f_{\max} t} \qquad (3.3.7)$$

After envelope detection, the spectrum will be the self-convolution of the rectangular input spectrum:

$$S(f) = S(0)(1 - |f|/f_{max}), \quad |f| < f_{max} \qquad (3.3.8)$$

and the correlation function will be the square of (3.3.7). If we characterize the correlation time as the value at which this $(\sin x)/x$ function drops to a null, then

$$t_c = 1/f_{max} = \lambda/2\omega_a L_x \qquad (3.3.8)$$

If the scattering sources are not uniformly distributed, then some equivalent L_{nx} can be defined to characterize the target fluctuation, for use in these equations [3.7, p. 174].

3.4 SPATIAL DISTRIBUTION OF CROSS SECTION

Radar measurement of target position and velocity is based on the assumption that some point on the target may be defined as the position reference. Targets which are small with respect to the radar wavelength present no problem in this regard, since they are seen as small point sources in all coordinates. Larger targets, however, show significant shifts in all coordinates, including doppler frequency, as a function of changing aspect angles. As with target amplitude or RCS, these shifting apparent positions are best modeled statistically [3.8].

Target Glint

Glint is defined [3.1, p. 391] as

The inherent random component of error in measurement of position or doppler frequency of a complex target due to interference of the reflections from different elements of the target.

Although the interference between signal components which leads to glint will also cause amplitude fluctuations and a related scintillation error in some tracking systems, the two effects are quite different. The scintillation error is a function of the sequential sampling processes used in some measurement systems, and must be considered independently of glint. It is sometimes erroneously assumed that glint merely represents the wandering of the dominant reflection (or radar center of gravity) over the extent of the target, but in fact the glint may lead to a radar pointing angle far beyond the physical span of the target. The pdf of glint error in any radar coordinate is basically described by the Student distribution with two degrees of freedom:

$$dP_\epsilon = \frac{\mu}{2(1 + \mu^2\epsilon^2)^{3/2}} \qquad (3.4.1)$$

where ϵ is the error normalized to the span of the target and μ determines spread of the pdf (larger values of μ giving narrower spreads). This distribution has infinite variance and peak values, although in the practical radar case these will be limited by circuit considerations. It has been shown [3.8, pp. 14, 51] that this pdf applies to all coordinates and all extended targets and that the distribution of scatterers across the target merely changes the parameter μ. For example, for the two-point target $\mu = 1$, for uniform distribution $\mu = \sqrt{3}$, and for targets with greater concentrations of scatterers near the center $\mu > \sqrt{3}$. A physical interpretation of μ is that it is the ratio of target span L to twice the radius of gyration of the scattering voltage sources.

Because the Student distribution cannot be characterized by a standard deviation, it is convenient to match to its main lobe a normal distribution, leading to an equivalent glint standard deviation for uniform scatterers over a target span L given by

$$\sigma_g = L/3 \qquad (3.4.2)$$

The probability that the error will lie beyond the target span, for a normal distribution with this standard deviation, is 0.134, which agrees with the exact calculation for the Student distribution with uniform scatterers. If the scatterer distribution is triangular, the effective value of L will be reduced by $\sqrt{2}$, or the factor in the denominator may be increased to 4.2. For most aircraft, if L is the wingspan or fuselage length, σ_g will be $L/4$ to $L/6$.

The Two-Element Target

It is convenient to illustrate glint effects by considering the two-element target, in which two spheres or other point sources are rigidly connected with spacing L, and are rotated about the line of site. Let the signal amplitude of source 1 be unity, and that of source 2 be k. If the axis connecting the sources makes an angle α with the line of sight, the received signal voltage will be

$$E_s = \sqrt{1 + k^2 + 2k \cos\phi} \qquad (3.4.3)$$

where $\phi = (4\pi L/\lambda) \sin\alpha$ is the phase angle between the two signal components. The glint error with respect to the center point of the target pair, and normalized to the projected target span $L \cos\alpha$ can be written

$$\delta = (1/2)\frac{1 - k^2}{1 + k^2 + 2k\cos\phi} \tag{3.4.4}$$

This is plotted in Figure 3.4.1 as a function of phase angle ϕ for different values of k. The extreme values of glint can be found for $\phi = 0°$ and $180°$. For $k \to 1$, the tracking point lies at the center of the pair, $\delta \approx 0$, for all phase angles except $\phi \to 180°$, at which angle it jumps abruptly to the large peak value $\sigma = (1 + k)/2(1 - k)$. For small k, the tracking point will oscillate sinusoidally about the position of the stronger source, $\delta \approx -k \cos\phi$. An important result is that the average tracking point will lie at the stronger source, with deviations occurring in both directions from this source. As the second target becomes the stronger, the average tracking point shifts abruptly to it, and the interference from the first target causes deviations on both sides of the second.

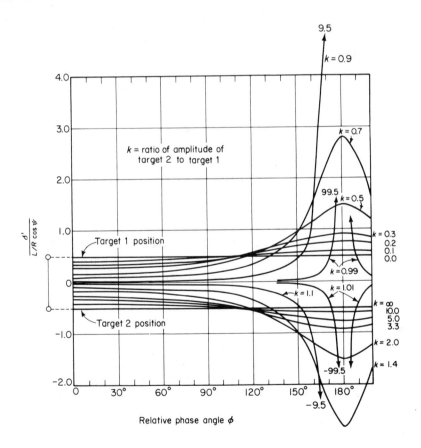

Figure 3.4.1 Glint for the two-element target.

Angle, Range, and Doppler Glint

In angle measurement or tracking, the glint error ϵ will be normalized to the angle L_x/R subtended by the target span at range R. The deviations described in the previous section can be attributed to the ripples in the phase fronts arriving at the radar antenna [3.9], as illustrated in Figure 3.4.2. Abrupt discontinuities in these phase fronts, and corresponding peaks of glint, will occur when the target signal goes through deep fades, between lobes in the RCS pattern. The basic rms glint error in radians will then be $\sigma_{\theta g} \approx L_x/3R$. This value, of course, will be modified by the processing steps and dynamic response of the tracker, to the extent that the circuits may not follow the sharp peaks of glint when the target fades. The value also depends, to some extent, on the beamwidth λ/D of the radar antenna, relative to the subtended angle L_x/R of the target. As the target extent approaches the radar beamwidth, the edges of the target will no longer receive full illumination or weighting by the antenna pattern, and their contributions to glint will be reduced. The condition for target extent approaching beamwidth is equivalent to the antenna width approaching that of the RCS lobe arriving at the radar, providing smoothing of the phase front over the antenna. This effect also applies to doppler glint. It should be remembered, however, that the condition for this glint reduction also corresponds to peak errors approaching the beamwidth, if the antenna is permitted to follow the error signal, and loss of target may result in spite of the reduced value of normalized glint.

In range, the glint error has the same form, with peaks also occuring during signal fades. The rms glint error will be $\sigma_{rg} \approx L_r/3$, where L_r is measured radially, along the radar beam. When the range resolution cell of the radar waveform and signal processing approaches the radial length of the target, glint noise error may be reduced in a way similar to the beamwidth effect in angle. In doppler, the effect is the same, but the spectral span of the target is $f_{max} = 2\omega_a L_x/\lambda$, as given in (3.3.6). The rms glint error is $\sigma_{fg} \approx 2\omega_a L_x/3\lambda$.

Glint Spectrum

The same interference between target scattering sources that causes glint is also the source of amplitude fluctuation, for which the spectral width and correlation time were given in Section 3.3. The radar measurement process, however, introduces an inherent nonlinearity into the error signal channel, causing sharp spikes in the error output and broadening

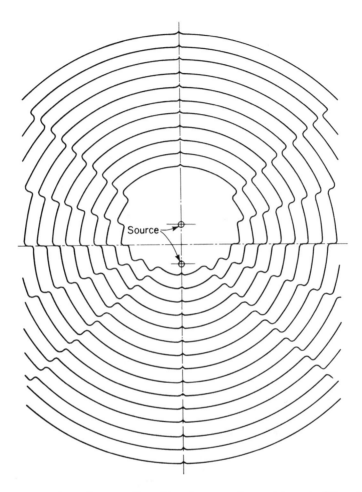

Figure 3.4.2 Phase front radiated by the two-element target. (Courtesy of D.D. Howard, Naval Research Laboratory.)

the spectrum of glint error. This effect is illustrated in Figure 3.4.3 for targets rotating at a uniform rate and with random oscillation. The half-power widths of these glint spectra are about 1.7 times as wide as the corresponding fluctuation spectra. Accordingly, the expressions for fluctuation spectrum and correlation function, (3.3.6) to (3.3.8), must be modified before they are applied to glint. The glint spectrum at the error detector output for uniform target rotation will be approximately triangular with a maximum width $3.4f_{max}$, and the correlation time will be

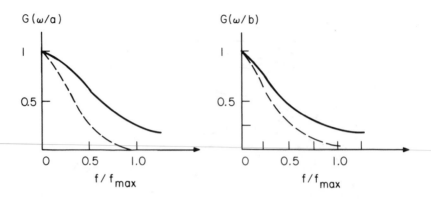

(a) Spectra of glint (solid line) and scintillation (dashed line) for uniformly rotating target.

(b) Spectra of glint (solid line) and scintillation (dashed line) for target with random oscillation.

Figure 3.4.3 Spectra of glint and fluctuation [3.8].

$$t_c = 1/1.7f_{max} = \lambda/3.4\omega_a L_x \tag{3.4.5}$$

When the target rotates randomly, the rms rate σ_a will replace the constant ω_a in calculating an effective f_{max} for these relationships.

3.5 BISTATIC CROSS SECTION

A bistatic radar is one which uses antennas at different locations for transmission and reception. The wave arrives at the target from one direction, and leaves toward the receiver in a different direction, and hence the monostatic RCS is not a measure of the echo signal. In the case of simple objects, the bistatic RCS can be found by considering the target as an antenna, on which the phase is established by the incident wave, and calculating the radiation lobe structure of that antenna over the directions covering the receiving site. For bistatic angles (the angle between the transmitting and receiving paths) of less than about 90°, the monostatic-bistatic hypothesis may be used: the bistatic RCS is the same as the monostatic RCS measured at the bisecting angle. This does not apply, of course, to special shapes like the corner reflector.

For complex objects, the statistics of the bistatic and monostatic RCS are similar. This follows from the monostatic-bistatic hypothesis, at bistatic angles of less than 90°, and can also be applied at larger angles, because the details of the lobe structure change more rapidly than their statistical measures. As a result, for targets such as aircraft, missiles, ships, and land vehicles, the bistatic RCS need only be evaluated separately at angles near 180°, or for targets which include significant corner reflector sources.

Forward Scatter RCS

At bistatic angles approaching 180°, a second scattering phenomenon becomes important in establishing the RCS. The wavefront, immediately after passing any object, contains a hole or shadow of the object. If we were to compare the resulting fields at some distance beyond the object with those that would have existed without the object, the difference would contain the field radiated by the shadow, as an antenna with negative illumination (180° phase) relative to the undisturbed plane wavefront. The result is a forward scatter lobe, shown in Figure 3.5.1. This lobe has a peak RCS given by

$$\sigma_f = 4\pi A^2/\lambda^2 \tag{3.5.1}$$

where A is the area of the shadow, the total projected area of the object as viewed by the transmitter. The lobe pattern is the antenna pattern of a uniformly illuminated flat antenna in the shape of the shadow. At the lobe peak, along the path of the original transmission, this component of RCS can be very large, but will have the much stronger direct transmission as an interference source. The interesting feature of bistatic forward scatter is that a receiver slightly off this line, not receiving significant direct signal from the transmitter, may still receive a significant scattered signal from the target. In the example shown in the figure, where the lobe width is 4°, the receiver can be 2° off the direct path and still get the benefit of a 2000 m² target.

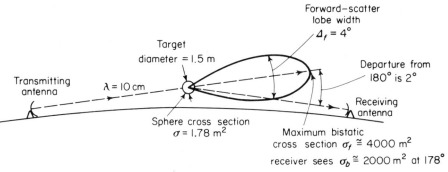

Figure 3.5.1 Forward-scatter RCS lobe.

Extent of Bistatic Enhancement

Because the forward scatter RCS is independent of the shape and surface absorption of the target, the bistatic enhancement of RCS can become important in cases where the monostatic RCS is very low, relative to the physical size of the target. An aircraft may have a projected area of 100 m^2, producing a peak forward lobe of 2×10^6 m^2 at L-band. For a projection which is 20×5 m, the lobe will be 0.023×0.09 radians at its first nulls, with sidelobes spreading over the entire forward hemisphere. Significant signal will be received, even in lobes which are 60 dB below the main lobe. For a rectangular shadow, these sidelobes will follow the $(\sin x)/x$ pattern

$$\sigma_f(\theta) = \sigma_f(0)\ (\sin^2 \pi\theta L/\lambda)/(\pi\theta L/\lambda)^2 \qquad (3.5.2)$$

Outside the main-lobe region, the average sidelobe level about the angle θ will be

$$\overline{\sigma_f(\theta)} = \sigma_f(0)/2(\pi\theta L/\lambda)^2 = 2H^2/\pi\theta^2 \qquad (3.5.3)$$

where H is the height of the shadow, θ is measured in the azimuth plane, and the receiver is at the elevation of the peak lobe, offset only in azimuth. If there is an elevation offset ϕ as well, placing the receiver outside the plane of principal azimuth sidelobes, the average will become

$$\overline{\sigma_f(\theta, \phi)} = \lambda^2/\pi^3(\theta^2 + \phi^2) = (\lambda/\pi\beta)^2/\pi \qquad (3.5.4)$$

where β is the angle away from forward scatter, or π minus the bistatic angle. In our L-band example, this gives an average of $0.0017/\beta^2$ m^2. The enhancement is thus very small, unless the receiver is in one of the principal planes of the lobe pattern. For our L-band example, the principal azimuth sidelobes will have an elevation width $\lambda/H = 0.046$ radians or 2.6°.

Analysis based on a rectangular shadow can be misleading, since most targets are tapered at both ends and will cast a more elliptical shadow. The lobe structure will then follow the Bessel function $2J_1(x)/x$, rather than $(\sin x)/x$. For a circular shadow of diameter D, the average level outside the main lobe will be

$$\overline{\sigma_f(\theta)} = 0.32D\lambda/\theta^3 \qquad (3.5.5)$$

For an elliptical shadow with major axis a and minor axis b,

$$\overline{\sigma_f(\theta)} = 0.64\lambda b^2/a\theta^3 \qquad (3.5.6)$$

$$\overline{\sigma_f(\theta)} = 0.64\lambda a^2/b\phi^3 \qquad (3.5.7)$$

where a is the shadow width in the θ plane and b is the width in the ϕ plane. For an angle in a slant plane, we can use (3.5.6) with an angle θ' given by

$$\theta' = \sqrt{\theta^2 + \phi^2(b^2/a^2)} \qquad (3.5.8)$$

In our example, with $a = 20$ m and $b = 5$ m, and an offset $\theta = \phi = 0.35$ radians (slant angle 0.5 radians), we have $\theta' = 0.36$ radians and $\sigma_f(\theta') = 4$ m^2. This represents a significant enhancement, greater than would be calculated in the slant plane for a rectangular shadow, but less than in the principal planes of a rectangular shadow. There will be difficulty in resolving the target echo from the radar transmission in this region of bistatic enhancement, if the transmitter is within line of sight of the receiver.

3.6 RADAR CLUTTER

Radar clutter is defined as unwanted echoes, typically from the ground, sea, rain or other precipitation, chaff, birds, insects, and aurora [3.1, p. 151]. In some cases, land vehicles may be added to this list. This is the definition of clutter for a typical air surveillance radar, but the distinction between clutter and targets in a radar system depends on the purpose of the radar. A ground-mapping radar, for instance, has the earth's surface as a target. Radar meteorologists regard weather sources as the target, and aircraft constitute a potential clutter source. Clutter may be divided into sources distributed over a surface (land or sea), within a volume (weather or chaff), or concentrated at discrete points (structures, birds, or vehicles). The methods of analyzing the radar response to these three types of clutter will differ.

Whatever the definition for a particular radar application, a number of parameters must be specified if the clutter is to be included in analysis of a radar's performance:

Clutter RCS or density of RCS (reflectivity) over the surface or volume;
Numbers of discrete sources;
Spatial extent and distribution of sources;
Velocity extent and distribution (spectrum);
Wavelength dependence of RCS;

Amplitude distribution (pdf);
Spatial correlation of amplitudes;
Polarization properties.

In addition, it is important to recognize that the propagation factor F_c along the path from radar to clutter will play an important role in establishing the clutter power received by the radar. In many cases, the factor F_c^4 is embedded in the measurement of RCS or its density, and must be removed if the paths to be considered differ from those used in the measurement program.

The procedure for inclusion of distributed clutter in a radar analysis was outlined in Section 1.5. The volume V_c or surface area A_c within the radar resolution cell is found from (1.5.1) or (1.5.12) and multiplied by the reflectivity (or RCS density) η_v or σ^0 to obtain the clutter RCS in each cell. For discrete sources, the RCS is taken directly from a model. The pattern-propagation factor F_c^4 for the clutter is then included in the radar equation to find clutter power or C/S ratio. The radar antenna pattern and resolution considerations will be discussed in Chapters 4 and 5, and here we shall be concerned with the reflectivity, the clutter propagation factor, and the other parameters listed above. From the hundreds of papers, many chapters, and several books which have been published on clutter, only a brief summary can be presented here, with references to the most important background literature.

Surface Clutter

The most convenient starting point for modeling of surface clutter is the *constant-γ* model, in which the surface reflectivity is modeled as

$$\sigma^0 = \gamma \sin\psi \qquad\qquad (3.6.1)$$

where ψ is the grazing angle at the surface and γ is a parameter describing the scattering effectiveness of the surface. This model is in excellent agreement with most measurements, at angles high enough that $F_c \approx 1$ and not too close to 90°. At low grazing angles, the measurements fall below the model because of propagation factor. Near zenith, the measurements rise because of quasispecular reflections from the surface facets.

The propagation factor for paths between the radar antenna at height h_r and the clutter sources on the undulating, random surface of the earth can be written, for low grazing angles, as [3.10, p. 219]:

$$F_c \approx \psi/\psi_c = \psi\,(4\pi\sigma_h/\lambda) = R_1/R \qquad\qquad (3.6.2)$$

where σ_h is the rms surface height deviation from the average height, R is range, and R_1 is the range at which ψ_c is reached. Thus, below the critical grazing angle $\psi_c = \lambda/4\pi\sigma_h$, or beyond range R_1, the measured value $\sigma^0 F_c^4$ should vary as ψ^5 rather than ψ (using the small-angle assumption $\sin\psi \approx \psi$). This relationship may be modified by diffraction for large values of σ_h.

Near vertical incidence, the reflectivity resulting from small features on the surface is supplemented by quasispecular reflectivity σ_f from randomly tilted facets of the surface:

$$\sigma_f^0 = (\rho_0/\beta_0)^2 \exp(-\beta^2/\beta_0^2) \tag{3.6.3}$$

where ρ is the surface reflection coefficient, $\beta = (\pi/2) - \psi$ is the angle from vertical, and β_0 is $\sqrt{2}$ times the rms slope of the surface. This leads to a large increase in σ^0 near vertical incidence. For example, if $\rho = 0.8$ and $\beta_0 = 0.05$ radians, or $3°$, σ^0 rises to $+24$ dB at zenith. Most measured data will not show such a large peak value, because the beamwidth of the measuring antenna observes the surface over an angle approaching or exceeding β_0, averaging σ^0 over regions removed from the actual peak. The convolution of the measuring beamwidth θ_3 with the surface slopes leads to a measured reflectivity:

$$\sigma^0 = [\rho^2/(\beta_0^2 + 0.36\theta_3^2)] \exp[-(\beta^2/(\beta_0^2 + 0.36\theta_3^2)] \tag{3.6.4}$$

For example, if a $10°$ beamwidth is used, the zenith reflectivity for $\beta_0 = 3°$ will be reduced to $+17$ dB.

Sea Clutter

When the preceding model is applied to sea clutter, averaging all wind directions, it is found that γ depends on the Beaufort wind scale K_B and radar wavelength, such that

$$10 \log\gamma = 6K_B - 10 \log\lambda - 64 \tag{3.6.5}$$

where wavelength is in meters. This relationship is plotted as solid lines in Figure 3.6.1, for a medium sea condition (Beaufort scale 4 to 5 wind). The rms surface deviation σ_h for this wind state depends on the time and space over which the wind has acted on the sea, but for steady-state conditions we can make the approximations:

$$\sigma_h \approx K_B^3/300 \tag{3.6.6}$$

$$\psi_c \approx 24\lambda/K_B^3 \tag{3.6.7}$$

$$\beta_0 \approx 0.04 \text{ radians}$$

The reason for the near constant value of rms slope is that the wavelength and wave height both increase with wind speed.

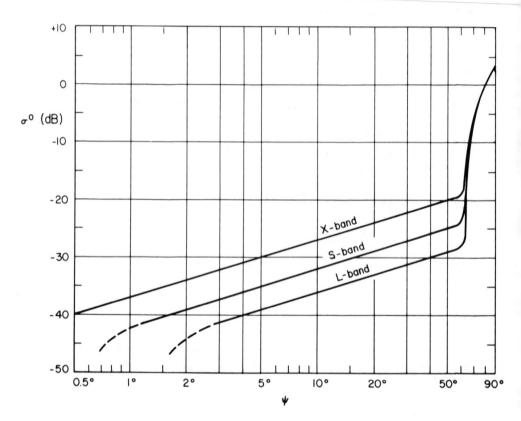

Figure 3.6.1 Reflectivity of sea clutter.

The range extent of sea clutter is determined primarily by the radar antenna height, but also by atmospheric conditions, which can cause superrefraction and ducting. The horizon range can be expressed as a function of height h_r and effective earth's radius ka.

$$R_h = \sqrt{2kah_r} \approx 4130\sqrt{h_r} \qquad (3.6.8)$$

where $a = 6.5 \times 10^6$ m and the approximation applies to $k = 4/3$, with h_r and R_h given in meters. Even before the horizon is reached, the declining $\sigma^0 F_c^4$ will usually bring the clutter level below the noise level.

The velocity spectrum of sea clutter [3.11, pp. 243–247] with a wind velocity v_w will be approximately Gaussian, with a standard deviation $\sigma_v \approx v_w/8$ and a mean (downwind) between $v_w/8$ and $v_w/4$. After envelope detection, the standard deviation will be greater by $\sqrt{2}$. The amplitude distribution will be approximately Rayleigh for grazing angles above ψ_c, and also at low angles for large resolution cells, in which many wave tops contribute to the clutter power. As resolution increases, the distribution departs from the Rayleigh, corresponding to large regions in which clutter is very low, punctuated by sharp peaks at the larger wave tops. This phenomenon is especially noted when horizontal polarization is used. Under these conditions, the pdf has been modeled as log-normal [3.12]. The correlation function, which is roughly sinusoidal at the sea wavelength, for higher grazing angles, becomes peaked at the wavelength and multiples of the wavelength for lower angles.

Land Clutter

The reflectivity of land clutter is much more difficult to characterize than that of the sea. The constant-γ model is still followed at the higher grazing angles, with values of γ between -10 and -15 dB widely applicable to land covered by crops, bushes, and trees. Desert, grassland, and marshy terrain is more likely to have γ near -20 dB, while urban or mountainous regions may have $\gamma \rightarrow -5$ dB. These values are almost independent of wavelength and polarization, and apply to modeling of the mean clutter reflectivity.

128   *TARGETS AND INTERFERENCE*

At low grazing angles, as apply to ground-based radars, propagation considerations become dominant. One study [3.13] showed that practically all of the variations in measured reflectivity of vegetation-covered terrain could be attributed to the propagation factor, so that use of a constant $\sigma^0 = -30$ dB could closely reproduce the measured mean and pdf, at least in the microwave region. In fact, the definition or calculation of a grazing angle becomes quite indefinite for a ground-based radar looking out over typical terrain. Another difficulty is that large regions of the surface tend to be shadowed completely, and the measured data are necessarily restricted to regions in which clutter is visible to the radar. The true mean reflectivity is then the measured mean multiplied by the fraction of terrain over which the data has been measured, often less then 10%.

A simple modeling approach [3.14] is to assume a homogeneous surface with given σ_h and $\gamma \approx -14$ dB, and to assume the radar antenna to be located on a local $2\sigma_h$ point in the terrain, so that its height above the mean is $h_r' = h_r + 2\sigma_h$. A single curve for σ^0 then results, which may be modified for very flat and for mountainous terrain, and for propagation factor at different wavelengths (Figure 3.6.2). In range, there are three distinct clutter regions, depending on the propagation mode. For $R < R_1$, $F_c \approx 1$. Between R_1 and R_h, reflection-interference phenomena dominate the path, and F_c is given by (3.6.2). Beyond the horizon, a choice is made between smooth-sphere and knife-edge diffraction (see Chapter 6), either one leading to a rapid reduction in F_c with range. The effective reflectivity as a function of range, for different antenna heights, may then be plotted as in Figure 3.6.3. The horizon range is computed from (3.6.8), and propagation beyond the horizon is computed by using smooth-sphere or knife-edge diffraction approximations. This captures the average effects of propagation, as well as the average reduction in grazing angle with increased range. Observed variation in land clutter reflectivity with wavelength is explained primarily by the propagation factor.

The spatial extent of land clutter depends not only on antenna height, siting, and propagation conditions, but on terrain roughness. Mountains rising to 5000 m altitude will normally be visible at ranges near 300 km, but with atmospheric ducting can appear at 500 to 600 km range. Clutter from normal, rolling terrain ($\sigma_h = 15$ m) will be visible some 15 km beyond the radar's horizon range, but will be extended to hundreds of km under ducting conditions. The frequency spectrum of land clutter is narrow, having an approximately Gaussian shape with $\sigma_v < 0.5$ m/s for vegetation in windy conditions.

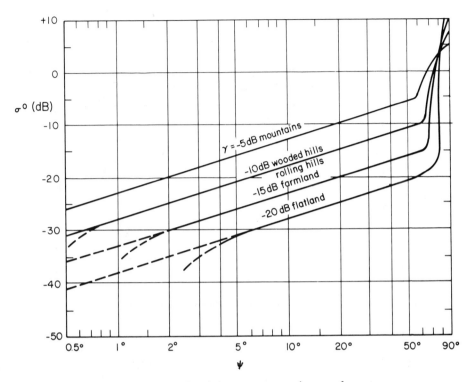

Figure 3.6.2 Land clutter reflectivity *versus* grazing angle.

Clutter Amplitude Distribution

A major analytical problem with land clutter is its amplitude distribution. Knowledge of the mean clutter is not adequate for prediction of
its effect on the radar, unless the pdf is known. At grazing angles well
above the critical angle, this pdf may be assumed Rayleigh, but at and
below ψ_c it spreads to a log-normal or Weibull form, the latter given by

$$dP_\sigma = (\sigma^{1/a}/a\alpha\sigma) \exp(-\sigma^{1/a}/\alpha)\, d\sigma \tag{3.6.9}$$

where α determines the mean and a gives the spread of the pdf:

$$\bar{\sigma} = \alpha^a \Gamma(1 + a) \tag{3.6.10}$$

$$\sigma_{50} = (\alpha \ln 2)^a \tag{3.6.11}$$

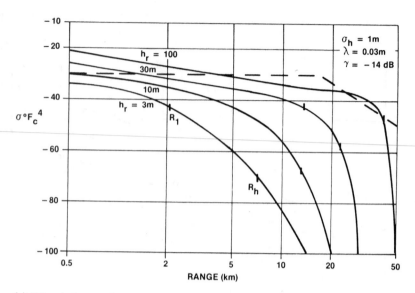

(a) X-band clutter reflectivity *versus* range, for different radar heights over level terrain.

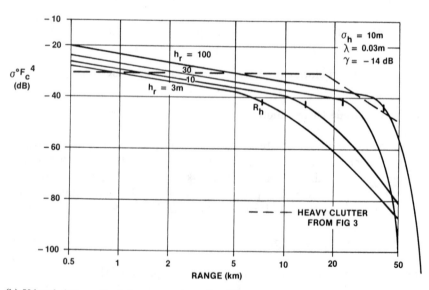

(b) X-band clutter reflectivity *versus* range, for different radar antenna heights over rolling or hilly terrain.

Figure 3.6.3 Predicted average land reflectivity *versus* grazing angle for different values of antenna height and surface roughness.

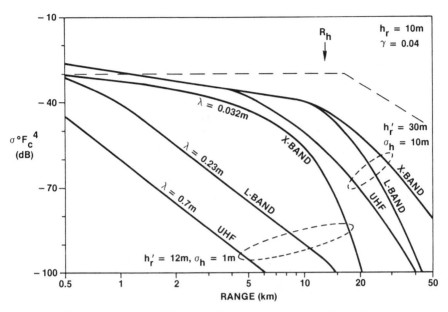

(c) Comparison for different bands over rough and smooth terrain.

Figure 3.6.3 (cont'd)

where Γ is the Gamma function. Note that this pdf, written in terms of σ, actually applies to the effective value σF_c^4. For $a = 1$, the exponential pdf of power (Rayleigh amplitude) is generated, while, for other values of a, the ath root of amplitude is Rayleigh distributed. On a log scale, the spread of the pdf becomes directly proportional to a, while the mean moves farther above the median (Figure 3.6.4). The log-normal and Weibull distributions may both be characterized by the ratio of mean to median, plotted in Figure 2.5.4. Ratios between 10 and 20 dB are not uncommon for land clutter viewed at low grazing angles by high-resolution radars, and these can be compared with 1.5 dB for the Rayleigh distribution.

The broad spread of the land clutter pdf is largely the result of the propagation factor between the radar and individual patches of the terrain, with the peak value of the pdf (e.g., the value exceeded with probability 10^{-4}) holding almost constant for small resolution cells to the longest clutter range. The simple model in which F_c varies smoothly as a function of range cannot capture the actual variability of clutter from cell to cell,

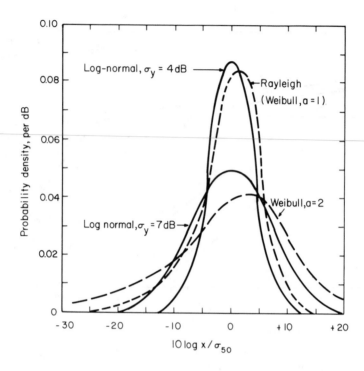

Figure 3.6.4 Weibull and log-normal distributions plotted on logarithmic scale.

but gives only the average clutter *versus* range. For $F_c \to 0$, the average results from the contribution of a smaller and smaller fraction of the total cells, corresponding to increase Weibull a or log-normal σ_y. A way of modeling this effect is to relate F_c, as calculated from the simple model, to the Weibull parameter a, as shown in Figure 3.6.5 for large and small resolution cells. Using a model of this form, the relative performance of a system can be estimated as a function of resolution cell size and clutter attenuation. Remaining to be determined are the correlation properties of clutter under different conditions, which will determine the effectiveness of different CFAR approaches.

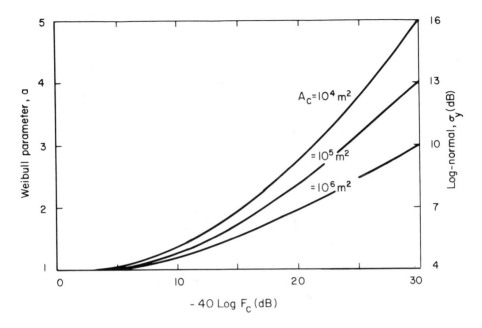

Figure 3.6.5 Variation in spread of clutter pdf with average propagation factor.

Volume Clutter

The reflectivity of volume clutter is described by a factor η_v, expressed in m^2/m^3. Using (3.2.1), with adjustments for the dielectric constant of the droplets, this factor can be expressed, for rain at $\lambda > 0.02$ m, as

$$\eta_v = 5.7 \times 10^{-14} \, r^{1.6}/\lambda^4 \qquad (3.6.12)$$

where r is the rainfall rate in mm/h. For dry snow at these wavelengths,

$$\eta_v = 1.2 \times 10^{-13} \, r^2/\lambda^4 \qquad (3.6.13)$$

where r is the rate in terms of water content of the snow. For shorter wavelengths [3.16], the scattering properties depart from the pure Rayleigh RCS prediction, as shown in Figure 3.6.6. This reflectivity is for linearly

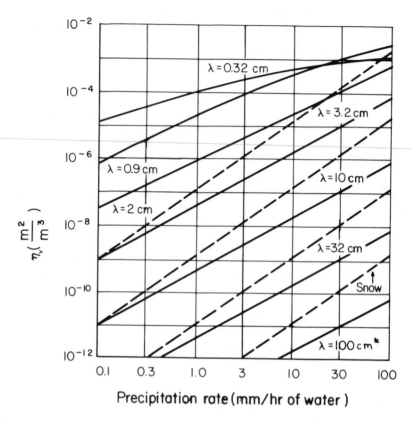

Figure 3.6.6 Volume reflectivity of rain and snow.

polarized radar, or circularly polarized systems using opposite sense for transmission and reception. For same-sense circular polarization, cancellation up to about 20 dB can be expected (somewhat more at low rainfall rates, less at high rates). The data are necessarily approximations, since different types of rain have different drop size distributions.

Chaff

The RCS of an individual dipole of chaff has been given in Section 3.2 as $0.88\lambda^2$, when viewed broadside. When averaged over all aspect angles, this drops to $0.15\lambda^2$. Early data on chaff gave a total RCS, as a function of its weight W in kg,

$$\sigma = 6600\, W_{kg}/f_{GHz} = 22{,}000\, \lambda W \text{ m}^2 \tag{3.6.14}$$

for chaff dipoles cut to resonance in a specific band. More recent data [3.15, p. 187] indicate that more modern chaff can achieve this RCS over a two-octave band, with λ in (3.6.14) taken as the geometric mean value (Figure 3.6.7). For $\lambda > 0.3$ m, resonant dipoles usually give way to long streamers called rope, achieving somewhat lower RCS for its weight, but being less subject to breakage. Chaff normally falls with a motion which randomizes the orientation of individual dipoles, and hence is not very sensitive to the radar polarization.

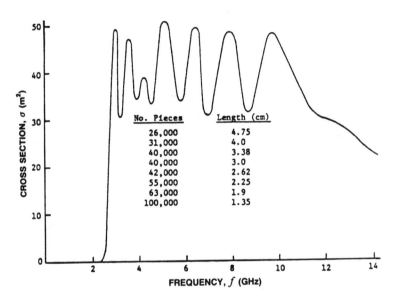

Figure 3.6.7 Radar cross section of a chaff package [3.15].

The density of a chaff cloud is thus entirely dependent on the weight of the chaff dispensed into a given volume. For example, in order to achieve the same reflectivity as 3 mm/h rain at X-band, $\eta_v = 3 \times 10^{-7}$ m^2/m^3, or 30 m^2/km^3, chaff must be dispensed at a rate of 0.045 kg/km^3. With uniform distribution, there would be one dipole per 17 m cube, and an average mass density of 45 g/km^3. This compares with 3 mm/h rain, which typically has 2×10^8 g/km^3.

Spatial and Velocity Extent of Volume Clutter

The spatial extent of rain is limited by its top altitude, usually 3 to 7 km, but extending up to 10 km in thunderstorms. Widespread rain, below 5 km altitude, can extend to the horizon in all directions: about 300 km range. Heavier rainfall will be more limited horizontally, but may extend to greater altitudes. Chaff is limited by the operational altitude of the dispensing aircraft, normally to about 12 km (for which the horizon range is 450 km). The horizontal extent of chaff depends only on the routes flown by the aircraft and the time since dispensing. Winds will disperse the cloud over considerable horizontal distances.

The velocity spectra of rain and chaff are dependent on the wind conditions, and in particular on the wind shear in altitude. For volume clutter filling the elevation beam, and subject to a wind shear k_{sh} in m/s per km altitude, Nathanson [3.11, p. 207] gives the standard deviation of the velocity spectrum as

$$\sigma_v = 0.3 k_{sh} R \theta_e \cos\alpha \qquad (3.6.15)$$

where R is in km, α is the azimuth angle between the beam and the wind direction, and the elevation beamwidth θ_e is in radians (the constant in this equation has been adjusted for our use of the one-way 3-dB beamwidth). Values of k_{sh} between 2 and 4 m/s per km are common, with a total velocity change up to 40 m/s between the surface and the highest clutter. The mean clutter velocity, with constant wind shear, will be equal to the wind radial velocity at the center of the beam.

The velocity spectrum of volume clutter competing with a particular target will depend not only on the wind and beamwidth, but also on the presence of range ambiguities in the radar response. For example, Figure 3.6.8 shows the spectrum that results from a narrow beam which penetrates an extended cloud of clutter at four separate range ambiguities. Although the clutter in each of these is narrowed by the beam, the total spread is the same as will be seen by a broad-beam radar.

The vertical velocity component of rain, near the surface, reaches several m/s, but is not usually a significant contributor to the spectral spread, even for beams at high elevation. Chaff has a very small vertical velocity, given in [3.15, p. 183] as near 0.3 m/s at sea level. This implies a rate near 0.6 m/s at 12 km altitude, and a fall time of eight hours to the surface from that altitude.

The amplitude distribution of volume clutter is normally assumed to be Rayleigh. However, both rain and chaff have local variations, leading to peaks well above those which would be predicted from the Rayleigh distribution. A Rayleigh distribution with spatially varying mean, correlated over perhaps hundreds of meters, might be an appropriate model.

Discrete Clutter Sources

Of the several types of discrete clutter, buildings (including towers and water tanks) present the largest RCS, and birds are the most extensive

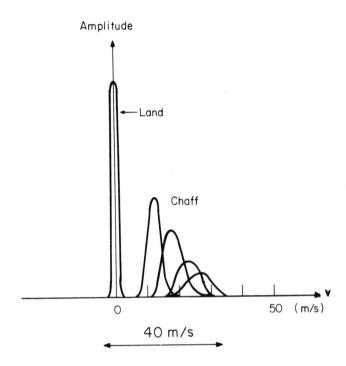

Figure 3.6.8 Spectrum of range-ambiguous chaff clutter.

in space and velocity. Table 3.2 shows different models of discrete land clutter taken from the literature, along with a recommended model for general use. These large, rigid sources can be regarded as having very small spatial extent (typically only one or a few meters) and spectra of essentially zero width. They establish the dynamic range at the input to the receivers of most systems, and place special requirements on methods of controlling false alarms.

Birds, although much smaller in RCS, present serious problems for radars which must detect small targets or targets at low altitude. The bird RCS for mid-microwave bands can be represented by a log-normal pdf, with a median of -30 dBm^2 and a standard deviation of 6 dB. This implies that only 0.13% of the bird population will exceed -12 dBm^2. However, densities of birds are such that there may be 10^5 birds within 50 km of the radar, of which 130 will then exceed -12 dBm^2 Variation with radar wavelength and type of bird is discussed in [3.21]. Migrating birds will have an exponential distribution in altitude, with a scale height near 700 m. Most local birds will be at lower altitudes. The most damaging feature of the bird population, in modern doppler radar systems, is its velocity distribution. The birds will move at the wind velocity, $\pm$ 15 m/s air speed. Bird RCS is not sensitive to polarization, nor are the larger bird RCS

Table 3.2 Models of Discrete Point Clutter

	RRE	Ward	Mitre Rural	Metro	Suggested
Density (per km^2):					
$\sigma_c = 10^2 \ m^2$	3.5	1.8			2
$= 10^3$	0.8	.36	.02	2	0.5
$= 10^4$	0.15	.18	.002	.2	0.2
Resulting $\overline{\sigma^0}$ (dB)	−26	−26	−44	−24	−26
Number of points per 1° beam, 0–5 km:					
$\sigma_c = 10^2 \ m^2$	.7	.35			0.4
$= 10^3$	.16	.07	.004	.4	0.1
$= 10^4$	.03	.035	.0004	.04	0.04
	[3.17]	[3.18]	[3.19]	[3.20]	

values sensitive to wavelength for $\lambda < 0.3$ m. Small birds will become resonant near that wavelength, and will be in the Rayleigh region at longer wavelengths. Larger birds will resonate in the UHF bands, but all will be in the Rayleigh region at VHF and below.

A further problem with birds is that they often fly at altitudes high enough to provide free-space propagation, $F_c = 1$, or even in reflection lobes with $F_c \rightarrow 2$. Thus, the fact that a bird RCS is -12 dBm^2 does not prevent it from appearing at the radar as $\sigma F_c^4 = 0$ dBm^2. This, plus the presence of multiple birds in a flock, makes it impossible to reject birds on the basis of signal thresholding. This same factor applies to insects, for which the median of the log-normal distribution may be -70 dBm$^2 \pm 10$ dB in microwave bands. With $\sigma_y = 6$ dB, the mean will be -66 dBm^2, but clouds of thousands of insects will increase this to -30 dBm^2. The propagation factor, averaging $+8$ dB, places these also within the detection capabilities of modern systems.

Land vehicles become a clutter problem for many air surveillance radars because their RCS is comparable to aircraft, and their velocities lie in the passband of MTI and many doppler systems. Here, at least, the propagation factor almost always favors the desired aircraft target, which necessarily lies at higher altitude than the land vehicle. However, when there are dozens or hundreds of land vehicles within line of sight of the radar, numbers of false alarms are inevitable unless velocity filtering removes targets below 30 m/s.

3.7 JAMMING

Noise Jamming

The most common type of ECM is noise jamming, which is intended to prevent detection of the echo signal by raising the receiver noise level. The effect on the radar is to increase the effective input temperature from T_s, determined by the receiver and the natural environment, to a level $T_i = T_s + T_j$, where the jammer temperature is given by

$$T_j = \frac{P_j G_j A_r F_j^2}{4\pi k B_j R_j^2 L_{aj}} \tag{3.7.1}$$

where $P_j G_j$ is the effective radiated power (ERP) of the jammer, F_j is the pattern-propagation factor of the radar receiving antenna in the jammer direction, B_j is the bandwidth of the jammer spectrum, R_j is the jammer range, and L_{aj} is the one-way propagation loss from the jammer. When more than one jammer is present, a temperature is calculated for each, and these temperatures are all added to T_s to find T_i.

The radar detection range on echo targets, in the presence of noise jamming from remote platforms (stand-off jamming, SOJ), is best evaluated by calculating the increase in input temperature from (3.7.1). When the jammer range decreases along with the target (either for escort jamming, ESJ, or self-screening jamming, SSJ), a *burn-through range* may be calculated, at which the radar echo will become visible. This calculation is presented in Section 1.4, (1.4.4), and will not be discussed here.

In *barrage noise jamming,* the bandwidth B_j extends over the entire tuning band of the radar, typically 10% of the radar frequency. The *spot noise jammer* concentrates its power on a band that is little wider than the instantaneous signal bandwidth of the radar, achieving higher noise density at the risk of missing the frequency at which the radar is actually receiving. Spot jammers, along with other specialized jammers, require intercept receivers to control their frequency, and these receivers must operate during quiet (look-through) periods in which no jamming is emitted. When several tunable radars are within line-of-sight of the jamming platform, in the same band, barrage jamming will normally be used.

In addition to describing the jamming in terms of ERP and bandwidth, the quality of the noise should be considered. Direct amplification of a wideband noise source produces jamming with a nearly Gaussian amplitude distribution, but with average power well below the capability of the jammer output amplifier. By increasing the jamming level relative to the saturation level of the amplifier, higher average output power is obtained at the expense of clipping the noise peaks. This clipped noise is less effective in masking targets and creating false alarms in the radar. Many jammers produce broadband noise by sweeping rapidly in frequency over the band, while simultaneously applying a noise FM function to randomize the jamming signal. Full saturated output is obtained with this method. Again, depending on the timing of the sweep and the extent of the noise FM, the effectiveness on the radar may be degraded relative to true random noise. Whichever method is used, the ERP of the jammer should be reduced by a noise quality factor.

Typical ECM transmitters operate with average powers between 100 W and a few kW. The ERP is increased by using directional antennas, to the extent this is possible on the type of vehicle employed. For example, the SOJ can use fairly large antennas (e.g., $D = 0.5$ m), if the beamwidths are consistent with the coverage required. For simultaneous jamming of several radars, the azimuth beamwidth must be about 60°, and this requirement combined with the limited vertical aperture size normally limits the antenna gain to 10–15 dB. The corresponding ERP levels for SOJs are typically 30 to 100 kW. For the SSJ, antenna gains nearer 5 dB are used, resulting in ERP levels of 1–10 kW. These values are usually taken as the effective values after reduction for noise quality.

The noise jammer can mask target echoes, but in the process the jammer permits the radar to obtain an accurate reading of jammer angle (azimuth, for 2D search radars; azimuth and elevation, for 3D and tracking radars). The angle location procedure in search radars results in a *jammer strobe,* while in trackers angle location provides *track-on-jam* data. The jammer objective is met if the jammer is masking other targets (SOJ or ESJ), or if the using system is unable to operate without range and doppler information on the jamming platform. Since most missile guidance systems are designed to work in the track-on-jam mode, noise jamming is not effective as an SSJ technique.

Deception Jamming

The deception jammer uses specialized waveforms to produce multiple false targets or a target at an angle other than the jammer angle. The jammer waveform in many cases is an amplified, or repeated, version of the incident radar signal. Against search radars, the multiple false targets are intended to saturate the data system at the radar output. Against tracking and guidance radars, false targets delay acquisition of the actual target, and prevent accurate fire control or guidance on the jamming platform. In most cases, the deception jammer acts to exacerbate a problem naturally present in the radar. This may be the effect of target fluctuation, depolarization of target echoes, or target glint. Information about the effect of deception jamming on trackers is covered in Chapter 10.

Decoys

Decoys are a specialized form of deception jamming in which a signal source is established outside the jammer platform. Decoys may be passive, reflecting the radar transmission, or active, as repeaters or noise jammers. The basic purpose is to dilute the effectiveness of a defensive system, to delay acquisition of data on real targets, or to protect those targets by diverting missiles or gunfire to the decoy. As with other deception jamming techniques, the decoy acts against a natural response of the radar. An example of this technique is the so-called *ground-bounce jammer,* which directs its emission at the surface in such a way as to produce an exaggerated multipath error. Because of the similarity of these effects to natural multipath reflections of target signal, the subject will be discussed further, and quantitative analysis given, in connection with that subject in Chapter 11.

REFERENCES

[3.1] IEEE Standard Dictionary of Electrical and Electronic Terms, *ANSI/IEEE Std. 100-1984,* 1984.
[3.2] E.F. Knott, J.F. Shaeffer, and M.T. Tuley, *Radar Cross Section,* Artech House, 1985.
[3.3] N.C. Currie (ed.), *Techniques of Radar Reflectivity Measurement,* Artech House, 1984.
[3.4] P. Swerling, Probability of detection for fluctuating targets, *RAND Corp. Research Memo. RM-1217,* March 17, 1954; reprinted in *IRE Trans.* **IT-6,** No. 2, April 1960.
[3.5] J.D. Wilson, Probability of detecting aircraft targets, *IEEE Trans.* **AES-8,** No. 6, November 1972, pp. 757–761.
[3.6] W. Weinstock, Target cross section models for radar systems analysis, doctoral dissertation, University of Pennsylvania, Philadelphia, 1964.
[3.7] D.K. Barton and H.R. Ward, *Handbook of Radar Measurement,* Prentice-Hall, 1969; Artech House, 1984.

[3.8]   R.V. Ostrovityanov and F.A. Basalov, *Statistical Theory of Extended Radar Targets,* Soviet Radio Publishing House, 1982; Artech House, 1985.

[3.9]   D.D. Howard, Radar target angular scintillation in tracking and guidance based on echo signal phase-front distortion, *Proc. NEC* **15,** 1959, pp. 840–849.

[3.10] M.W. Long, *Radar Reflectivity of Land and Sea,* Artech House, 1983.

[3.11] F.E. Nathanson, *Radar Design Principles,* McGraw-Hill, 1969.

[3.12] G.V. Trunk, Radar properties of non-Rayleigh sea clutter, *IEEE Trans.* **AES-8,** No. 2, March 1972, pp. 196–204.

[3.13] S. Ayasli, Propagation effects on radar ground clutter, *Proc. IEEE National Radar Conference,* 1985.

[3.14] D.K. Barton, Land clutter models for radar design and analysis, *Proc. IEEE* **73,** No. 2, February 1985, pp. 198–204.

[3.15] D.C. Schleher, *Introduction to Electronic Warfare,* Artech House, 1986.

[3.16] R.K.Crane, Microwave scattering parameters for New England rain, *MIT Lincoln Laboratory Tech. Rep. 426,* October 3, 1966.

[3.17] A.K. Edgar, E.J. Dodsworth, and M.P. Warden, The design of a modern surveillance radar, *IEE Conf. Pub. No. 105, Radar-73,* October 1973, pp. 8–13.

[3.18] H.R. Ward, A model environment for search radar evaluation, *IEEE Eascon Record,* 1971, pp. 164–171.

[3.19] W.J. McEvoy, Clutter measurements program: Operations in Western Massachusetts, *Mitre Corp. Rep. MTR-2074,* March 1972, DDC Doc. AD742297.

[3.20] W.J. McEvoy, Clutter measurements program: Operations in the metropolitan Boston area, *Mitre Corp. Rep. MTR-2085,* March 1972, DDC Doc. AD742298.

[3.21] G.E. Pollon, Distributions of radar angels, *IEEE Trans.* **AES-8,** No. 6, November 1972, pp. 721–727.

Chapter 4

Radar Antennas

4.1 FOUR-COORDINATE RADAR RESPONSE

Radar operates in four-dimensional space: two angles (often specified as azimuth and elevation), range (measured as time delay), and frequency (including the doppler shift caused by target radial velocity). A target is said to be resolved if its signal is separated at the radar output from those of other targets, in at least one of these coordinates. For example, a surveillance radar may scan its beam in azimuth and elevation, detecting a target in a particular range gate and doppler filter. A second target signal may be resolved if it lies in a different azimuth or elevation beam position, or in a different range gate or doppler filter. A useful criterion for resolution is that the position of the desired target should be measurable by the radar with only small errors caused by the other target. As shown in Figure 4.1.1, the relative phase and amplitude of the second target affect the resolution process. However, for two targets of equal amplitude and arbitrary constant phase, resolution is normally possible when they are separated by approximately the half-power beamwidth or the half-power width of the processed pulse or doppler filter.

Resolution is thus determined by the relative response of the radar to targets separated from the target to which the radar is directed, or matched. The antenna and receiver, at a given time, are configured to produce a maximum response on a target at particular angles, range delay, and frequency. The radar will be designed to respond with reduced gain to targets at other locations. This *response function* can be expressed as a surface in a five-dimensional coordinate system: relative output voltage, as a function of the target displacement from the matched point in the four radar coordinates: $\chi(\theta, \phi, t_d, f_d)$. The parameters of the radar range equation determine this voltage at the matched point, at which χ is normalized to unity.

143

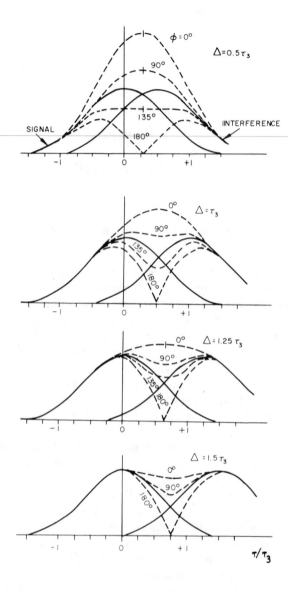

Figure 4.1.1 Resolution of signals from closely spaced target returns in time.

Separable Angle Response

In most cases, the angles response of the radar can be considered independent of the delay-doppler response:

$$\chi(\theta, \phi, t_d, f_d) = \chi(\theta, \phi)\chi(t_d, f_d) \qquad (4.1.1)$$

The angle response $\chi(\theta, \phi)$ is simply the two-way voltage gain pattern of the antenna, which has been represented in the radar equation by a pattern component $f^2(\theta, \phi)$ of the pattern-propagation factor F^2. In cases where the transmitting and receiving antennas have different patterns, the two-way voltage gain pattern will be the product of the two one-way patterns:

$$\chi(\theta, \phi) = f_t(\theta, \phi)f_r(\theta, \phi) \qquad (4.1.2)$$

It is these one-way gain patterns, or their squares that represent power gain patterns, which will normally be specified or measured to characterize the resolution properties of the antenna system. The power gain pattern $G(\theta, \phi) = G(0, 0)f^2(\theta, \phi) = G_m f^2(\theta, \phi)$, is often plotted on a decibel scale to permit sidelobe levels to be more readily seen.

In the case of ground-based antenna systems, we often speak of their antenna patterns in terms of azimuth and elevation coordinates. These can be related to the spherical coordinate system best used to make antenna calculations, as shown in Figure 4.1.2. The antenna aperture is located in the x-y plane, and the broadside beam will be directed along z. The angle θ is the deviation from broadside, while ϕ indicates the direction of this deviation. Placing the aperture in the vertical plane (Figure 4.1.3), with its width w along the x axis and height h in the y axis, and z pointing north, we can see that the azimuth angle $A = \theta \cos\phi$, while elevation $E = \theta \sin\phi$.

A typical plot of the power pattern in decibels for a pencil-beam tracking antenna is shown in Figure 4.1.4. Other ways of plotting antenna patterns are as contours (Figure 4.1.5a), or as gain *versus* angle along one of the principal planes (e.g., azimuth or elevation), as in Figure 4.1.5b. Note that the voltage plots preserve the reversals in phase of the pattern as the sidelobes are traversed.

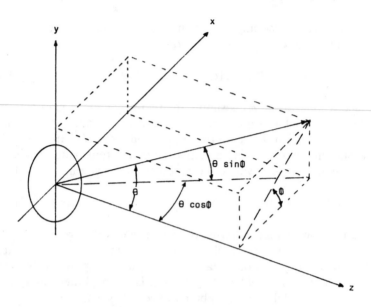

Figure 4.1.2 Antenna pattern coordinates.

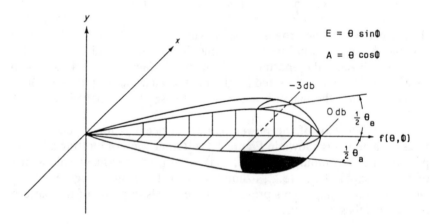

Figure 4.1.3 Coordinates for horizontal beam.

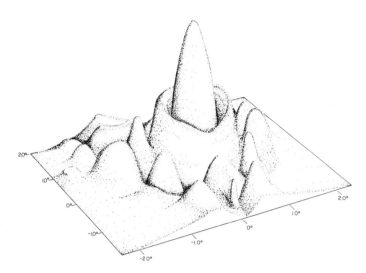

Figure 4.1.4 Power gain for two angular coordinates [4.1].

In general, voltage gain of an antenna is a complex quantity, with magnitude and phase angle measured relative to the on-axis voltage. Many types of antenna, however, produce real patterns with phase angles essentially zero over the main lobe and oscillating between 180° and zero over the sidelobes. Gain is also a function of polarization, $f_t(\theta, \phi)$ being defined for the designed transmitting polarization and $f_r(\theta, \phi)$ for reception of that polarization. Polarization diversity systems require separate functions for each polarization used.

Reciprocity of Antenna Patterns

An important concept in antenna theory is the principle of reciprocity. Simply stated, the patterns of a given antenna for receiving and transmitting will be the same. We may measure the pattern by radiating from the antenna and measuring field strength as a function of angle, at any convenient long range from the antenna, or by radiating from a small source at that range and measuring the received voltage at the antenna terminal, as a function of the same offset angle from the source. In the

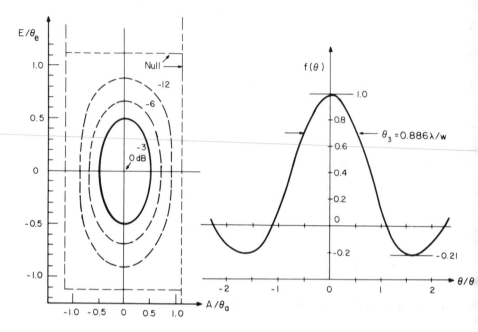

Figure 4.1.5 Antenna pattern plots: (a) contours in two angular coordinates; (b) principal-plane voltage pattern.

first case, we may consider that the antenna aperture has been *illuminated*
by the transmitter through its feed system, causing it to radiate a beam
into space. If the aperture is composed of discrete elements, these are said
to be *excited* by the transmitter through the feed system. In the second
case, the aperture is placed in a field of uniform intensity, created by the
small remote source, and components of this field are *weighted* by the feed
system in forming the received beam pattern. In either case, the amplitude
and phase of signal components corresponding to different portions of the
aperture are controlled by the feed system. For simplicity, we will refer
to this process as *illumination* for all types of antennas in both transmitting
and receiving modes. Uniform illumination, with phase matched to a plane
wave normal to the desired beam direction, gives maximum gain in that
direction.

Reciprocity is violated in actual antennas only when a ferromagnetic
device such as a circulator or ferrite phase shifter is included in the circuit,
or when a nonlinear device such as a gas-tube duplexer is included.

4.2   ANTENNAS AND ARRAYS

Antenna Fundamentals

The earliest radar antennas, and some of the most recent, use arrays of half-wave dipole elements (Figure 4.2.1). The pattern of a dipole is a toroid surrounding the dipole axis, and the gain in any direction normal to this axis is approximately $\pi/2$. When the dipole is mounted $\lambda/2$ in front of a ground plane, the rear half of the toroid is reflected into the front hemisphere, reinforcing the forward lobe, narrowing its width in the θ coordinate, and increasing its gain to approximately π.

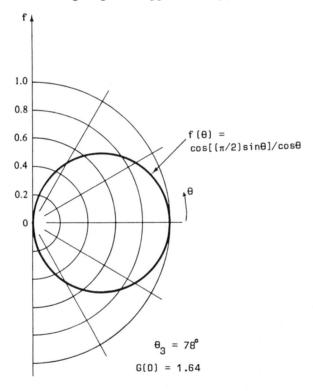

Figure 4.2.1 Radiation pattern of half-wave dipole. (After Elliott [4.2].)

It is convenient to consider any antenna as an array of $T = n_a \times n_e$ elements, each $\lambda/2$ square with gain π. Such elements can be thought of as small dipoles, spaced $\lambda/2$ apart in both directions and mounted in front of a continuous ground plane. Elements other than dipoles may be used, or the antenna may actually be a continuous reflector or lens illuminated by a horn feed, in which case the elements will be merely small areas on the surface, $\lambda/2$ square. If the elements are illuminated uniformly and in phase, the antenna gain will be

$$G_0 = G(0, 0) = \pi T \tag{4.2.1}$$

where the z axis ($\theta = 0$) is normal to the ground plane. For a rectangular aperture of area $A = w \times h$, there will be $n_a = 2w/\lambda$ columns and $n_e = 2h/\lambda$ rows of elements, giving $T = 4A/\lambda^2$ total elements. The corresponding gain is

$$G_0 = \pi T = 4\pi A/\lambda^2 \tag{4.2.2}$$

The beam pattern in this case will be described by the product of two $(\sin x)/x$ functions:

$$f(\theta, \phi) = \frac{\sin[(\pi w/\lambda) \sin\theta \, \cos\phi]}{(\pi w/\lambda) \sin\theta \, \cos\phi} \frac{\sin[(\pi h/\lambda) \sin\theta \, \sin\phi]}{(\pi h/\lambda) \sin\theta \, \sin\phi} \tag{4.2.3}$$

In this case, not only is the antenna pattern separable from the range-doppler response, but the azimuth and elevation patterns are separable from each other. This response was shown in Figure 4.1.5, with the angle scale normalized to the 3-dB beamwidth. The first nulls, beside the main lobe, will occur at $\sin A = \sin\theta \, \cos\phi = \lambda/w$ and at $\sin E = \sin\theta \, \sin\phi = \lambda/h$, and other nulls will occur at integral multiples of these angles. Solving (4.2.3) for the half-power beamwidths, we find

$$\begin{aligned} \sin\theta_a \approx \theta_a &= 0.886 \, \lambda/w \\ \sin\theta_e \approx \theta_e &= 0.886 \, \lambda/h \end{aligned} \quad \text{(uniform illumination)} \tag{4.2.4}$$

Considering the azimuth coordinate only (elevation will be similar with h replacing w), and making the small-angle asumption $\sin\theta \approx \theta$, the pair of first sidelobes, at $\theta = \pm 1.4\lambda/w$, $\theta/\theta_a = 1.6$, will have an amplitude -0.21, or -13.6 dB from the main-lobe peak. The sidelobe peak amplitudes will fall off inversely with $(\pi w/\lambda) \sin\theta$, and will not fall below -30 dB until the tenth sidelobe at $\theta = 10.5\lambda/w = 11.9\theta_a$. The relationship between number of elements and beamwidth is

$$\begin{aligned} T &= 3.14/\theta_a\theta_e \quad \text{(in radians)} \\ &= 10{,}300/\theta_a\theta_e \quad \text{(in degrees)} \end{aligned} \quad \text{(uniform illumination)} \tag{4.2.5}$$

The gain and the beamwidth are related by

$$G_0 = 9.84/\theta_a\theta_e \quad \text{(in radians)}$$
$$= 32{,}300/\theta_a\theta_e \quad \text{(in degrees)} \quad \text{(uniform illumination)} \quad (4.2.6)$$

The constant 9.84 corresponds to an adjustment factor $L_n = 1.28 = 1.1$ dB in the gain-beamwidth relationship, (1.3.1). We will see how these rather fundamental relationships among gain, beamwidth, and number of elements (or elementary areas) change as the aperture illumination is tapered for lower sidelobes, or as the aperture itself is made elliptical to match the illumination produced by a horn feed.

Tapered Aperture Illumination

The high sidelobes produced by the uniformly illuminated antenna are not often acceptable, and are reduced by tapering the illumination as a function of distances x and y from the center of the aperture. The relative illumination amplitude (normalized to unity at the center) is described by a function $g(x, y)$. For uniform illumination of a rectangular aperture,

$$g(x, y) = 1, \ |x| < w/2, \ |y| < h/2$$

For any given illumination function, the antenna pattern is given by the two-dimensional Fourier transform of that function:

$$f(\theta, \phi) = \frac{1}{A'} \int_{-h/2}^{h/2} \int_{-w/2}^{w/2} g(x, y) \exp[j(2\pi/\lambda)(x \sin\theta \cos\phi$$

$$+ \ y \sin\theta \sin\phi)] \ dx \ dy \qquad (4.2.7)$$

where $A' = \int\int g(x, y) \ dx \ dy$ is an effective area which normalizes $f(0, 0) = 1$. The gain of the tapered aperture can be expressed as

$$G_m = \frac{4\pi |\int\int g(x, y) \ dx \ dy|^2}{\lambda^2 \int\int g^2(x, y) \ dx \ dy} = G_0\eta_a \qquad (4.2.8)$$

For patterns not extending too far from the z axis, $\sin\theta \approx \theta$, and we may express the azimuth and elevation patterns as

$$f(A, E) \leftrightarrow g(x, y)$$

where $\leftrightarrow$ represents the Fourier transform.

In a rectangular array, the illumination and pattern functions are usually separable into functions in the two principal planes:

$$g(x, y) = g(x)g(y), \quad f(A, E) = f(A)f(E)$$

Setting $\phi = 0$ so that θ represents azimuth angle A, we may write

$$f(\theta) = \frac{1}{A_x} \int_{-w/2}^{w/2} g(x) \exp [j(2\pi/\lambda)x \sin\theta] \, dx \qquad (4.2.9)$$

where $A_x = \int g(x) \, dx$ normalizes the pattern. Similar expressions apply to elevation with $\phi = \pi/2$, y replacing g, and h replacing w. Figure 4.2.2 illustrates the Fourier transform relationship in one plane. Substitution of $g(x) = g(y) = 1$ yields (4.2.3), while for tapered illumination the beamwidth and efficiency will vary with level of the first sidelobe as shown in Figure 4.2.3. In a rectangular aperture with separable illuminations, the aperture efficiency η_a will be

$$\eta_a = G_m/G_0 = G_m\lambda^2/4\pi A = \eta_x\eta_y \qquad (4.2.10)$$

Thus, an aperture with cosine taper in both coordinates will be described by the following parameters:

$$\eta_a = 0.64, \text{ or } -1.9 \text{ dB}$$

$$\theta_a = 1.19\lambda/w, \ \theta_e = 1.19 \ \lambda/h$$

First sidelobes $G_s/G_m = -23$ dB

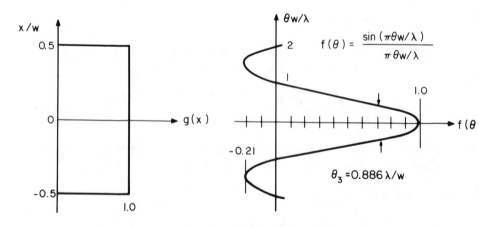

Figure 4.2.2 Fourier transform relationship between aperture illumination and beam pattern.

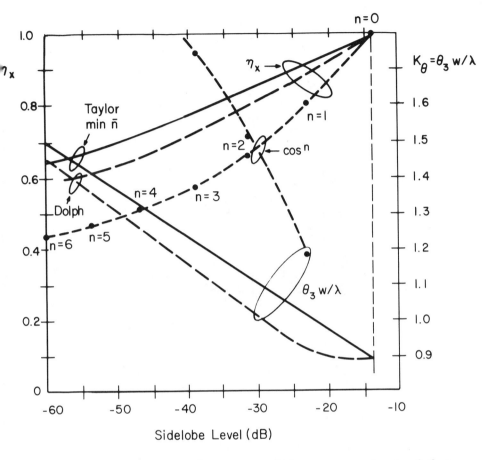

Figure 4.2.3 Beamwidth and aperture efficiency *versus* level of first sidelobe.

Table 4.1 lists the first sidelobe ratio, the beamwidth constant, and the efficiency for some common taper functions used in linear and rectangular arrays. More distant sidelobes will have amplitudes varying reciprocally with the square of the angle θ, or, more accurately, with the quantity $[(2.4\theta/\theta_a)^2 - 1)]$. Only the first sidelobe will exceed -30 dB, for cosine illumination.

The relationship between gain and beam solid angle remains almost constant, for all practical illuminations:

$$G_m = 11.3/\theta_a\theta_e \quad \text{(in radians)}$$
$$ = 37,100/\theta_a\theta_e \quad \text{(in degrees)} \quad \text{(tapered illumination)} \quad (4.2.11)$$

Table 4.1 Antenna Parameters for Tapered Rectangular Apertures

Illumination function, $g(x)$	Sidelobe G_s/G_m (dB)	Beamwidth $\theta_3 w/\lambda$	Efficiency η_x (dB)	Efficiency η_x (ratio)	Efficiency η_x^2 (dB)	Efficiency η_x^2 (ratio)
$\cos(\pi x/w)$	−23	1.19	−0.9	0.80	−1.8	0.64
$\cos^2(\pi x/w)$	−32	1.44	−1.8	0.66	−3.6	0.44
$1 - (2x/w)^2$	−21	1.18	−0.8	0.83	−1.6	0.69
$[1 - (2x/w)]^2$	−27.5	1.37	−1.6	0.69	−3.2	0.48
$1 - |2x/w|$	−27.1	1.27	−1.3	0.74	−2.6	0.55
Taylor, $n = 3$	−25	1.05	−0.5	0.90	−1.0	0.81
$n = 4$	−30	1.12	−0.7	0.85	−1.4	0.72
$n = 5$	−35	1.18	−0.9	0.80	−1.9	0.64
$n = 6$	−40	1.25	−1.2	0.76	−2.4	0.58

This corresponds to $L_n = 1.1 = 0.5$ dB in (1.3.1). Note that the adjustment L_n is less than for uniform illumination, indicating that the use of taper increases the beamwidth more rapidly than it decreases the gain. This is the result of transferring power from the sidelobes into the mainlobe. The constant 37,100 may be compared with the frequently used relationship $G_m = 25,000/\theta_a\theta_e$ for reflector antennas, corresponding to $L_n = 1.65$ or 2.2 dB, which is indicative of the greater loss of illumination power in spillover and blockage lobes. The area required in the aperture, for a given gain or beamwidth product, varies as $1/\eta_a = 1/\eta_x\eta_y$, as does the number of $\lambda/2$ spaced elements covering this area.

Elliptical and Circular Apertures

When low-sidelobe tapers are used for $g(x)$ and $g(y)$, the combined $g(x, y)$ near the corners of the aperture is very small, and those elements or portions of the aperture can be omitted with little effect on performance. The result is an elliptical aperture with gain and pattern that are essentially the same as those of the rectangle having the same width w and height h. The area of the ellipse and the number of elements are $\pi/4$ times those of the rectangle, and hence its aperture efficiency η_a will be greater by about $4/\pi = 1.05$ dB.

Consider a circular aperture of diameter D formed by omitting the corners from a square, $w = h = D$, and designed with 40-dB Taylor weighting in both coordinates (Figure 4.2.4). The original square aperture gave

$$\theta_3 = 1.25\lambda/w, \quad \eta_x = \eta_y = 0.763, \quad \eta_a = 0.582$$

The circle, with illumination adjusted slightly for -40 dB circular sidelobes, has

$$\theta_3 = 1.29\lambda/D, \quad \eta_a = 0.706, \quad \eta_a\pi/4 = 0.555$$

Considering the reduced aperture area of the circle, the circle is 0.84 dB more efficient than the square, with a beamwidth only three percent greater. The solid angle occupied by the principal sidelobes will, on the other hand, be greater than for the rectangular aperture, but in many cases the actual levels outside the main beam will be set by random errors in the illumination rather than by the intended $g(x, y)$. The relationship between gain and beamwidth is essentially the same as for the rectangular aperture. Table 4.2 compares parameters for common circular taper functions with those of rectangular arrays.

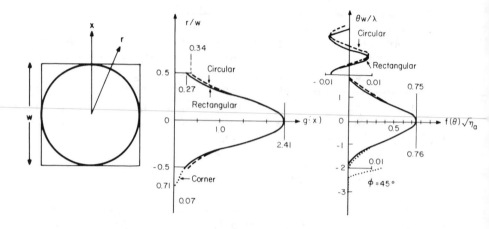

Figure 4.2.4 Circular aperture derived from square.

Table 4.2 Typical Parameters for Rectangular and Elliptical Antennas
(Taylor Illumination Functions)

Sidelobe Level	Parameter	Rectangular Aperture	Elliptical Aperture
−25	K_θ	1.05	1.12
	η_a	0.81	0.91
−35	K_θ	1.18	1.23
	η_a	0.65	0.77
−45	K_θ	1.30	1.33
	η_a	0.53	0.65

NOTE: $K_\theta = \theta_3 w/\lambda$ or $\theta_3 D/\lambda$. In all cases, $L_n = 4\pi/\theta_3^2 = 1.11$.

Antenna Sidelobes

Ideally, the sidelobe levels of an antenna would be determined by the design illumination function $g(x, y)$ in accordance with (4.2.7). However, in practice, the sidelobes will be higher because of errors in generating this illumination. In reflector antennas, these errors result also from spillover, currents on the edge of the reflector, blockage of the aperture by the feed structure, and the shape of the reflector surface. In arrays, there will be errors in the phase and amplitude of the illumination of individual elements, resulting from both the feed network and the phase shifters, as well as edge current effects and mechanical deviations in location of the elements.

While it has long been customary to specify the sidelobe characteristics of antennas in terms of the peak lobes (usually the first pair beside the main lobe), it is often the average sidelobe levels (rms voltage) over a sector removed from the main lobe that will be important (Figure 4.2.5). This far sidelobe level can be calculated as the average gain of an additive pattern established by the aperture when illuminated by the error from the intended illumination function. For example, if the phase errors of small elements ($\lambda/2$) of the aperture are independent, and have an rms value σ_ϕ radians ($\sigma_\phi \ll 1$), a broad, random error pattern will be produced for which the power gain, averaged over the entire sphere, is $G_s = \sigma_\phi^2$ relative to isotropic gain. In the forward hemisphere, the average level will be about twice the value. Near broadside to the array, where the gain of each element is maximum ($\approx \pi$), the gain relative to the main-lobe gain γ_m is given by

$$G_s/G_m = \pi\sigma_\phi^2/G_0\eta_a = \sigma_\phi^2\lambda^2/A\eta_a \qquad (4.2.12)$$

When the phase errors are not random over the array, but correlated in rows, columns, or groups (e.g., subarrays), the average sidelobe level remains unchanged, but particular lobes will be much higher than the average.

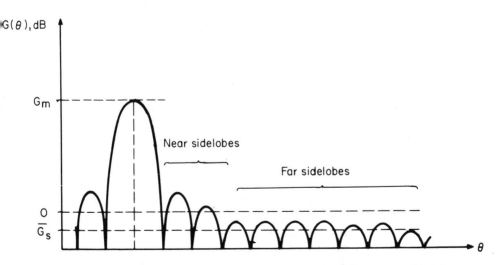

Figure 4.2.5 Typical antenna pattern showing peak and average sidelobe levels.

Reflector Antennas

A paraboloid of revolution, illuminated by a feed at the focal point, reflects a plane wave along its axis (Figure 4.2.6). If we measure the field in a plane just beyond the feed, it will appear as if it had been radiated from a planar array with the illumination $g(x, y)$ of the reflector, but with three important differences:

(a) A portion of the aperture, shadowed by the feed and its support structure, will not contribute to the field, producing a broad pattern of *blockage lobes* in the forward hemisphere;
(b) The power intercepted by the feed structure will be scattered over a wide angle, mostly in the rear hemisphere occupied by the reflector;
(c) Additional power, radiated by the feed, will miss the reflector and produce a second, direct *spillover* pattern in the rear hemisphere.

The far-field antenna pattern will thus consist of the sum of the intended pattern $f(\theta, \phi) \leftrightarrow g(x, y)$ plus the three unintended components: blockage, feed scattering, and spillover. In practice, the illumination taper will establish the level only of the principal sidelobes, blockage will control the level in most of the forward hemisphere, while feed scattering and spillover will control the level in the rear hemisphere (which may also have components caused by reflector edge effects: diffraction and currents circulating around the edges onto the rear surface). The lobes caused by direct feed spillover can be calculated readily from the pattern of the horn, but the other components usually require measurement.

Aperture Blockage

The blockage pattern is sometimes estimated as the radiation pattern of an aperture area A_b represented by the optical shadow area of the feed structure, negatively phased and subtracted from the basic reflector pattern (Figure 4.2.7). Applying (4.2.7) for the unblocked reflector pattern, and assuming that the blocked area is fully illuminated, $g(x,y) = 1$ over A_b, the on-axis voltage for the blockage pattern (normalized to the voltage of the unblocked main pattern) is

$$f_b(0, 0) = -A_b/A' = -A_b/\iint g(x,y) \, dxdy \approx -2A_b/A \qquad (4.2.13)$$

The blockage lobe amplitude (not its power) is proportional to the blockage area because both the power intercepted by the blockage and the power gain of the resulting pattern are proportional to this area. Values of A'/A for different illumination functions may be found from tabulations in Appendix A of [4.3]. The typical value of 1/2 used in the final approximation of (4.2.13) applies to circular apertures with sidelobes near -25

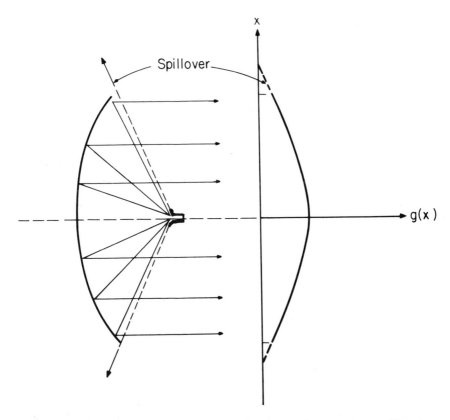

Figure 4.2.6 Parabolic reflector and equivalent planar aperture illumination.

dB, and this value will decrease with lower-sidelobe tapers. The half-power width of the blockage lobe will be approximately λ/w_b, where w_b is the width of the blockage area. This lobe will extend over the original main lobe and its first several sidelobes, reducing the main-lobe amplitude and those of even sidelobes while increasing the odd sidelobes.

The blocked main-lobe pattern will have an amplitude given by

$$f'(0, 0) = 1 + f_b(0, 0) \approx 1 - 2A_b/A \qquad (4.2.14)$$

The first sidelobe, at an angle $\theta_s \approx 1.6\theta_3$, will be negative, with $f(\theta_s, 0) \approx -0.056$. After blockage, its magnitude will increase to

$$f'(\theta_s, 0) = f(\theta_s, 0) + f_b(\theta_s, 0) \approx -0.056 - 2A_b/A \qquad (4.2.15)$$

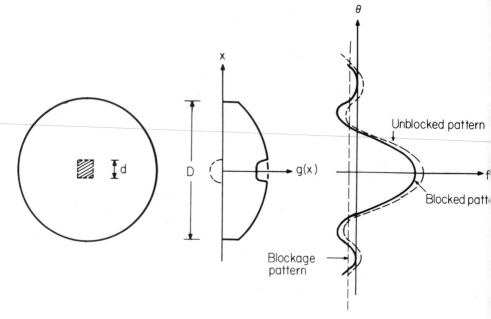

Figure 4.2.7 Blockage lobe effects.

For example, if the feed blocks a circle with diameter $d = D/10$ at the center of a circular aperture, the blockage lobe will have a relative amplitude f_b $(0,0) = 2/10^2 = 0.02 = -34$ dB. The main-lobe voltage gain will be reduced to $0.98 = -0.2$ dB, and a typical first sidelobe level will be increased from $0.056 = -25$ dB to $-0.076 = -22.4$ dB.

In many cases, the feed structure extends a considerable distance along the reflector axis, increasing its effective blockage through end-fire coupling phenomena. Use of the geometric shadow to estimate blockage in these cases will give optimistic results. Structure which supports the feed also must be taken into consideration in establishing blockage.

Cassegrain Reflector Antennas

The Cassegrain antenna (Figure 4.2.8a) is the most commonly used of the folded-optics type of reflector system. The hyperboloidal subreflector of diameter d intercepts the rays coming from the main paraboloidal reflector, focusing them on the feed. Waveguide runs from the feed to the transmitter and receiver can be made shorter than in front-fed designs, and the supporting structure for the subreflector can be lighter, with less

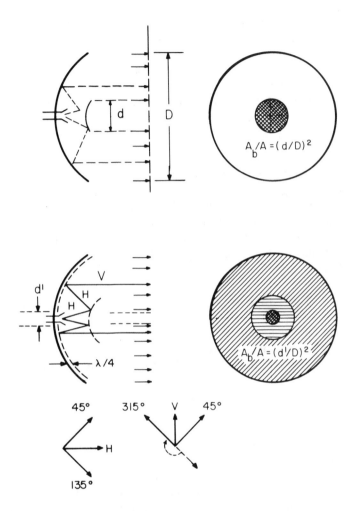

Figure 4.2.8 Cassegrain reflector configurations: (a) solid subreflector system; (b) polarization-twist Cassegrain system.

blockage. The subreflector is typically allowed to have a diameter about 1/10 that of the main reflector, giving blockage lobe effects as calculated for the earlier example.

The subreflector can be increased in size, without blockage, if the polarization-twist Cassegrain configuration is used (Figure 4.2.8b). In this case, the subreflector is composed of thin wires embedded in a plastic surface which establishes the hyperboloidal shape. In a typical tracking antenna, these wires will be horizontal, and the feed will be horizontally polarized. The horizontally polarized illumination will reach the surface of the main reflector, where it is rotated to vertical by a grid of wires oriented at 45° angle and located $\lambda/4$ in front of the solid reflector surface. The radiation into space is then vertically polarized, passing through the subreflector without blockage. The return signal goes through the process in reverse, appearing horizontally polarized at the feed. The blockage area can be limited to that occupied by the feed itself.

Cosecant-Squared Reflector Antennas

In two-dimensional search radar, the desired coverage pattern is often shaped as in Figure 4.2.9a, with the long-range lobe covering angles up to θ_1 and a broadened main-lobe skirt extending to θ_2. To produce a detection contour at constant altitude from θ_1 to θ_2, the gain of the antenna should follow the equation:

$$G\,(\theta) \,=\, G\,(\theta_1)\,\csc^2\theta/\csc^2\theta_1, \quad \theta_1 < \theta < \theta_2 \tag{4.2.16}$$

The resulting *cosecant-squared antenna* pattern will have a peak gain in the region $\theta < \theta_1$, which is reduced, relative to that of a simple fan beam, by diversion of energy into the upper elevations. By integrating the upper energy term, we find

$$G_{cs} \,=\, G_m/L_{cs} \,=\, G_m/(2-\theta_1\cot\theta_2) \tag{4.2.17}$$

Aperture efficiency is thus reduced by at most 3 dB (usually 2 dB) in order to obtain the upper coverage.

In reflector antennas fed by a simple horn, the energy is diverted into the $\csc^2$ sector by distortion of the parabolic surface (typically involving the lowest 1/3 of the reflector, Figure 4.2.9b). Multiple-horn feeds can be used, with power apportioned to produce the desired pattern.

Efficiency of Reflector Antennas

The efficiency of the reflector antenna system may be written as the product of factors expressing effects of the illumination function, blockage, spillover, and surface tolerances:

$$n_a \,=\, n_i n_b n_s n_t \tag{4.2.18}$$

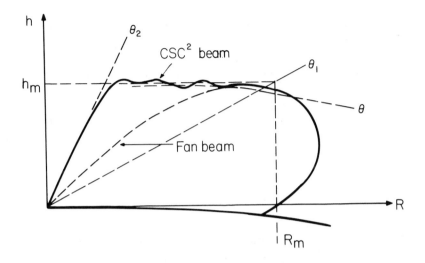

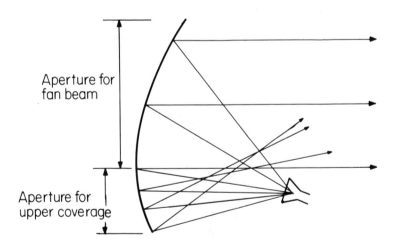

Figure 4.2.9 Coverage pattern and reflector shape for csc² search radar antenna.

Typical values of n_a lie between 0.5 and 0.63, or -3.0 to -2.0dB, with illumination taper loss near 1.5 dB and the balance of the loss divided between blockage and spillover. Loss caused by surface tolerance is normally small, except for antennas with very large D/λ (beamwidths less than about 0.5°). In most designs, it is necessary to strike a balance between the illumination efficiency, blockage, and spillover, because the large horn structures necessary to achieve control of illumination and small spillover will cause more blockage at the center of the aperture. Offset-fed paraboloids and polarization-twist Cassegrain designs provide solutions to this problem.

4.3 PHASED ARRAYS

The phased array is an electronically steerable antenna, in which each element or row of elements has its phase shift controlled by a computer command. The feed system and radiating elements are similar to those in fixed arrays, except that element spacing must generally be closer, to suppress grating lobes at maximum scan angles. The main classifications of phased arrays are shown in Table 4.3.

Table 4.3 Classification of Phased Arrays

Scan coordinates*	Elevation only (mechanical in azimuth)
	Azimuth and elevation
Steering method	Phase scan
	Frequency scan
	Switched beam-forming matrix
Feed method	Constrained (corporate)
	Space (optical)
Amplifier location	At feed input
	At rows or columns
	At subarrays
	At individual element
Amplifier function	Receiving only
	Transmitting and receiving

*Azimuth-elevation coordinates may be replaced by direction cosines, yaw-pitch, or any other selected pair of angles.

Element and Array Factors

At an angle θ from broadside, the gain of a phased array depends on the product of two factors:

$$f(\theta) = f_a(\theta)f_e(\theta) \qquad (4.3.1)$$

where $f_a(\theta)$ is the array pattern factor, depending on the complex illumination $g(x, y)$, and $f_e(\theta)$ is the element pattern factor. The array pattern factor is given by (4.2.7) and (4.2.9), and its effect on aperture efficiency, beamwidth, and sidelobe level has already been considered. When the antenna is an electronically scanned array, however, the phase gradient across the aperture can be controlled to steer the beam over a broad region centered on the broadside direction.

Phase Scanning

In the previous section, $g(x, y)$ was considered to be a real, symmetric function, producing a real pattern with its beam normal to the aperture plane. In the phased array, we usually find that $g(x, y)$ can be regarded as the product of a real, symmetric illumination function and a linear phase function that steers the beam. In Figure 4.3.1, we can see that the phase advance of the wave arriving at an element at x, for a steering angle θ in the plane containing x, is given by

$$\phi(x, \theta) = (2\pi x/\lambda) \sin\theta \qquad (4.3.2)$$

If the steering is to be in a slant plane, the phase $\phi(x, y, \theta, \phi)$ is calculated with x replaced by $x\cos\phi + y\sin\phi$. This phase is normally applied modulo 2π, so that the phase shifter need only be controllable over that range. The phase delay to be inserted between elements is

$$\Delta\phi = (2\pi d/\lambda) \sin\theta \qquad (4.3.3)$$

the delay being greater for increasing x. Referring to (4.2.7), this phase term in $g(x, y)$ serves to cancel, at each point on the aperture, the corresponding phase advance caused by the off-broadside angle θ, and hence to match the antenna to signals propagating to or from that angle in space. As a result, the full array gain $f_a = 1$ is brought to bear at that angle.

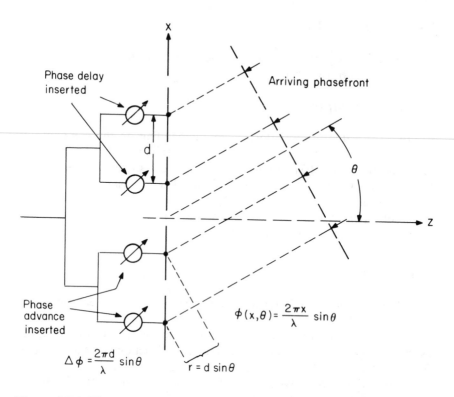

Figure 4.3.1 Phase relationships for beam steering.

Ideally, $f_e^2(\theta) = \cos\theta$, so that the array gain is proportional to the aperture area projected normal to the beam direction. The short dipole would have $f(\theta) = \cos\theta$, while the half-wave dipole would produce an even narrower pattern. Hence, radiating elements based on the dipole will normally have additional structure associated with each element to spread the pattern over the desired scan angle. In order to obtain proper matching and to control mutual coupling between elements as the beam scans, it is often necessary to configure the radiating elements such that

$$f_e^2(\theta) = \cos^{3/2}\theta \tag{4.3.4}$$

A compromise element matching can give a relationship approaching $\cos\theta$, but with a fixed loss of about one dB relative to the usual $G_e(0) = \pi = 5$ dB.

Dipole radiators, crossed dipoles (turnstiles), or open-ended wave-guides are often used in two-coordinate steered arrays. When mechanical rotation of the array, or frequency scanning, is employed for one coordinate, slotted wavegide radiators are often used instead.

The spacing of elements is determined by the need to restrict power to the desired main lobe, suppressing the grating lobe while scanning over the entire field. The peak of the grating lobe appears in real space (at 90° from broadside) when the spacing d between elements is

$$d = \lambda/(1 + |\sin\theta|) \tag{4.3.5}$$

Thus, if the maximum scan angle $\theta_m = 60°$, $d < \lambda/1.866 = 0.54\lambda$ is required to suppress grating lobes. Since the radiated lobe is the product of array pattern and element pattern, and the latter is essentially zero at 90°, spacings of 0.6 to 0.65λ are generally permissible without significant grating lobe problems. Arrangement of the elements in a triangular grid helps suppress grating lobes without raising the number of elements in the array too close to the value $4A/\lambda^2$, which would result from half-wavelength spacing in both coordinates. For wide-angle scanning arrays, the number of elements can be estimated as

$$T = n_a \times n_e = (1.4w/\lambda)(1.4h/\lambda) = 2A/\lambda^2 \tag{4.3.6}$$

Expressing T in terms of the product of the beamwidths, for a tapered illumination with $\theta_3 = K_\theta\lambda/w \approx 1.25$, we find

$$
\begin{aligned}
T &= 2K_\theta^2/\theta_a\theta_e \approx 3.14/\theta_a\theta_e \quad \text{(in radians)} \\
&= 10{,}300/\theta_a\theta_e \qquad\qquad \text{(in degrees)}
\end{aligned}
\tag{4.3.7}
$$

which is the same result as given in (4.2.5) for uniform illumination with $\lambda/2$ spacing.

For scanning over more limited sectors ($\theta_m < 20°$), directional elements may be used; subarrays fed from a single phase shifter, large horns, Yagis or other end-fed directional elements. Control of grating lobes is made difficult by both the closer spacing of the lobes, at intervals such that $\sin\theta_i - \sin\theta_0 = i\lambda/d$, and the effects of control quantization over the smaller number of elements.

Frequency Scanning

The scanning of a planar array to an angle θ in the plane containing the x axis requires inserting of equal phase increments $\Delta\phi$ between elements, as given by (4.3.3). A simple way of controlling phase shift is to locate feed slots with spacing s along a waveguide (Figure 4.3.2), and to steer the beam by changing radar frequency. The cut-off wavelength of a rectangular waveguide is $\lambda_c = 2a$, where a is the larger dimension of the guide. Shorter wavelengths will propagate in the guide, but as $\lambda \to \lambda_c$ the phase velocity increases, following the relationship:

$$c/c_g = \lambda/\lambda_g = \sqrt{1 - (\lambda/\lambda_c)^2} \qquad (4.3.8)$$

The phase increment between slots spaced a distance s along the guide will be

$$\Delta\phi = -2\pi s/\lambda_g \qquad (4.3.9)$$

where the negative sign implies a phase delay in the waveguide. For a straight length of waveguide $d = s$. Alternate slots will normally be given an additional phase shift π by their location or angle relative to the waveguide wall, so that the incremental phase delay will be

$$\Delta\phi = (2\pi s/\lambda_g) - \pi \qquad (4.3.10)$$

To place the beam at broadside, for a frequency $f_0 = c/\lambda_0$, we set $\Delta\phi = 0$, and the slot spacing is

$$d = \lambda_{g_0}/2 = a/\sqrt{(2a/\lambda_0)^2 - 1}$$

Then, at other frequencies, the beam will squint by an angle

$$\theta = \sin^{-1}\left[(\lambda/\lambda_g) - (\lambda/\lambda_{g_0})\right] \qquad (4.3.11)$$

The derivatives of squint angle with wavelength and frequency are

$$\begin{aligned}
d\theta/d\lambda &= (1/\lambda\,\cos\theta)(\sin\theta - \lambda_g/\lambda) \\
d\theta/df &= (1/f\,\cos\theta)[(\lambda_g/\lambda) - \sin\theta]
\end{aligned} \qquad (4.3.12)$$

For example, using standard WR-340 S-band waveguide ($a = 0.086$ m) with a center frequency of 3.0 GHz ($\lambda = 0.1$ m), the guide wavelength would be 0.123 m and the slots would be spaced $d = 0.0615$ m apart for a broadside beam. At other frequencies, the beam would squint to an angle such that

$$\sin\theta \approx \theta = \sqrt{1 - (\lambda/0.172)^2} - \lambda/0.123 \approx 1.23\,\Delta f/f_0$$

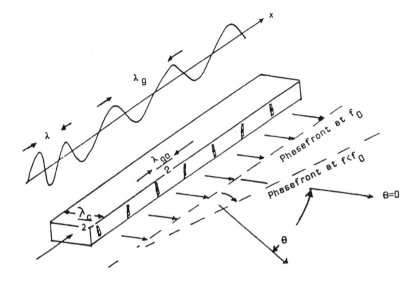

(a) Squinted beam from slotted waveguide.

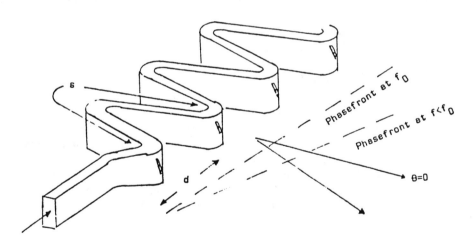

(b) Serpentine waveguide from increased squint.

Figure 4.3.2 Frequency scanning waveguide antennas.

One percent frequency increase would shift the beam by 0.0123 radians or 0.7°.

In order to scan over wider angles (e.g., 45° within the usual 5% tuning band), the waveguide feed is formed into serpentine or toroidal loops (Figures 4.3.2b) to give $s = n\lambda_{g_0} \gg d$, where n is an integer for identical slots, or an integer $+ 1/2$ for out-of-phase slots. The spacing d of the radiating elements is selected to give the desired scan angle without grating lobes. The frequency scan equation now becomes

$$\theta = \sin^{-1} \{[s/d] [(\lambda/\lambda_g) - (\lambda/\lambda_{g_0})]\} \tag{4.3.13}$$

the derivatives of scan angle with respect to λ and f are now multiplied by the ratio s/d, to give

$$d\theta/d\lambda = (1/\lambda \cos\theta) (\sin\theta - s\lambda_g/\lambda d)$$
$$d\theta/df = (1/f \cos\theta) [(s\lambda_g/\lambda d) - \sin\theta] \tag{4.3.14}$$

Using the waveguide of the previous example, but with a serpentine loop to give $s = 12d$, the steering sensitivity becomes 0.148 radians or 8.5° per one percent frequency change, near broadside, which is sufficient to give the required scan coverage.

As the beam scans from broadside (a line perpendicular to the array face), its width varies according to

$$\theta_3 (\theta) = \theta_3 (0) \sec\theta \tag{4.3.15}$$

When slotted waveguides are used for the rows of an array, coupled to a serpentine for elevation steering, the beam actually scans around a cone that has its apex aligned with the slotted waveguides, and a half-angle of $\tan^{-1} s/d$, rather than in a plane (as it would for $s/d \to \infty$). The array is often tilted by an angle that retains a nearly vertical scan at the horizon and at the highest elevation. No severe problems are encountered as long as the maximum elevation does not exceed about 20°.

Phase Error Effects

Electronic scanning at a constant frequency is performed by using digitally controlled phase shifters at each element (or each row, for one-dimensional scanning). Equations (4.3.2) and (4.3.3) apply, with ϕ calculated for each element position x,y. The digital-to-phase conversion process, carried out to m-bit precision, introduces a phase error as shown in Figure 4.3.3, for which the standard deviation is

$$\sigma_\phi = \frac{2\pi}{2^m \sqrt{12}}$$ (4.3.16)

This phase quantization error causes three distinct effects on the array radar system:

(a) *Quantization loss:*

$$L_q = 1 + \frac{\pi^2}{3 \times 2^{2m}}$$ (4.3.17)

(b) *Increased sidelobes:* The peak sidelobes for scanning in a cardinal planes, or for a linear array, are

$$G_i/G_m = (1/i2^m)^2$$ (4.3.18)

where i is the sidelobe number (Figure 4.3.3b).
The average sidelobes for independent phase errors are

$$\overline{G}_s/G_m = \frac{L_q - 1}{T\eta_a} = \frac{\pi^2}{3 \times 2^{2m}T\eta_\alpha}$$ (4.3.19)

(c) *Beam-steering error:* The peak angle error for scanning in a cardinal plane, or for a linear array is

$$\epsilon_\theta = \frac{1.2\lambda}{2^m w \cos\theta} \sqrt{G_d(\theta)/G_0} \approx (\theta_3/2^m) \sqrt{G_d(\theta)/G_0}$$ (4.3.20)

where θ is the scan angle and $G_d(\theta)$ is the normalized pattern derivative, or monopulse pattern, for this angle [4.3].
The average angle error for independent phase errors is

$$\sigma_\theta = \frac{\theta_3 \sigma_\phi}{1.61 \sqrt{T}}$$ (4.3.21)

For example, when scanning in a cardinal plane or with a linear array, controlled by three-bit phase shifters, we have

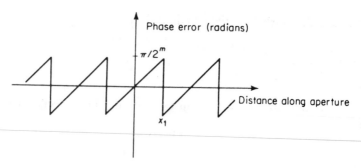

(a) Phase quantization error.

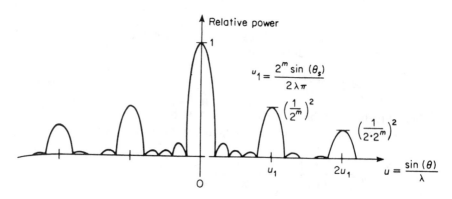

(b) Antenna pattern with phase quantization error.

Figure 4.3.3 Effect of phase quantization on array [4.3].

$$\sigma_\phi = \pi/8 \sqrt{3} = 0.23 \text{ radian}$$
$$L_q = 1 + 14.3/64 = 0.22 \text{ dB}$$
$$G(u_1)/G_m = 1/64 = -18 \text{ dB}$$
$$G(u_2)/G_m = 1/256 = -24 \text{ dB}$$
$$\epsilon_\theta = \pm \theta_3/8 \text{ at } \theta = \pm \theta_3/8 \text{ (no beam motion has yet occurred)}$$

In a typical two-coordinate scanning array of $T = 2000$ elements, $\theta_3 = 0.04$ radian $= 2.3°$ at broadside. When scanned in an intercardinal plane such as to randomize the quantizing errors, we have (for $m = 3$):

$$L_q = 0.22 \text{ dB}$$

$$\overline{G}/G_m = \pi^2/3 \times 64 \times 2000 \times 0.7 = 3.7 \times 10^{-5} = -44.4 \text{ dB}$$

$$\sigma_\theta = 0.04 \times 0.23/1.61 \cos\theta \sqrt{2000} = 0.125 \times 10^{-3}/\cos\theta \text{ (radians)}$$

(corresponding to 0.125 mr near broadside, or 0.25 mr near 60° scan).

Practical phase shifter and feed designs introduce additional random errors, often equal to the value of σ_ϕ caused by quantization. The loss in dB is then doubled, as are the sidelobe powers and the variance of the pointing error.

In corporate-fed arrays with regular element spacings, phase commands are usually computed for each row and column, and the appropriate row and column values are added together to obtain the command for a given element. In optically fed planar arrays, the commands may be obtained in the same way, if fixed phase shifts are applied in the electrical paths to focus the array on the feed (removing the spherical-to-planar phase term caused by array geometry, see Figure 4.4.2). A better procedure from the standpoint of performance is to apply the focusing correction in digital form at each element, adding the correction to the row and column commands before truncation of the command to m bits. This causes the phase errors to be almost entirely random at all steering angles, avoiding the peak errors and peak quantization sidelobes which otherwise occur.

As with frequency scanned arrays, there is a scan loss in gain, controlled by the element pattern factor according to (4.3.4), and a broadening of the beam according to (4.3.15).

Thinned Arrays

To reduce the number of radiating and phase shifting elements (and the number of amplifiers, in an active array), dummy elements are sometimes used at random locations in the array. A fraction of active elements F_a can be distributed in a random way so as to create the desired average taper of illumination, while avoiding the generation of large sidelobes in particular directions.

The total number of active elements required to establish given beamwidths in the thinned array is proportional to F_a:

$$T_f \approx 3.14 F_a/\theta_a\theta_e \quad \text{(in radians)}$$
$$= 10,300 F_a/\theta_a\theta_e \quad \text{(in degrees)} \tag{4.3.22}$$

The narrow beam formed by a thinned array is advantageous in resolving closely spaced targets, achieving high tracking accuracy, and reducing the clutter within the main beam. However, three major disadvantages attach to the use of thinning.

(a) The gain is reduced, when compared with the fully filled, uniformly illuminated array, but not when compared to an array with the same tapered illumination:

$$G_f(0) = G_0 F_a = \pi T F_a = \pi T_f \tag{4.3.23}$$

where T is the total number of element positions in the array.

(b) The presence of inactive elements causes an increase in the average sidelobe level, corresponding to radiation of power $(1 - F_a)$ in a random pattern with an average gain pattern equal to that of the individual element. This average sidelobe level, relative to main-lobe gain at broadside, is then

$$\overline{G}_s/G_m = (1 - F_a) f_e^2 / \pi T F_a \tag{4.3.24}$$

where $f_e(\theta)$ is the element pattern factor. Highly thinned arrays have average sidelobes approaching isotropic level, and can only be used when T is very large and when jamming is absent.

(c) In the case of a search radar, or a tracker in the acquisition mode, the reduced gain will apply for both transmitting and receiving, but the gain reduction is not accompanied by the increase in beamwidth that would be experienced by a smaller antenna with the same gain. Hence, the effective A_r in (1.3.3) must be reduced by F_a^2, rather than by F_a. This rapid reduction in search efficiency rules out the use of thinning in cases where volume scanning is an important radar function.

Phase Shifters

The element phase can be controlled by diode switching of RF paths or by variation in waveguide phase velocity through ferromagnetic action. The control input in either case is digital, but, with the ferrite, a D/A conversion process is used to control current or residual magnetization (through control of $\int i\, dt$). Neither approach is inherently superior in all respects, as both have their advantages and disadvantages (see Table 4.4).

Table 4.4 Phase Shifter Characteristics

Characteristic	Diode	Ferrite
Controllable element	Switched paths	Adjustable phase velocity
Control input	Inherently digital	Inherently analog, via D/A converter
Switching speed	Fast	Slower than diodes
Power limitation	Surface density	Volume density
Power rating	Lower	Higher
Reciprocity	Reciprocal	Nonreciprocal, or reciprocal with higher loss
Loss	Higher	Lower
Control current	Continuous	Latching or continuous
Cost	Lower	Higher
Size	Smaller	Larger
Weight	Lower	Higher

While the qualitative characteristics listed for a particular type of phase shifter will apply generally, specialized designs can overcome a disadvantageous characteristic, often at the expense of another. Still, the ferrite phase shifter has been preferred in high-power radars without element amplifiers, while diodes are generally used in lower power systems, receiving arrays, and active amplifier arrays (where they appear at the inputs to the final power amplifiers and the outputs of the low-noise receiving amplifiers).

Array Feed Systems

Two basic methods exist for coupling transmitter power to (and collecting receiver power from) the elements of an array:

(a) the *constrained feed*, in which the signal is confined to waveguide stripline, or coaxial structures with discrete branches connecting to the elements;

(b) the *space feed*, in which the signal is radiated from a horn and propagates as a wave through air or a continuous dielectric medium to reach the elements.

The general characteristics of the two types are shown in Table 4.5.

It was at one time believed that constrained feeds were required to produce the precise illumination tapers of ultra-low-sidelobe antennas (see

Table 4.5 Feed System Characteristics

Constrained (Corporate)	Space (Optical)
Closed circuits	Radiation through space
Coupler control of illumination	Illumination determined by horns in focal plane
No spillover	No divider loss
No rear-face reflections	No divider mismatch
Complex coupler networks	No networks unless many horns are coupled
Flat panels possible	Conical volume required
Monopulse networks very complex	Monopulse horn clusters easily implemented
Multiple-beam networks very complex	Multiple-beam feeds easily implemented

Section 4.4 below). However, advances in feed horns and networks, including the focal-plane array feed (or transform feed) have shown that space-fed arrays and even parabolic reflector antennas can be designed with ultra-low sidelobes. Hence, no difference in sidelobe performance is listed in the table.

Constrained feeds are classified as either series or parallel networks. The series feed is simpler, and may be implemented as a slotted waveguide or with a number of two-port couplers in series, each diverting the power required for its associated element or group of elements (Figure 4.3.4). The usual series feeds introduce squinting of the beam (if end-fed) or defocusing (if center-fed) as a function of frequency. Parallel feed networks (Figure 4.3.5) split the input power into two or more fractions, and continue to subdivide each of these until the levels for each element are reached. Either type of feed may be designed for equal path lengths to each element, providing a very wide bandwidth, but this approach is more common in parallel feeds.

A type of feed often used, for economy and ease of fabrication, is the slotted waveguide. This is similar to the frequency scanning structure discussed earlier in this section, and indeed some squint (in an end-fed line) or defocusing (in a center-fed line) is inevitable with this approach. The individual element phase shifters can compensate for this at a specific frequency, but the instantaneous signal bandwidth may be affected. It is the ability to control phase and amplitude very precisely by milling of the waveguide slots that makes this feed very attractive, especially for lines that radiate directly into space (as in arrays with phase scanning in only one coordinate).

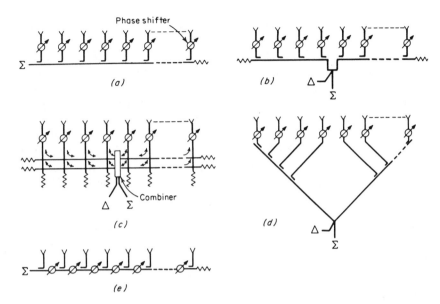

Figure 4.3.4 Series feed networks [4.4]. (Copyright 1970 by McGraw-Hill Book Co. Reprinted with permission.)

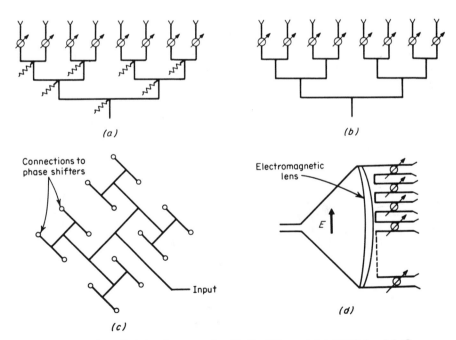

Figure 4.3.5 Parallel feed networks [4.4]. (Copyright 1970 by McGraw-Hill Book Co. Reprinted with permission.)

The space-fed array may be implemented either as a reflector or a lens (Figure 4.3.6). The lens may be a planar array of phase shifting elements, or a passive lens of the Rotman type (Figure 4.3.7). The Rotman lens is a one-dimensional divider, and so to produce narrow beams in both dimensions it is necessary to stack a number of lenses on top of each other, connecting their outputs with a second stack of lenses, side by side, to form the two-dimensional beam matrix (Figure 4.3.8). The space-fed planar array is a two-dimensional system that can form a pencil beam. Additional beams may be formed simultaneously, within a limited angle sector, by adding more feed horns in the focal plane of the array, in which case the entire beam cluster is scanned together by the phase shifters.

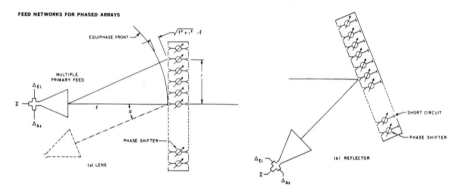

Figure 4.3.6 Space-fed arrays [4.4]. (Copyright 1970 by McGraw-Hill Book Co. Reprinted with permission.)

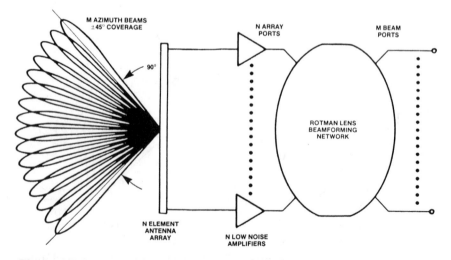

Figure 4.3.7 Array with Rotman lens feed [4.5].

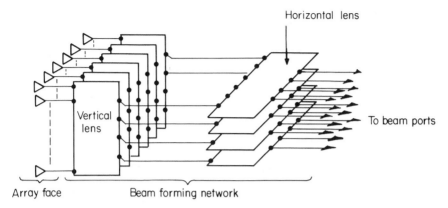

Figure 4.3.8 Two-dimensional Rotman lens feed [4.6].

Space feeds have a major advantage in cost, as their power division occurs in a bulk dielectric medium (usually air). The array elements must now couple efficiently into this medium on the feed side, as well as into space on the opposite side of the array face (in a reflector array, a short circuit is placed at one end of the phase shifters, and coupling to the feed and to beam space is on the other end). A disadvantage of this approach is that the feed horn structure, which launches the wave into the medium, must establish the desired illumination function for low sidelobe operation, without excessive spillover. Further advantages are the ease with which smooth functions may be produced (especially for monopulse operation), and with which multiple beam ports may be established for monopulse, stacked-beam, and other applications. In this respect, the contrast with constrained feeds is significant.

Amplifier Arrays

Regardless of the feed arrangement, it is possible to place the final transmitter amplifier and the first receiving amplifier at the array element. In constrained feed systems, these amplifiers may be placed, instead, at any level in the dividing network: at subarrays, or at row or column level, as well as at the individual elements. An advantage of placing amplifiers at the element (or at rows in a one-dimensional scanning array) is that the phase shifter may be placed on the feed side of the amplifiers, reducing its power rating and avoiding loss in output power and receiving noise factor. Only the final duplexer (a switch or circulator device) and a short cable or guide to the radiating element will then appear in the loss budget for L_t and L_r. In a typical solid-state modular array, separate feed networks are used for transmitting and receiving, providing uniform illumination for efficient transmitting, while tapering the receive illumination for low side-lobes. A common phase shifter at each element amplifier is switched between the two feeds, while the radiating element is connected to the amplifiers through a circulator (Figure 4.3.9).

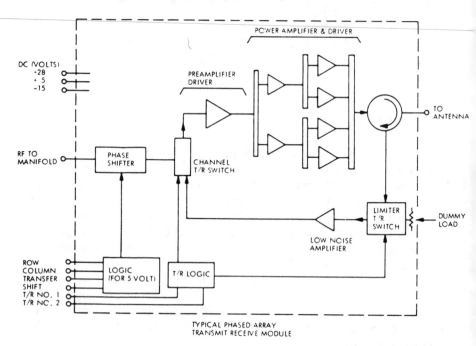

Figure 4.3.9 Typical phased array transmit-receive (T/R) module [4.7].

The total power of the amplifier array is limited only by the available prime power, the RF power that can be generated within the volume associated with each element, the heat that can be dissipated from this volume, and cost considerations. For microwave arrays, the available volume dictates solid-state power amplifiers, packaged in modules having a cross section of approximately $(\lambda/2)^2$. The limiting technical factor will then generally be heat dissipation, since an RF power P_{av} generated at efficiency η_t requires the dissipation of heat

$$P_{\text{heat}} = P_{av}[(1/\eta_t) - 1] \tag{4.3.25}$$

Practical heat-transfer technology limits P_{heat} to about 2 kW per m^2 of antenna aperture, when solid-state modules are embedded in the array. This gives

$$P_{av} < 2000A\eta_t/(1 - \eta_t) \approx 1000T\lambda^2\eta_t/(1 - \eta_t) \qquad (4.3.26)$$

At X-band, with $\eta_t = 0.17$, the average transmitter power in watts, for an active array, is limited to about $T/5$. This limitation may be compared to that of the passive phase-shifting array in which the phase shifter loss is 1 dB. The calculations may be made by setting $\eta_t = 1/L = 0.8$ in (4.3.26), giving $P_{av} = 2T$ at X-band.

Beam-Forming Matrices

To avoid use of controllable phase shifters or frequency scanning, an electronically steerable array can be based on use of a beam-forming matrix and a switch. Examples are the (analog) Rotman lens (Figure 4.3.7), and the (discrete) Butler and Blass matrices (Figure 4.3.10). In each case, an array with aperture $A = T\lambda^2/2$ can form T possible beams in space, which are selectable by switching. For transmission at high power level, the problem of switch implementation can be extreme. Use of amplifiers between the matrix and the radiating elements can reduce this problem, but even in this configuration the fast microwave crossbar switch is a difficult and expensive component to obtain, as compared with phase shifters at each amplifier.

With a beam matrix, search and tracking operations can be carried out by sequentially switching the transmitter and receiver among the beams, or by illuminating the entire sector with a transmitting pattern and then receiving and processing all beams in parallel to detect and track targets. This subject will be discussed further in Section 4.5. Beam matrices used only for receiving may be implemented at IF (Figure 4.3.11) or in digital form (Figure 4.3.12), as well as in the RF networks already described. In the IF matrix, amplifiers for each element produce outputs at 0°, 90°, 180°, and 270° phase, from which a resistive network can select signals of arbitrary gain and phase. The number of parallel beams formed in such a network is limited only by the gain of the preceding amplifiers and the required isolation between beam channels. This approach is most appropriate for a stack of beams formed in one dimension. The equivalent digital approach can readily form two-dimensional matrices of beams, limited only by processor throughput. The required throughput will be

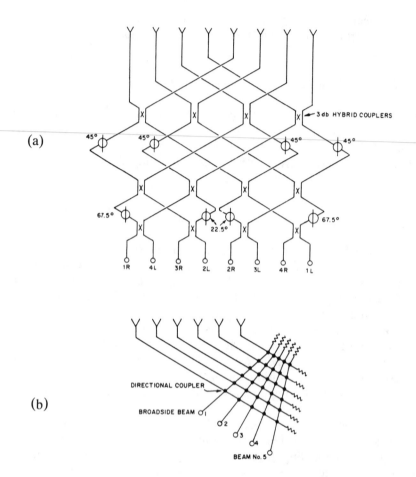

Figure 4.3.10 RF beam-forming matrices: (a) Butler matrix; (b) Blass matrix [4.4]. (Copyright 1970 by McGraw-Hill Book Co. Reprinted with permission.)

high in any case, because the M differently weighted complex sums of N element outputs must be obtained at a rate that is at least equal to the signal bandwidth. An advantage of forming the beams after amplification is that aribtrary beam shapes and spacings can be generated without concern for dissipative loss in RF networks. Low-sidelobe beams may be generated by combining several adjacent elementary beams with suitable weights.

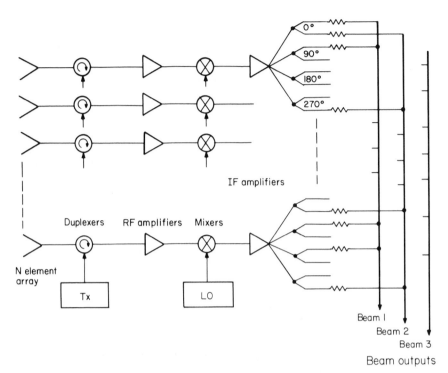

Figure 4.3.11 IF beam-forming matrix.

Phase and Amplitude Error Effects

Errors in the phase and amplitude of the feed network or the phase shifters will increase the losses, sidelobes, and beam pointing errors of the array system. The equations already developed for phase quantization errors may be used, but with phase and amplitude errors assumed random and characterized by their standard deviations σ_ϕ radians and σ_a/a. The following expressions are written for phase errors, and for amplitude errors the term σ_ϕ may be replaced by σ_a/a.

$$L_\phi = 1 + \sigma_\phi^2 \tag{4.3.27}$$
$$10 \log L_\phi = 4.3\,\sigma_\phi^2$$

$$\overline{G}_s = \pi\sigma_\phi^2 \tag{4.3.28}$$
$$\overline{G}_s/G_m = \sigma_\phi^2/T\eta_a$$

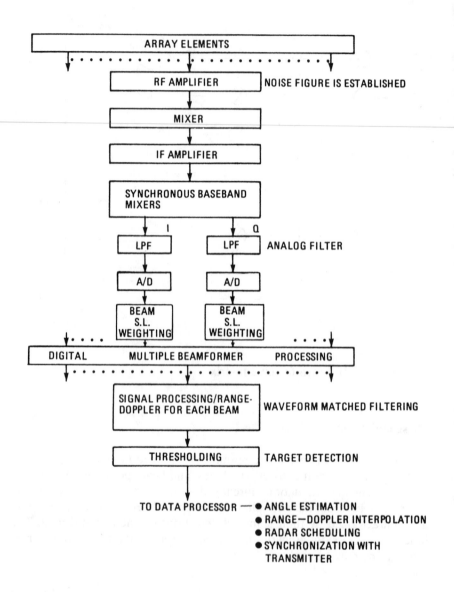

Figure 4.3.12 Digital beam-forming system [4.8].

For large arrays with reasonable errors (e.g., $T = 5000$, $\sigma_\phi = 0.3$, $\sigma_a/a = 0.2$, or 2 dB), the losses are small (0.5 dB) and the error sidelobes are lower than the principal sidelobes from the illumination function (-46 dB). However, in linear arrays ($T = 100$) or where ultra-low sidelobes are required, the error tolerances must be very strict. Active amplifiers in the array introduce additional opportunities for errors in both the phase and amplitude of the illumination function, requiring monitoring and correction circuits, if high performance is to be maintained.

Array Bandwidth

The signal bandwidth, which will be passed by a phased array without distortion, is dependent on both the feed design and the scan angle of the array. At broadside, the array can be characterized by the impulse response of the feed network: the sum of the element weights, each delayed by the electrical path length from the common antenna terminal. The received and processed signal waveform will be convolved twice with this feed impulse response, once upon transmission and again upon reception. Examples are shown in Figure 4.3.13. The intentionally elongated frequency-scan series feed has the longest impulse response, while the space-fed array and equal-line-length constrained feed have short impulse responses (approaching zero in the latter case). A space-fed array with focal length L will have an impulse response spread over a time $D^2/8Lc$, with the delayed components reduced by the illumination taper.

When scanned off broadside, an additional array delay for each element appears in series with that of the feed. In many cases, it is this array delay that will determine the instantaneous signal bandwidth of the radar. The array impulse response alone has the shape of the aperture illumination function. For scanning in the plane containing x,

$$h_a(t) = (x/c)(\sin\theta)g'(x) \tag{4.3.29}$$

where $g'(x)$ is the aperture illumination collapsed onto the x axis. The same relationship will apply with y substituted for x, when scanning in the plane containing y. Thus, for a circular array scanned at $\theta = 30°$, the one-way impulse response will spread over a time $D/2c$, tapered at the ends according to the illumination function.

In the end-fed series feed, this array delay may add to or subtract from that of the feed network, depending on the direction of scan:

$$h_a(t) = (x/c)[(\lambda_g/\lambda) + \sin\theta]g'(x) \tag{4.3.30}$$

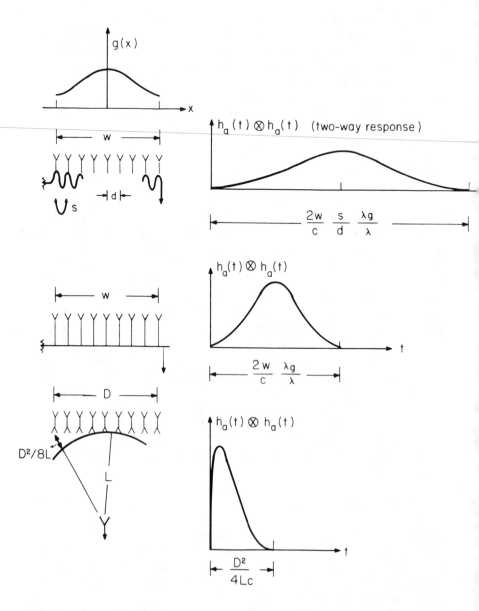

Figure 4.3.13 Impulse responses for arrays at broadside: (a) frequency-scanned array; (b) end-fed array; (c) space-fed array.

where λ_g is the wavelength in the guide, typically 20 to 30% greater than in space (see Eq. (4.3.8)). In frequency scanning arrays with serpentine feeds, the term (λ_g/λ) is multiplied by ratio s/d from Figure 4.3.2b.

The two-way impulse response of the array is given by the convolution of the transmitting and receiving array responses. When the same antenna is used for transmitting and receiving, its impulse response is convolved with itself, doubling the total time spread, but generally increasing the effective width only by $\sqrt{2}$. As the response approaches Gaussian shape with standard deviation σ_t, an equivalent 3-dB bandwidth can be found as

$$B_a = 0.265/\sigma_t \qquad (4.3.31)$$

If we assume that the feed is broadband, relative to the array at angles off boresight, we can use (4.3.31), and the relationship between the illumination function and the broadside beamwidth, to find the one-way bandwidth of the array, at an operating frequency f:

$$B_a = \theta_3 f/\sin\theta = cK_\theta/D \sin\theta \qquad (4.3.32)$$

for array diameter D. Thus, regardless of the illumination function, the fractional bandwith B_a/f of an array will be given by its beamwidth θ_3 divided by the sine of the scan angle, and the bandwidth for two-way operation will be 0.7 times this value. Any array producing a one-degree beam ($\theta_3 = 0.018$ radians, one-way) will have a two-way bandwidth equal to 1.2% of the center frequency, divided by the sine of the scan angle.

Wider bandwidths can be obtained by using time-delay steering techniques, in which elements or subarrays are fed through adjustable time-delay networks which compensate for $h_a(t)$ and for the internal feed impulse response.

The effect of limited array bandwith on SNR, for a system with a processed signal bandwidth B_3 is to introduce an array loss given by

$$L_z = 1 + 2B_3^2/B_a^2 \qquad (4.3.33)$$

This loss will be 1 dB when the array bandwidth is $B_a = 2.8B_3$. For example, an array diameter $D = 3.4$ m will generate a two-degree beam at S-band. When scanned to $\theta = 30°$, the array bandwidth $B_a = 200$ MHz, and signals with $B_3 = 70$ MHz (e.g., transmitted bandwidth of 100 MHz, received with a 100-MHz matched filter) will suffer a loss of 1 dB. The output pulse will be broadened by a factor of

$$\tau_x/\tau_3 = \sqrt{1 + 2B_3^2/B_a^2} \qquad (4.3.34)$$

relative to its width with a wideband antenna.

4.4   ULTRA-LOW-SIDELOBE ANTENNAS

Definitions

There are no established standards defining low-sidelobe, very-low-sidelobe, and ultra-low-sidelobe antennas, but Table 4.6 may be used as a guide.

Table 4.6 Definitions of Sidelobe Levels

Antenna Description	dB below G_m		dB below isotropic
	Peak	Average*	Average*
Normal	> -25	> -30	> -3
Low-sidelobe	-25 to -35	-35 to -45	-3 to -10
Very-low-sidelobe	-35 to -45	-45 to -55	-10 to -20
Ultra-low-sidelobe (ULSA)	< -45	< -55	< -20

*Averages apply to the worst 30° sector starting 10° from the main lobe.

A very-high-gain antenna is bound to have low average sidelobes over most solid angles, since the average level outside the main beam must be less than $(1 - \eta_a)/G_m$ relative to the main beam, or $1 - \eta_a$ relative to isotropic. Hence, to fall within a given description, all three sidelobe criteria must be met.

Using (4.3.28), the random phase errors needed to obtain -20 dBi average sidelobes must not exceed 0.05 radians = 3° rms, if amplitude errors are absent. If phase and amplitude errors are allowed to have equal contributions, the requirement is $\sigma_\phi < 0.04$ radians = 2.2°, and $\sigma_a/a < 1.04 = 0.33$ dB. These tolerances are difficult to meet in electronically scanned arrays. An early example of an ultra-low-sidelobe array was the AWACS array, which used slotted waveguide radiating lines (Figure 4.4.1). This and the subsequent array design for the AN/TPS-70 3D radar have sidelobes so low that the manufacturer cannot disclose the numbers, for security reasons. Neither of these antennas, however, uses two-coordinate electronic scanning. Both rotate mechanically in azimuth, the AWACS having electronic elevation scan, and AN/TPS-70 having an elevation beam-forming matrix. For such arrays, the phase and amplitude errors introduced by the elevation network or phase shifters are identical across the rows of elements, producing much larger peak sidelobes in the elevation plane. It is only in the azimuth plane and slant planes that ULSA performance can be expected.

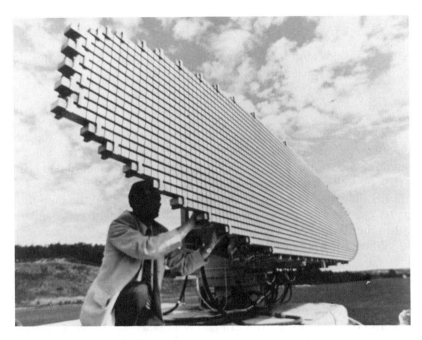

Figure 4.4.1 AWACS waveguide ULSA. (Photo courtesy of Westing-house Electric Co.)

Design of Scanning ULSA Systems

Arrays using phase shifters at each element, for two-coordinate scanning, are more difficult to design for ULSA performance. Random phase and amplitude errors can originate in the feed network, the phase shifters, or the radiating elements themselves. A major share of the total error must be assigned to the phase shifters, imposing very tight tolerances on the feed and the radiators. Consider an error budget to achieve -20 dBi average sidelobes, for which eight equal allocations are made, for the following error sources: phase shift insertion phase, phase quantization, phase response to control, feed network phase, phase shifter insertion amplitude, phase shifter amplitude change, feed amplitude, and radiator amplitude. The four phase errors will each be 0.02 radian $= 1.1°$ rms, and the amplitude errors will be 2% $= 0.17$ dB. A six-bit phase shifter will not quite meet the requirement (4.3.16).

The requirement for peak sidelobes at least 45 dB down from the main lobe is also difficult to meet, unless the array is very large. Consider a 4000 element array, for which the gain G_0 for uniform illumination would be (4.2.1):

$$G_0 = \pi T = 12,500$$

Figure 4.2.3 indicated that the aperture efficiency, for a Taylor function, would be 0.7 for each coordinate, or 0.5 overall:

$$G_m = \pi T \eta_a = 6250 = +38 \text{ dB}$$

The average sidelobes are then −58 dB from the main lobe, and peak sidelobes must be below −7 dBi, or +13 dB from the average. Since the random sidelobes are Rayleigh distributed, this requirement should be met in regions well off the main lobe, but the first few sidelobes, where the −45 dB illumination lobes are not much above the peak random lobes, can be expected to vary above and below their design value unless a 50-dB taper is applied.

When a corporate feed is used to illuminate an array $n_a \times n_e = 64 \times 64$ elements, the receiver will normally connect to the elevation network, from which will branch 64 networks. The 1.1° allowance for random errors in the row networks is difficult to meet, but the tolerance on the elevation feed is eight times tighter, since it contributes errors which are correlated across each row. In addition, when the array is steered in one of the cardinal planes, the peak sidelobes will be increased by the periodic phase quantization error (4.3.18). It is the practical impossibility of meeting requirements for low correlated errors that leads the designer to orient the array with the first feed in the elevation plane, where sidelobe requirements can normally be relaxed. Then, when the array is steered out of a cardinal plane, the azimuth and slant-plane sidelobes can be held below specified levels.

In a space feed, the phase error problem is significantly relaxed because correlated errors are avoided by the system geometry. There is no elevation column feed to introduce errors correlated over rows of elements. In addition, the phase command to a given element will be the sum of three terms: a row command, a column command, and a focusing correction. The last term is proportional to the square of the distance of the element from the center of the array, and this has the effect of randomizing the quantizing error (Figure 4.4.2). As a result, in a space-fed array, sidelobes from sources other than the illumination function will be random, following (4.3.19).

The illumination function produced by simple horns in a space-fed array system does not produce ULSA performance for two reasons: the approximate cosine shape of the horn function does not match the complex curvature of the Taylor of Dolph function (Figure 4.4.3), and the spillover (Figure 4.2.6) may radiate into space or generate reflections from the

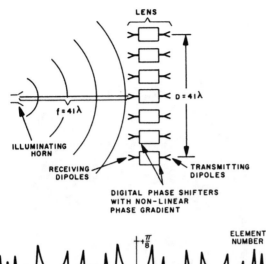

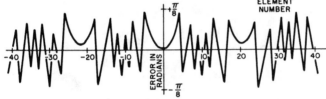

Figure 4.4.2 Randomization of phase error in space-fed array [4.9].

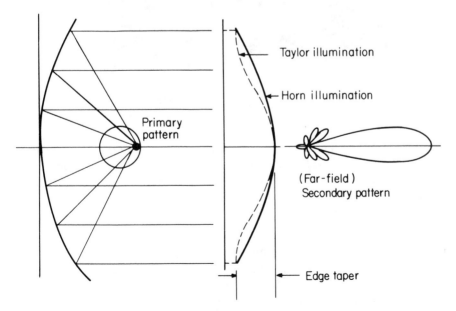

Figure 4.4.3 Horn illumination compared to Taylor function (after Schrank [4.10].)

Figure 4.4.4 Focal-plane array feed using 44 horns [4.11].

structure surrounding the array elements. More elaborate feeds (e.g., corrugated horns or focal-plane arrays of horns) are required to generate the required illumination. One example is the 44-horn feed shown in Figure 4.4.4. The focal-plane array can also generate monopulse patterns, as discussed in the next section.

Ultra-Low-Sidelobe Reflector Antennas

The conventional reflector antenna system, using a horn feed and doubly curved paraboloid, is designed to give a reasonable compromise between efficiency and sidelobe level. Reflector antennas often achieve between 50 and 60% efficiency, with first sidelobes about -25 dB from the main-lobe peak, and spillover lobes near −35 dB. Two design approaches have been used to approach ULSA performance.

(a) *Parabolic cylinder with line feed.* In this design, a constrained line-feed is used to produce the low-sidelobe illumination function (usually in the azimuth plane). The feed illuminates a parabolic cylinder, which is made somewhat wider than the feed to minimize spillover (Figure 4.4.5). The parabola is fed from a position offset in elevation to avoid blockage. The feed itself may use an equal-line-length series configuration (Figure 4.3.4d), and the design challenge is to fold this to meet mechanical requirements.

Figure 4.4.5 Line-fed parabolic cylinder antenna. (Photo courtesy of Marconi Radar Systems, Ltd.)

(b) *Doubly curved reflector with complex feed.* A precision reflector surface is illuminated by an oversized feed (e.g., corrugated horn or array as in (Figure 4.4.4). The feed must be offset to avoid blockage, and must be designed to approximate the 45-dB Taylor or similar illumination function. An example is shown in Figure 4.4.6. Spillover lobes near 120° are about −15 dBi or −53 dB from the main lobe. The first sidelobes are replaced, in this design, by shoulders on the main lobe at about −40 dB, and the first separated sidelobes are at −42 dB. Another example is the Gregorian configuration, Figure 4.4.7, with first sidelobes again near

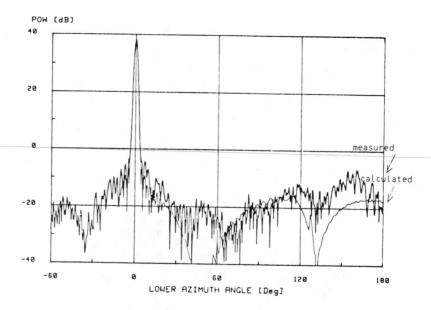

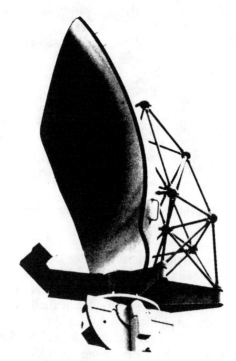

Figure 4.4.6 ULSA reflector antenna [4.12]: (a) azimuth beam pattern;
(b) antenna on test rotor. (Courtesy of L.M. Ericsson Co.)

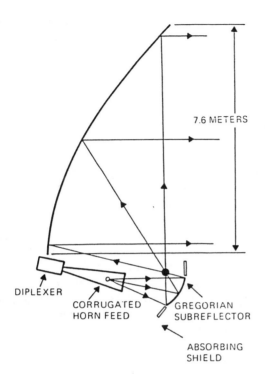

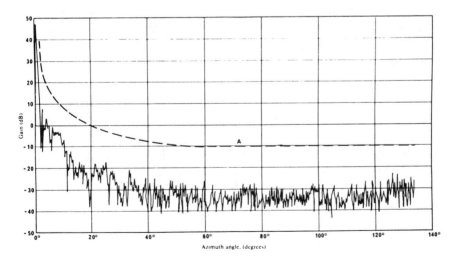

Figure 4.4.7 ULSA reflector antenna using Gregorian configuration [4.10].

-40 dB and average sidelobes below -30 dBi, -78 dB with respect to the mainlobe. Designers of satellite communications antennas have gone beyond their radar counterparts in reflector ULSA systems.

4.5   MULTIPLE-BEAM ANTENNAS

Stacked-Beam Systems

In order to obtain adequate time-on-target for MTI or doppler processing, high-performance 3D radars must cover the entire elevation search sector in parallel with a *stacked beam* configuration, rather than scanning sequentially with a single pencil beam. In the 1945–1975 period, such radars would use an array of feed horns in front of the doubly curved reflector (Figure 4.5.1). In the lower portion of the elevation sector, each horn would provide a single beam, illuminating the entire reflector. At higher elevations, outputs of two or more horns would be added to produce broader beams, corresponding approximately to the csc^2 receiving gain requirement. The transmitter would couple through duplexers into each beam, producing a single fan beam with csc^2 extension. A basic problem with this class of antenna is the blockage area of the vertically extended feed structure, which causes large azimuth sidelobes. Elevation sidelobes are also a problem because of the blockage and the off-axis location of most of the horns.

More modern stacked-beam antennas use planar arrays, in which n_e rows of n_a elements each are fed from an elevation beamforming matrix. The antenna is still scanned mechanically in azimuth. The row feeds and radiating elements can consist of slotted waveguides (Figure 4.5.2), or corporate feed networks (Figure 4.3.5) with radiating dipoles. The elevation beam-forming matrix can be at RF, using a modified Butler matrix (Figure 4.3.10) or a Rotman lens (Figure 4.3.7), at IF using a resistive matrix (Figure 4.3.11), or in the form of digital processing (Figure 4.3.12). Since only the single elevation matrix is used, the elevation sidelobes will not be very low unless great precision is maintained in coupling to the n_e rows of radiating elements. On the other hand, the one matrix can be fabricated with considerable care, since the one matrix will not be a major expense.

Figure 4.5.1 AN/TPS-43 stacked-beam antenna. (Photo courtesy of Westinghouse Electric Co.)

In some systems, and necessarily in those using IF or digital beam forming, amplifiers are used prior to the matrix. Placing amplifiers before an RF matrix makes it easier to design the matrix, since moderate losses can be tolerated without affecting the system noise temperature. However, the gain and phase stability of these amplifiers is critical in establishing the elevation sidelobe levels. In the digital beamformer, these amplifiers consist of the entire receiver, providing outputs through phase detectors and A/D converters. This makes it more difficult to maintain the phase and gain stability required for low elevation sidelobes. In addition, whenever amplifiers are used, the failure of one row amplifier will introduce large elevation sidelobes, even though it may have small effect on the gain of each elevation beam channel.

Figure 4.5.2 AN/TPS-70 stacked-beam array. (Photo courtesy of West-inghouse Electric Co.)

Monopulse Antennas

The antenna system for the conventional monopulse radar must form three simultaneous beams: a sum beam and two difference beams, one for azimuth and one for elevation (for 3D radars scanning mechanically in azimuth, only the sum and elevation beams are necessary). Early monopulse trackers used a cluster of four horns, displaced about the focal point of a reflector or lens antenna. Each of these horns could be associated with a pencil beam in space, squinted about the axis of the antenna (Figure 4.5.3). Outputs of these four horns were combined to produce the three patterns:

$$\Sigma = A + B + C + D$$

$$\Delta_a = A + D - (B + C)$$

$$\Delta_e = A + B - (C + D)$$

A basic problem in such antennas was the compromise between sidelobe levels (especially in the difference pattern) and efficiency in the Σ and Δ patterns. If the width of the four-horn assembly was made correct for an efficient, low-sidelobe Σ pattern, the Δ illumination was spread too widely beyond the aperture (Figure 4.5.4), the high edge illumination producing large spillover lobes and high sidelobes. Increasing the width of the feed corrected the problem in the Δ patterns at the expense of under-illumination and inefficiency in the Σ pattern.

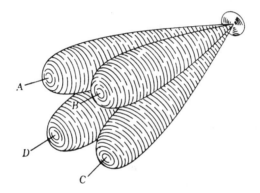

Figure 4.5.3 Beams from antenna with four-horn feed [4.14].

More advanced feed systems (Figure 4.5.5) used a 12-horn cluster or a stack of multilayer, multimode horns to permit separate optimization of Σ and Δ illuminations. By 1960, satisfactory design had been achieved for low sidelobes and high efficiency in all three patterns [4.13]. This class of monopulse feeds was subsequently applied to space-fed array antennas.

The generation of monopulse patterns in constrained feed arrays is a much more difficult problem, addressed by Lopez [4.15]. In a linear array (or row of a planar array), a hybrid junction placed at the center leads to a sharp discontinuity in the Δ illumination (Figure 4.5.6), which generates very high Δ sidelobes. The difference pattern of such an antenna is totally unsuited for operation in an environment of clutter, ECM, or multipath reflections. A partial solution can be obtained by further subdivision of the array into modules (subarrays) with additional couplers to

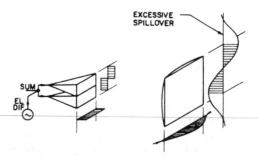

(a) Difference mode when feed is optimized for sum mode.

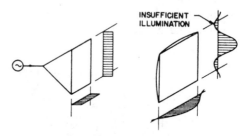

(b) Sum mode when feed is optimized for one difference mode.

Figure 4.5.4 Nonoptimum illuminations of four-horn feed [4.13].

the Δ channel (Figure 4.5.7). This creates a stepped approximation to the desired smooth illumination functions. However, the amplitude errors, relative to the smooth functions, are such as to produce unacceptably high sidelobes, unless large numbers of modules are used. In a series feed, the coupling to individual elements may be adjusted to match the desired Σ or Δ illumination function, but not both.

The two-dimensional array of subarrays uses the modular approach to produce a compromise between Σ and Δ illuminations. All elements of a subarray have the same illumination, and the subarray is coupled with different coefficients to the Σ and Δ channels. For example, consider an array with 64 subarrays, so that in each coordinate there will be eight-module approximations to the desired illumination functions. For a -45 dB Taylor illumination function, the approximation for the Σ channel is shown Figure 4.5.8. The rms error, normalized to the illumination at the center of the array, is

$$\sigma_a/a = 0.07$$

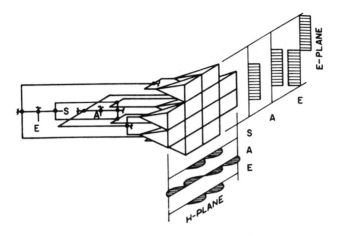

(a) Twelve-horn feed.

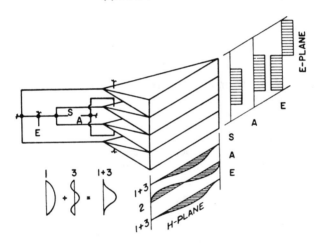

(b) Four-horn triple-mode feed.

Figure 4.5.5 Optimized monopulse feed designs [4.13].

If this error were random, it would produce average sidelobes (4.3.28, with σ_a/a replacing σ_ϕ):

$$\overline{G}_s = \pi(0.07)^2 = 0.015 = -18 \text{ dBi}$$

However, the amplitude error is periodic across the array and will produce peak lobes of much larger amplitude. In the case shown, the peak sidelobe

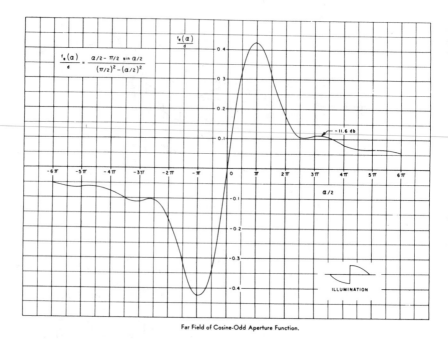

Far Field of Cosine-Odd Aperture Function.

Figure 4.5.6 Illumination and pattern of array with central hybrid junction [4.16].

will be about -40 dB with respect to the main lobe, or -3 dBi for a 4000-element array. In the Δ channel, the amplitude errors are much larger. The eight-module approximation to the -45-dB Taylor derivative pattern (which would have sidelobes near -35 dB, if implemented with a continuous illumination function) is shown in Figure 4.5.9. The rms error relative to the peak of the Δ illumination function is 0.14, and relative to the peak of the Σ function it is

$$\sigma_a/a = 0.125$$

The average sidelobe level would be

$$\overline{G}_s = \pi(0.125)^2 = 0.048 = -13 \text{ dBi}$$

and the peak sidelobes (for $T = 4000$ elements) would be near -37 dB from the Σ main lobe, or 0 dBi. These lobes can be reduced to approach the average value by staggering the modules in a planar array, but the sidelobe levels remain well above those for the design illumination, es-

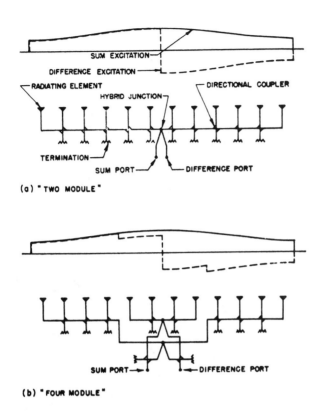

Figure 4.5.7 Four-module series feed [4.15].

pecially when other error sources are considered. The design of a monopulse feed network, coupling 64 subarrays into the three monopulse
outputs, is not a simple matter, and subdivision into more subarrays to
reduce sidelobes presents formidable problems.

A different approach to monopulse feed design for phased arrays is
the dual-ladder feed (Figure 4.5.10). Here the primary line has couplers
which produce the Σ illumination function. A secondary line is coupled
only to the Δ channel. The discontinuity introduced in the Δ illumination
by the center hybrid of the primary line is canceled by an opposite discontinuity in the secondary line, leading to the desired smooth functions

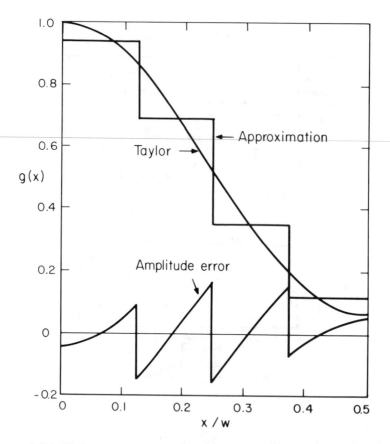

Figure 4.5.8 Eight-module approximation to −45-dB Taylor illumination
function.

in both channels. Each element will require two precision couplers, holding
their electrical tolerances over the operating band of the radar. This makes
it relatively costly to obtain low sidelobes in the Δ channel by this method,
as compared with the single-line feed, which produces low sidelobes only
in the Σ channel. In an amplifier array, however, the couplers and networks
can be relatively lossy, and the inherent errors of the amplifiers tend to
overshadow those of the networks, making this approach more acceptable.

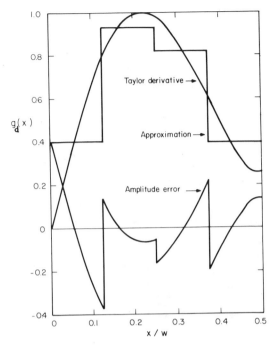

Figure 4.5.9 Eight-module approximation to -45-dB Taylor derivative illumination function.

Focal-Plane Array Feeds

The inherent simplicity of the space feed technique, especially for large arrays (thousands of elements), creates a strong incentive for solution of the sidelobe and spillover control problem. One approach, going beyond the twelve-horn or four-layer, multimode feeds of Figure 4.5.5, is to expand the feed into an array of many horns or waveguide ports, in the focal plane of the main array. An example is the 44-horn feed of Figure 4.4.4. The arrangement of couplers to support low-sidelobe monopulse patterns is complex, compared with conventional horn feeds, but not when compared to the networks for coupling to subarrays or elements in the aperture plane. The fundamental difference lies in the wideband, two-dimensional Fourier transform operation performed in the space between the focal plane and the aperture of the space-fed system.

The application of a complex focal-plane array to feed a large aperture is not limited to space-fed phased array antennas. The same approach, applied to an offset-fed parabolic reflector, can produce ULSA performance, monopulse patterns, stacked beams, or other desired patterns with reasonable cost and complexity.

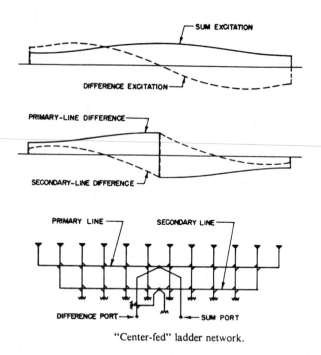

"Center-fed" ladder network.

Figure 4.5.10 The Lopez feed: a dual-ladder network for separate optimization of Σ and Δ excitation of a linear array.

Digital Beam Forming

The ultimate in flexibility of receiving arrays is provided by digital beam-forming systems (Figure 4.3.12). Each of T elements passes its signal through a low-noise receiver to $2T$ phase detectors into A/D converters, which operate at a rate equal to the signal bandwidth B on the I and Q baseband outputs. These outputs, having a digital word rate $2TB$ can be processed to yield $M \leq T$ orthogonal beams. These beams can be further combined with weighting to produce any number of overlapping beams, each with the full signal bandwidth and with pattern and sidelobe characteristics determined by T, by the number of bits in the digital words, and by errors in the element channels and digital weighting coefficients.

For example, consider the ULSA requirements discussed in Section 4.4, where each of eight error sources was budgeted a phase or amplitude error of 0.02. The total error from all phase components was 0.04 radians and from all amplitude components was $\sigma_a/a = 0.04$, producing average sidelobes of -20 dBi. A 4000-element array was considered. In a digital system, the specification would be for errors of 0.04 in each I and Q channel, considering all phase and amplitude sources from the radiating elements to the A/D converters. There are now no feed networks or phase shifters, so the entire budget can be allocated to the receivers and A/D converters, if the designer is willing to perform digital operations with enough bits to avoid build-up of error later in the system. If the receiver gains can be held to 2.8% = 0.24 dB rms, an equal allocation to the A/D converters requires five-bit performance, or probably six-bit design. The receivers can be allocated the entire phase error budget of 0.04 radians = 2.5°.

If the digital system operates with a signal beamwidth of 1 MHz, the requirement is for 4000 receivers feeding 8000 A/D converters operating at 1 MHz with five-bit accuracy. The computer then performs two-dimensional Fourier transforms from 4000 complex inputs to M complex outputs at a rate of one complete transform per microsecond, carrying enough bits to avoid build-up of errors, and the computer forms the final output beams by summing adjacent beams with proper weighting. Monopulse Σ and Δ channels, stacked beams, or any other combination of beams would be available. The cost in receiver implementation and digital throughput is a large price to pay for the resulting flexibility, even assuming that the specifications for accuracy and sidelobe levels can be met. For both these reasons, the more conventional types of array are preferable for most applications.

REFERENCES

[4.1]   D.D. Howard, Analysis of the 29-foot monopulse Cassegrainian antenna of the AN/FPQ-6 and AN/TPQ-18 precision tracking radars, *U.S. Naval Research Laboratory Memo. 1776,* June 1967, DDC Doc. AD 816772.

[4.2]   R.S. Elliott, *Antenna Theory and Design,* Prentice-Hall, 1981.

[4.3]   D.K. Barton and H.R. Ward, *Handbook of Radar Measurement,* Artech House, 1984.

[4.4]   T.C. Cheston and J. Frank, Array Antennas, Chapter 11 in *Radar Handbook* (M.I. Skolnik, ed.), McGraw-Hill, 1970.

[4.5]   L. Cardone, Ultra-wideband microwave beamforming technique, *Microwave J.,* April 1985, pp. 121–131.

[4.6]   D. Archer, Lens-fed multiple beam arrays, *Microwave J.,* October 1975, pp. 37–42.

[4.7]   B.C. Dodson, Jr., L-band solid state array overview, Chapter 19 in *Radar Technology* (E. Brookner, ed.), Artech House, 1977.

[4.8]   A.E. Ruvin and L. Weinberg, Digital multiple beamforming techniques for radar, *IEEE Eascon '78 Record,* pp. 152–163.

[4.9]   P.J. Kahrilas, HAPDAR—An operational phased array radar, *Proc. IEEE* **56,** No. 11, November 1968, pp. 1967–1975.

[4.10]   H.E. Schrank, Low sidelobe reflector antennas, *IEEE APS Newsletter,* April 1985, pp. 5–16.

[4.11]   N.S. Wong, R. Tang, and E.E. Barber, A multielement high power monopulse feed with low sidelobes and high aperture efficiency, *IEEE Trans.* **AP-22,** No. 3, May 1974, pp. 402–407.

[4.12]   E. Carlsson, *et al.,* Search radar reflector antenna with extremely low sidelobes, *Proc. Military Microwaves '82,* Microwave Exhibitions and Publishers, Ltd., Tunbridge Wells, England, October 1982, pp. 500–505.

[4.13]   P.W. Hannan, Optimum feeds for all three modes of a monopulse antenna, *IRE Trans.* **AP-9,** No. 5, September 1961, pp. 444–461.

[4.14]   D.R. Rhodes, *Introduction to Monopulse,* Artech House, 1980.

[4.15]   A.R. Lopez, Monopulse networks for series feeding an array antenna, *IEEE Trans.* **AP-16,** No. 4, July 1968, pp. 436–440.

[4.16]   G.M. Kirkpatrick, Final Engineering Report on Angular Accuracy Improvement, General Electric Company, August 1, 1952; reprinted in D.K. Barton, ed., *Radars,* Vol. 1, *Monopulse Radar,* Artech House, 1974.

Chapter 5

Waveforms and Signal Processing

5.1 THE AMBIGUITY FUNCTION

In Chapter 4 we introduced the concept of the four-dimensional radar resolution cell, which covers specific regions in azimuth, elevation, range (or time delay t_d) and radial velocity (or doppler frequency f_d). Most radar systems operate in such a way that these resolution properties in azimuth and elevation can be considered separately from those in t_d and f_d. The response of the radar in coordinates t_d and f_d can then be described, for a radar using a matched filter, by the *ambiguity function* $\chi_0(t_d, f_d)$. The corresponding response function with a mismatched filter is given by the *cross ambiguity function* $\chi(t_d, f_d)$. The radar transmitter in Figure 5.1.1 accepts from a waveform generator the modulation function $a(t)$, which in general is a complex function (having phase as well as amplitude variation). This is applied to a high-level RF carrier of frequency f_0, radiated from the antenna, and reflected back to the receiver from a target at range R and with radial velocity v_r. In the receiver, the RF echo signal is downconverted to a convenient IF, processed in a filter or correlator having frequency response $H(f)$ and impulse response $h(t)$, and envelope detected for presentation to an output device.

It is convenient to consider that the receiving system is adjusted for peak response to a target with particular values of delay and frequency, and to measure the response (and the ambiguity function) for values t_d and f_d with respect to this tuned point in the delay-frequency plane. This avoids the issue of realizability of filters having negative delay, and permits the waveform and filter functions to be written without inclusion of the carrier term $\exp(j2\pi f_0 t)$. We will assume that $a(t)$ corresponds to the rms level of the modulated carrier. It is also convenient to ignore the actual path losses and gains through the system, and to assume that the circuit

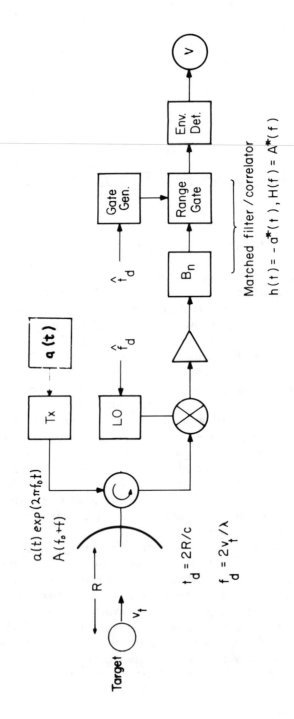

Figure 5.1.1 Transmission, reception, and processing of a radar waveform.

resistances are one ohm, so that the peak value of $a(t) = 1$ V at transmission, reception, and at the output of the matched receiver, with a corresponding signal power of one watt.

The receiver output will then depend on the waveform and the filter functions, in the following way:

$$\chi(t_d, f_d) = \int_{-\infty}^{\infty} H(f)A(f - f_d) \exp(j2\pi f t_d) \, df \qquad (5.1.1)$$

or

$$\chi(t_d, f_d) = \int_{-\infty}^{\infty} h(t_d - t)a(t) \exp(j2\pi f_d t) \, dt \qquad (5.1.2)$$

The Fourier transform relationship between time and frequency functions is similar to that between the antenna pattern and the aperture illumination function, but there are two major differences between waveform and antenna resolution limits.

(a) In the antenna, the two resolution coordinates are independent of each other because the Fourier transform relationship is to the illumination functions across the width and height of the aperture. In the waveform case, the two resolution coordinates are connected by the Fourier transform of time to frequency, and *vice versa*, so that the volume of the resolution cell (as defined by the response or ambiguity function) is constant.

(b) In the antenna, the field incident on the aperture from a target at long range is uniform, so that the pattern depends only on the aperture illumination function established by the feed system. In the waveform case, the signal and filter are described in general by different functions, and in the case of the matched filter these are complex conjugates of each other.

The theory of the matched filter traces to the work of North [5.1], who showed that it would produce the maximum possible SNR in the presence of white noise. The matched filter frequency response, $H(f)$, and impulse response, $h(t)$, are defined by

$$H(f) = (1/C_e)A^*(f), \text{ and } h(t) = (1/C_v)a^*(-t) \qquad (5.1.3)$$

where $*$ denotes the conjugate, and the constant $C_e = 1$ V-s is included to make the expression dimensionally correct. The maximum output signal-to-noise ratio from the matched filter is equal to the ratio of signal energy to noise spectral density at the input of the filter:

$$(S/N)_{mf} = E/N_0 \qquad (5.1.4)$$

When a radar processes its reflected echo through the matched filter, the output response is described by the function:

$$\chi_0(t_d, f_d) = \int_{-\infty}^{\infty} A^*(f)A(f - f_d) \exp(j2\pi f t_d) \, df \tag{5.1.5}$$

or

$$\chi_0(t_d, f_d) = \int_{-\infty}^{\infty} a^*(t - t_d)a(t) \exp(j2\pi f_d t) \, dt \tag{5.1.6}$$

The ambiguity function, as used by Woodward [5.2] is given by $|\chi_0(t_d, f_d)|^2$. An important property of the ambiguity function is that the volume under the surface $|\chi_0(t_d, f_d)|^2$ remains constant as the waveform is changed. The volume under $|\chi(t_d, f_d)|^2$ for mismatched filters is also constant, with a peak response lower than that of the matched filter, but with increased main-lobe width or sidelobe levels.

Response to a Rectangular Pulse

As a basis for discussion of radar waveforms, we can consider the rectangular pulse of width τ. Figure 5.1.2 shows this pulse and its spectrum, both of which are real functions with even symmetry. Any sharp limit on the extent of a function in time implies an infinite extent of the frequency spectrum, and in this case the spectrum has sidelobes of infinite extent following the function:

$$A(f) = C_v \tau \frac{\sin \pi \tau f}{\pi \tau f} \text{ (V/Hz)} \tag{5.1.7}$$

where $C_v = 1$ V is included to make the expression dimensionally correct. The energy of the pulse is

$$E = \int_{-\infty}^{\infty} a^2(t) \, dt = \int_{-\infty}^{\infty} A^2(f) \, df = \tau \text{ (W-s)} \tag{5.1.8}$$

The matched filter for this pulse has a rectangular impulse response of duration τ, and a frequency response identical to $A(f)$:

$$H(f) = \frac{\sin \pi \tau f}{\pi \tau f} \tag{5.1.9}$$

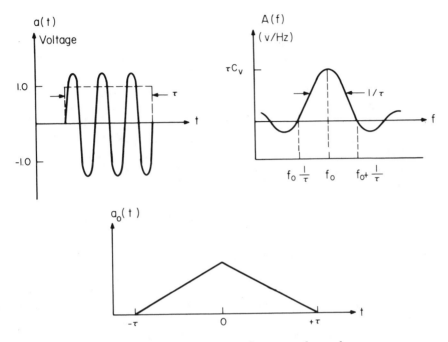

Figure 5.1.2 Waveform and spectrum of rectangular pulse.

The output spectrum is

$$A_o(f) = A(f)H(f) = C_V\tau \frac{\sin^2 \pi\tau f}{(\pi\tau f)^2} \text{ (V/Hz)}$$

and the peak output, for $f_d = 0$, $t_d = 0$ is

$$\chi_0(0, 0) = \int_{-\infty}^{\infty} C_V\tau \frac{\sin^2 \pi\tau f}{(\pi\tau f)^2} \, df = C_v = 1 \text{ V} \qquad (5.1.10)$$

In the assumed one-ohm circuit, the peak output power is one watt. If noise is present with density N_0(W/Hz), the noise power output will be

$$N = N_0 \int_{-\infty}^{\infty} |H(f)|^2 \, df = N_0 B_n = N_0\tau \text{ (W)} \qquad (5.1.11)$$

where B_n is the *effective noise bandwidth* of the filter. The SNR at the output peak is thus

$$C_v^2 \tau / N_0 = E/N_0$$

as required by matched filter theory. The output waveform, obtained by convolution of $a(t)$ with $h(t)$, or by Fourier transformation of $A(f)H(f)$, is the triangular function shown in Figure 5.1.2c.

If we consider mismatched filters, a possible family of responses would be those for rectangular bandpass filters (Figure 5.1.3). For rectangular bandpass, $B_n = B$, and with $B\tau = 6$, the output waveform will be almost rectangular, but with leading and trailing edges having durations $\tau_e \approx 1/B = \tau/6$. The peak output power (if we ignore the small ripples at the top of the pulse) remains 1 W, but the noise power is now $N_0 B_n = N_0 B = 6N_0/\tau$, giving $S/N = E/6N_0$, about 8 dB below that of the matched filter. At $B\tau = 2$, the maximum output voltage increases to 1.18 V, and $S/N = 1.18^2/2N_0 = 0.7E/N_0$, about 1.5 dB less than for the matched filter.

Measurement of the Ambiguity Function

The response $|\chi(t_d, f_d)|$ of a radar waveform and processor, to a target located anywhere in time delay or doppler frequency, can be measured directly. In Figure 5.1.4 the radar transmits $a(t)$ and receives, from a target at the range R_0 and velocity v_0 an echo with delay and doppler shift to which the radar is tuned:

$$t_d = (2R_0/c) - t_{d0} = 0, \quad f_d = (2v_0/\lambda) - f_{d0} = 0 \qquad (5.1.12)$$

After down-conversion and amplification to give $a(0) = 1$ V rms, this echo signal is applied to a processing circuit which in general consists of a filter $H_1(f)$, a range gate which multiplies the signal by $h_2(t)$, and an output filter $H_3(f)$. The output is envelope detected and measured by a voltmeter and a power meter. To measure the ambiguity function, the cascaded combination of filters and gates is configured to produce an overall frequency response $H(f) = (1/C_e)A^*(f)$. The reference target then gives an output voltage $E_o = 1$ V, power $P_o = 1$ W. As the target is moved to other values of range R and radial velocity v_r (assuming that the change in range is not sufficient to alter the path loss in the radar equation), the voltmeter readings will decrease, with E_o tracing out $|\chi_0(t_d, f_d)|$ and P_o giving the ambiguity function $|\chi_0(t_d, f_d)|^2$. For a rectangular pulse, the voltage response will be as shown in Figure 5.1.5.

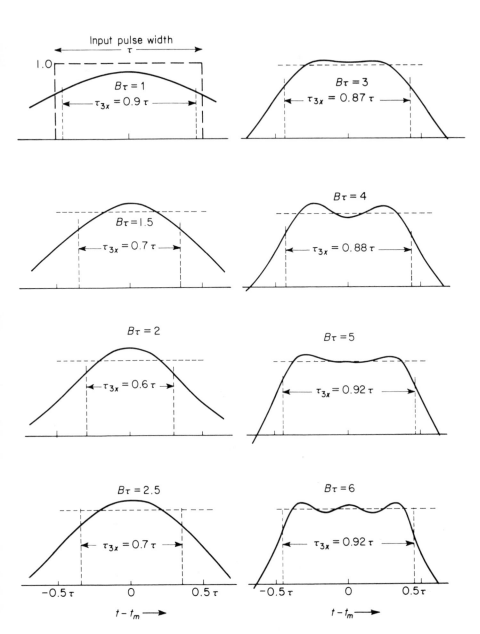

Figure 5.1.3 Waveforms for rectangular pulse with rectangular bandpass filter.

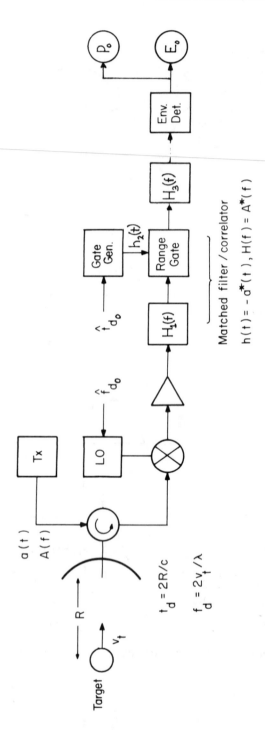

Figure 5.1.4 Measurement of the ambiguity function.

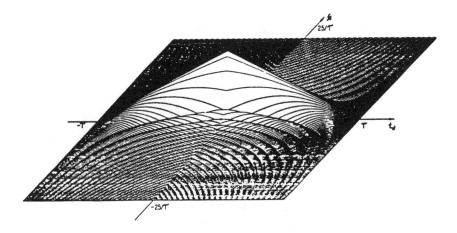

Figure 5.1.5 Matched response for rectangular pulse [5.3]. (Copyright 1969 by A.W. Rihaczek. Reprinted with permission.)

Implementation of the Matched Filter

The matched filter can be implemented in either of two ways:

(a) as a correlator, in which the input signal is multiplied by a replica of the signal expected at a particular time t_d, and the product is integrated over the duration of the signal, or

(b) as a filter, which is matched to the spectrum of a signal expected at a particular doppler shift f_d.

The filter, if implemented in the frequency domain, has the advantage that it passes signals arriving at all time delays, as long as their doppler shifts are not so large as to reduce the response significantly. On the other hand, for rectangular pulses, the rectangular gate needed for a correlator is easily generated, while it is not easy to provide a filter which matches the $(\sin x)/x$ spectrum. The choice between filter and correlator implementation then depends on the waveform and the region, in time and doppler, over which the receiver must process signals.

For a rectangular pulse, the matched filter is most conveniently implemented with a wideband filter $B_1 \gg 1/\tau$, a rectangular gate function $h_2(t) = a(t)$, and a narrowband output filter $B_3 \ll 1/\tau$, which performs integration over τ. The gate and narrowband filter constitute a matched correlator when the gate delay is equal to the signal delay, as in (5.1.12). A bank of parallel gates with different delay values $t_1 < t_d < t_2$, each followed by a narrowband filter and envelope detector, constitute a set of approximately matched filters for signals with delay lying between t_1 and t_2. If the gates are spaced closely enough in time, each possible signal in this interval will produce a matched response in one channel, with reduced response in a number of nearby channels. A sampling of such gate outputs, for a single target, traces out the function $|\chi_0(t_d, 0)|$.

A more complex implementation would use $H(f) = (\sin \pi \tau f)/\pi \tau f$ for the first filter, followed by range gates narrowed to near impulses ($\tau_g \ll \tau$), wideband output filters ($B_3 \gg 1/\tau$) and envelope detectors. In this case, the gates and output filters could be placed after a single envelope detector with no loss penalty, since the S/N ratio would have been determined by $H_1(f)$. By omitting the gates and output filters, the output can be displayed directly on a wideband oscilloscope, presenting $|\chi_0(t_d, 0)|$ for $0 < t_d < \infty$. If digital representation of this function were needed, the impulse gates would be applied after envelope detection as strobes to an A/D converter, sampling at a rate such that the peak amplitude of any signal would be obtained. For sampling carried out at a rate near $1/\tau$, there would be a *straddling loss* on signals with delay that failed to match the position of a strobe.

Depending on the waveform properties and the interval in which targets are to be detected, the most convenient implementation will be found either in the frequency domain (using a matched filter H_1) or the time domain (with a matched correlator function h_2 followed by a narrowband filter H_3). Use of the matched filter H_1 is advantageous in providing a continuous stream of time samples at its output, for detecting targets within a broad range interval. A correlator or gate approach is simpler for tracking a target of range that has already been determined.

Resolution Properties of a Simple Pulse

The rectangular pulse is often used because it permits the transmitter to operate as a saturated oscillator or amplifier, and to be turned off entirely during reception of signals. When processed through a matched filter, a well defined (point) target produces an output 2τ wide at its base, with a sharp central peak. A second, similar target separated by τ in time delay can produce any one of the response envelopes shown in Figure 5.1.6, depending on the relative signal phase ϕ. For all phase angles, the location and magnitude of the second peak is identifiable, and the two targets can be considered *resolved*.

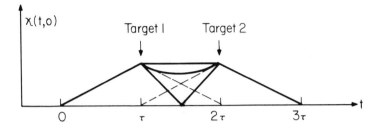

Figure 5.1.6 Resolution of two targets with matched filter.

If a wideband (mismatched) filter is used, the response to more closely spaced targets can be resolved, as shown in Figure 5.1.7. Here, the desired target S is assumed to be somewhat smaller than the interfering target I, and the receiver bandwidth $B_1 = 1/\tau_e \gg 1/\tau$. Using a narrow sampling gate, the response as the gate is passed across the combined signal will appear as in Figure 5.1.7e. It is possible to locate and measure the amplitude of both targets, regardless of signal phase, because they are separated by a delay equal to the sum of the gate width and the rise time of the leading edge of the received signal. Use of a differentiator on the output envelope (Figure 5.1.7f) provides resolution for delay separations of about $1/B_1$, as long as the SNR is adequate to support this type of processing.

In the frequency domain, the output spectrum from a matched filter has the form $(\sin^2 \pi \tau f)/(\pi \tau f)^2$, which provides resolution for targets separated by about $1/\tau$ in doppler shift. Thus, in each coordinate, resolution requires that the targets be separated by approximately the width of the output response at its -6 dB level. The product of the major and minor axes of this contour in the time-frequency plane, which we will call the *resolution area* of the waveform, is always equal to unity. As the pulsewidth changes, the shape of the contour changes, but the area remains the same. For mismatched filters, as B is increased and a differentiator is applied, the large spectral sidelobes of the rectangular pulse permit the effective signal bandwidth to be increased as well, improving the delay resolution at the expense of doppler resolution and SNR. The product of the axes of the resolution ellipse remains essentially unchanged at unity. Narrowing the filter bandwidth extends the time response without significant reduction in the size of the frequency resolution interval, showing that it is possible to expand the resolution area at the expense of reduced SNR.

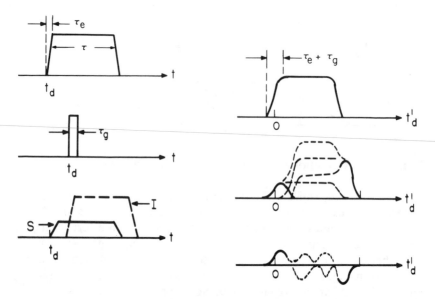

Figure 5.1.7 Resolution with mismatched filter.

If the restriction of saturated transmitters is removed, other pulse shapes such as the Gaussian pulse can be generated. Referring to the resolution properties shown in Figure 4.1.1 for a Gaussian waveform, we see that resolution is possible at any relative phase for signals separated by the $-6\,\text{dB}$ width of the output pulse. The resolution area of the resulting ellipse is 0.88 for a Gaussian pulse, and it remains between this value and unity for all simple pulse waveforms with matched and nearly matched filters.

5.2 PULSE COMPRESSION

Use of a real waveform $a(t)$ imposes two serious restrictions on the shape of the ambiguity function:

(a) The product of delay and doppler resolution remains near unity, as shown in the previous section;
(b) The pulse energy is directly proportional to the width of the resolution element in delay.

If complex $a(t)$ is permitted, both these restrictions can be evaded, and the resulting process is called *pulse compression*. The two basic approaches to pulse compression are to subdivide the transmitted pulse into short chips, to which a pseudorandom phase code is applied, or to introduce over the transmitted pulse a continuous phase modulation, such as a quadratic function. Phase-code pulse compression achieves a reduction in the cell width to $\tau_n = 1/B$ in delay, while it remains $B_o = 1/\tau$ in doppler, reducing the resolution area, at the -6 dB level, from unity to $1/B\tau \ll 1$. Continuous phase modulation, usually described as FM pulse compression, leaves the resolution area at unity, but achieves a delay resolution width, for signals near $f_d = 0$, of $\tau_n = 1/B$ rather than τ. In either case, the energy of the transmitted pulse increases with τ, which can be made as large as permitted by transmitter limitations and eclipsing of short-range targets. Because the total volume under the ambiguity function must be constant, the phase-coded system, with less volume in its main lobe, must have greater volume in sidelobes. The FM pulse compression waveform can concentrate the volume within its main lobe and achieve low sidelobes everywhere else in the time-frequency plane.

Phase-Coded Pulse Compression

A diagram of a simple phase-coded pulse compression waveform and its matched filter is shown in Figure 5.2.1. A Barker coded waveform has $m = \tau/\tau_n$ chips, which are either positive ($0°$ phase) or negative ($180°$ phase). The five-chip Barker code 11101 is illustrated. The matched filter consists of a $(\sin x)/x$ filter matched to the chip length τ_n (with bandwidth $B = 1/\tau_n$), followed by a tapped delay line having four delays τ_n. Taps 0, 1, 2, 3, 4 are weighted by the time-reversed code 10111 and summed before envelope detection. The output will be as shown, with $m - 1$ *time sidelobes* of unit amplitude C_v and a main lobe of amplitude mC_v, all having width τ_n. The ratio of transmitted pulsewidth to output width is $\tau/\tau_n = B\tau$, the *pulse compression ratio*. The noise components with power N_0B and with random phase will add in power to give a total power $mN_0B = mN_0/\tau_n$. The output SNR is then

$$S/N = \frac{C_v^2 m^2}{mN_0/\tau_n} = \frac{C_v^2 m\tau_n}{N_0} = C_v^2 \tau/N_0 = E/N_0 \qquad (5.2.1)$$

The relative sidelobe power level is $1/m^2 = 1/(\tau B)^2$.

The Barker code is the only code which has equal sidelobes at this low level, and this applies only along the zero-doppler axis. The longest Barker code is $m = 13$, and a plot of $|\chi_0(t_d, f_d)|$ for this code is shown in Figure 5.2.2. The low, equal-level sidelobes at $1/m^2 = -22$ dB cannot be seen in the figure because they are shadowed by larger sidelobes elsewhere in the doppler coordinate. A fundamental disadvantage of phase-coded

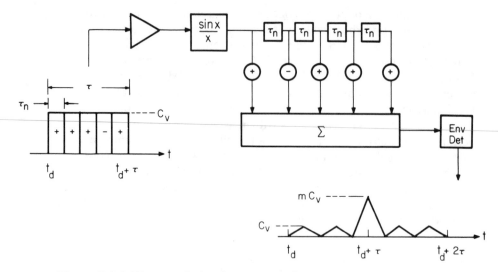

Figure 5.2.1 Phase-coded pulse compression.

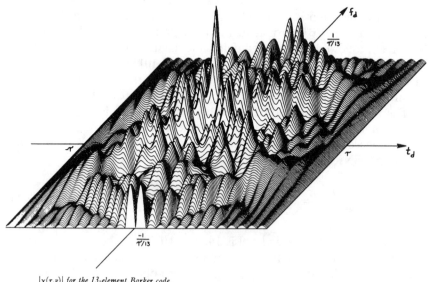

$|\chi(\tau,\nu)|$ *for the 13-element Barker code.*

Figure 5.2.2 Response of 13-chip Barker coded waveform [5.3]. (Copyright 1969 by A.W. Rihaczek. Reprinted with permission).

pulse compression is this sensitivity to doppler shift, such that the rapid decay of the main lobe (giving doppler resolution) is accompanied by an increase in sidelobes. The average sidelobe power over the time-frequency plane is $1/m$, rather than the $1/m^2$ which is obtained along the $f_d = 1$ axis.

For pulse compression ratios greater than 13, there are families of pseudorandom codes that can achieve peak sidelobe voltage along the axis $f_d = 0$, which approach $1/B\tau$ without restrictions on $B\tau$. There even exist so-called complementary sequence pairs, for which the sum of the two responses, processed separately, have zero sidelobes along this axis. In practice, however, to achieve separate processing without overlapping interference components, the two sequences must be transmitted in successive pulse repetition intervals, producing a strong sensitivity to target and clutter doppler shift. It is usually assumed that the long pseudorandom code will have widely distributed sidelobes at a power level $1/B\tau$ below the main-lobe peak, with a possible low-level trough at a specific doppler shift.

FM Pulse Compression

The basic FM pulse compression waveform is *chirp*, a linear FM (quadratic phase modulation) applied to a pulse of constant amplitude (Figure 5.2.3). The transmission occupies each frequency within B for an equal time, and hence the spectrum is uniform over B (although, in actuality, there are end effects that produce small ripples near the ends of the spectral band). The matched filter for an echo with $f_d = 0$ is a dispersive network which has the opposite ϕ *versus* t, and hence f *versus* t, function. Frequency components at $f_0 - B/2$ from the beginning of the pulse are delayed so that the last transmitted components at $f_0 + B/2$ can catch up, producing a large output at the end of the received pulse. As the leading edge of the received pulse enters the dispersive network or delay line, its upper spectral sidebands appear immediately at the output, and at intermediate times preceding the main output, producing time sidelobes. Similarly, lower frequency sidebands, generated at the trailing edge of the pulse, are delayed by amounts up to τ and appear as time sidelobes after the main output. The output waveform thus takes the form:

$$a_x(t) = a_0(\sin\pi B\tau)/\pi B\tau$$

where a_0 is increased by $\sqrt{B\tau}$ as compared with the received input pulse, and the -4 dB pulsewidth is $\tau_n = 1/B$. Noise of spectral density N_0 at the input appears with power $N_0 B$ at the output, giving an output power SNR,

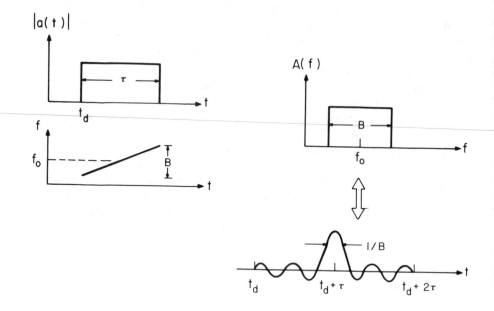

Figure 5.2.3 Linear FM (chirp) pulse compression.

at the peak of the main signal lobe,

$$S/N = a_0^2/N_0B = C_v^2B\tau/N_0B = C_v^2\tau/N_0 = E/N_0 \qquad (5.2.2)$$

To reduce the $(\sin x)/x$ sidelobes produced by the leading and trailing edges of the rectangular pulse, the frequency response $H(f)$ of the dispersive filter is deliberately mismatched so that $|H(f)|$ is tapered or *weighted* for $f \neq 0$. The effect on the output waveform $a_x(t)$ is analogous to that on the far-field pattern of a rectangular antenna with tapered illumination. As shown in Figure 5.2.4, the main-lobe peak is reduced in amplitude and broadened, while sidelobes are reduced substantially. For the 40-dB Taylor weighting shown, the loss in peak output is $L_m = 1.2$ dB, broadening is such that the -3 dB width is $1.25/B$, rather than $0.886/B$ for the unweighted pulse, and peak sidelobes are -40 dB with respect to the actual peak main lobe. The curves of Figure 4.2.3 may be used to relate efficiency $\eta = 1/L_m$ to broadening factor $K_\tau = \tau_n B$, when τ_n is measured at the -3 dB level.

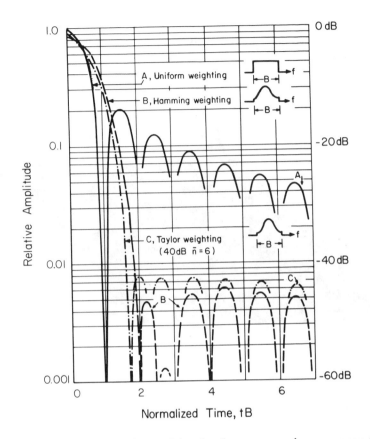

Figure 5.2.4 Weighted and unweighted pulse compression response (after Farnett [5.16]).

Ambiguity Function for Linear FM Pulse Compression

The response $|\chi(t_d, f_d)|$, which is the square root of the cross-ambiguity function, is shown in Figure 5.2.5, for a chirp pulse with Hamming weighting. Almost all of the volume in the function is concentrated in the ridge which runs diagonally across the rectangular response region $2\tau \times 2B$. The slope of this ridge is $-\tau/B$ Hz/s (for a chirp with positive slope). A target with doppler offset f_d from the tuned point will produce a peak output at time $\Delta = -\tau(f_d/B)$ from its true delay. Thus there is ambiguity between a target at $\Delta, 0$ and one at $0, f_d$. This does not constitute a major problem for $|f_d| \ll B$, since $|\Delta| \ll \tau$. For example, using $\tau = 50$ μs, $B = 2$ MHz at S-band, the doppler shift for a subsonic target ($v_r =$

300 m/s) will be f_d = 6 kHz and Δ = 0.15 µs. This is less than 1/3 the output pulsewidth, and it leads to an error of only 22 m in target location, if not corrected.

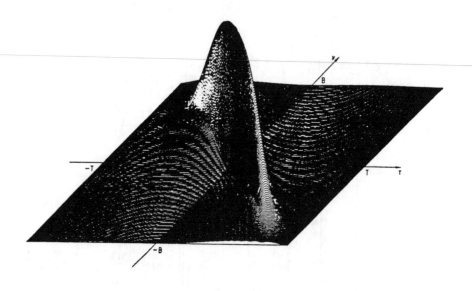

Figure 5.2.5 Response of chirp pulse with Hamming weighting [5.3]. (Copyright 1969 by A.W. Rihaczek. Reprinted with permission.)

Nonlinear Chirp Waveform

Several types of nonlinearity can be introduced in the chirp frequency sweep, with useful results.

(a) *Nonlinear FM.* By increasing the FM rate near the ends of the transmitted pulse and decreasing it near the center, the transmitted spectrum may be tapered so that the matched filter response has low sidelobes (Figure 5.2.6). For example, if the FM slope is made to vary inversely with the square root of the 40-dB Taylor weighting function, the energy spectrum will be weighted by this function. The application of the matched filter in the receiver will then produce the desired −40 dB sidelobe levels without matching loss.

(b) *Stepped FM.* The transmission can be stepped over the bandwidth B in discrete jumps (Figure 5.2.6a). The phase *versus* time plot (Figure

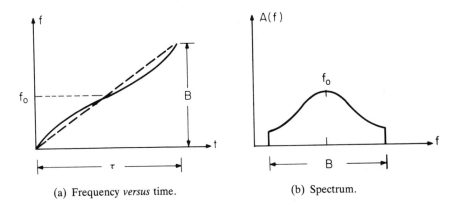

(a) Frequency *versus* time.   (b) Spectrum.

Figure 5.2.6 Nonlinear FM chirp.

5.2.6b) will follow straight line segments, which are an approximation of the linear FM characteristic. The presence of a saw-toothed difference between the steps and the linear function will generate grating lobes in the time response, but these lobes will be small if the magnitude of the step B/m is less than the reciprocal m/τ of the step duration, i.e., $m^2 > B\tau$. Apart from the grating lobes, the ambiguity function will be similar to that of linear chirp [5.4, p. 273].

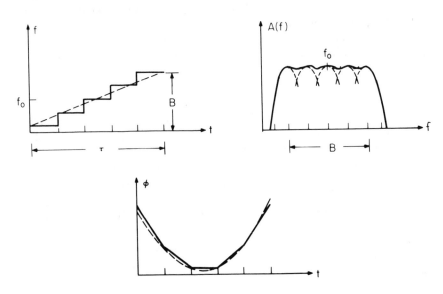

Figure 5.2.7 Stepped FM chirp.

(c) *Stepped-Phase FM.* One level further removed from linear FM is the signal in which the phase is stepped to approximate the linear phase segments of the stepped FM (Figure 5.2.8). This corresponds to the Frank polyphase code [5.4, p. 255], or to one of the related codes described by Lewis and Kretschmer [5.5, p. 15]. If phase quantization and frequency steps are made fine enough, the waveform and its ambiguity function become indistinguishable from linear chirp. With somewhat coarser steps, the peak sidelobes along the axis $f_d = 0$ are decreased to about -30 dB (Figure 5.2.8b). Both this waveform and stepped FM can be made to approximate a nonlinear FM function to provide reduced sidelobes without matching loss.

Doppler Tolerance of Pulse Compression Waveforms

In search radar applications, it is desirable to detect targets with a variety of velocities without having to use filters matched to each of many different doppler shifts. The linear FM waveform is well suited to this requirement, as shown in Figure 5.2.9. The peak response drops by less than 1 dB, for linear FM with a matched filter, when the doppler shift $|f_d| < B/10$ from the tuned frequency. At this point the sidelobes are increased only slightly relative to the peak response. For example, a linear FM pulse compression system at S-band with a bandwidth $B = 0.5$ MHz gives $\tau_n \approx 2$ μs and can tolerate target velocities up to 2500 m/s, regardless of the transmitted pulse length. The nonlinear FM waveforms become less tolerant as the departure from linear FM increases, but doppler deviations equal to several percent of the signal bandwidth are still permissible.

Phase-coded waveforms are much less tolerant of doppler shift, with significant loss in peak output and increases in sidelobe levels when the product $f_d\tau$ reaches about 0.2 (i.e., when the accumulated phase error $2\pi f_d\tau$ over the transmitted pulse reaches one radian). This means that this waveform's tolerance is less than that of linear FM by a factor of about $B\tau/2$. The problem is not severe unless long pulses are used. For example, if a 13-chip Barker code is used to compress a 26-μs pulse to 2 μs, the system tolerates a doppler shift of about 8 kHz, corresponding to 400 m/s at S-band. Longer phase-coded sequences much use receiver channels tuned to the velocities of desired targets. This sensitivity can be advantageous in rejecting an out-of-band target echo, converting it from a strong lobe to a series of sidelobes. However, in the case of continuous clutter, the average response will remain approximately the same, regardless of the receiver tuning.

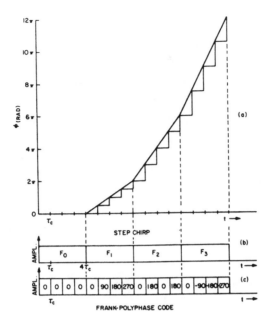

(a) Step-chirp and Frank polyphase code
relationships.

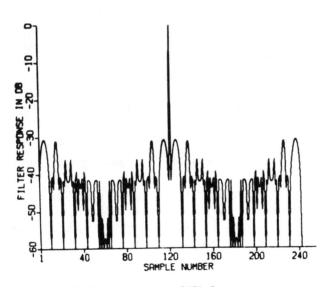

(b) Compressed pulse of 121-element
Frank-coded waveform.

Figure 5.2.8 Frank polyphase code [5.5].

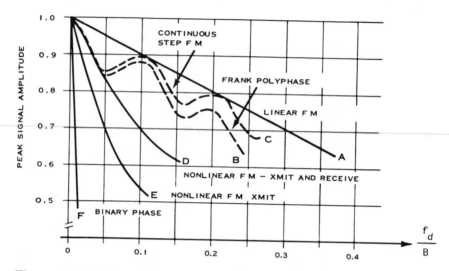

Figure 5.2.9 Peak response *versus* doppler shift for different pulse compression waveforms [5.6].

Summary of Pulse Compression Ambiguity Functions

The shapes of the main response lobes, and the regions occupied by large sidelobes for different pulses and pulse compression waveforms, are summarized in Figure 5.2.10. Pulses without phase modulation have lobes which vary from circular to elliptical, with resolution areas of unity and with the major axis elongated in delay for long pulses and in frequency for short pulses. Sidelobes appear only in the frequency coordinate, and the response drops to exactly zero beyond $\pm\tau$ in delay from the peak. The resolution area remains constant for these pulses.

Pulses with phase modulation fall into either of two classes:

(a) *Phase-coded*, with resolution area $1/B\tau$ and significant sidelobes over the entire area $4B\tau$;

(b) *FM (continuous phase)*, with an elongated main lobe (having a resolution area of unity) running along a diagonal. Although sidelobes also extend in this case over the area $4B\tau$, there is much less energy in the sidelobes than with phase-coded waveforms.

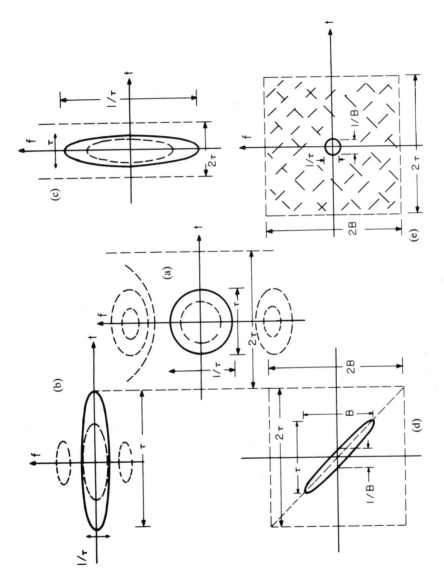

Figure 5.2.10 Resolution contours for single pulses.

5.3 MOVING-TARGET INDICATION

Spectra of Pulse Trains

The doppler resolution of single pulses is insufficient to permit rejection of fixed echoes from clutter while passing those from aircraft or land vehicles, unless very long pulses, with no phase modulation, are used. To develop such resolution, a train of pulses must be transmitted with consistent phase relationships (coherence). Consider the waveform obtained when an antenna scans its Gaussian beam across a fixed target (Figure 5.3.1a). The envelope of the received pulse train will be a broad time function established by the antenna pattern:

$$a_0(t/t_o) = \exp[-2.77(t/t_o)^2] \tag{5.3.1}$$

where $t_o = \theta_3/\omega$ is the time-on-target (measured within the one-way half-power beamwidth) and $t = 0$ corresponds to the target being centered in the beam. This envelope is sampled by individual pulses, with a repetition interval t_r and width τ. The sampling can be expressed as multiplication of the envelope by the convolution of the repetition function and the individual pulse waveform at carrier frequency f_0:

$$a(t) = a_0(t/t_o)a_1(t/t_r) \otimes a_2(t/\tau) \cos(2\pi f_0 t) \tag{5.3.2}$$

where $a_1(t/t_r)$ is an impulse at $t = it_r$, $i = 0, \pm1, \pm2, \ldots$, $a_2(t/\tau)$ is the pulse shape (nominally, a rectangular function $a_2 = 1$, $0 < t < \tau$), and $\otimes$ denotes convolution.

The Fourier transform of this sampled waveform gives the voltage spectrum (Figure 5.3.1b). This spectrum can be described as the product of two functions:

(a) the carrier line $\delta(f_0)$, convolved with the pulse spectral envelope $A_2(f\tau)$, given by (5.1.7);

(b) a fine line spectrum, which is the convolution of lines $A_1(ft_r)$, repeating at the pulse repetition frequency $f_r = 1/t_r$, and the spectral envelope $A_0(ft_o)$ produced by antenna scanning.

The resulting spectrum is

$$A(f) = [\delta(f_0) \otimes A_2(f\tau)][A_1(ft_r) \otimes A_0(ft_o)] \tag{5.3.3}$$

Note that multiplications in the time function are replaced by convolutions in the spectrum, and *vice versa*. As a result of the Fourier transform

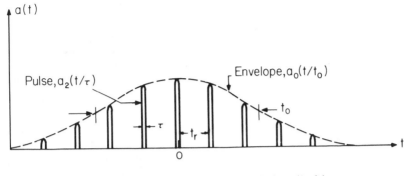

(a) Waveform, $a(t) = a_0(t/t_o)a_1(t/t_o) \otimes a_2(t/\tau) \cos(2\pi f_0 t)$

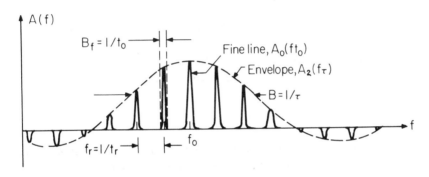

(b) Spectrum, $A(f) = [\delta(f_0) \otimes A_2(f\tau)][A_0(ft_o) \otimes A_1(ft_r)]$

Figure 5.3.1 Waveform and spectrum of pulse train modulated by scanning antenna.

relationship, the broad width B of the spectral envelope is the reciprocal of the short pulsewidth τ, the width B_f of the fine spectral lines is the reciprocal of the antenna time on target t_o, and the spacing f_r of spectral lines is the reciprocal of the pulse repetition interval t_r. If the amplitudes and phases (with respect to the carrier f_0) of every pulse in the transmitted train are the same, and if $a_0(t)$ is a smooth function with $t_o \gg t_r$, there will be regions in the received spectrum of a fixed target or clutter (Figure 5.3.3) where only receiver noise will be present. Echoes from a second, moving target in the beam will produce a spectrum in which the lines are offset from those of the fixed target by the doppler shift f_d. The presence of a moving target can then be detected, even if it is much weaker than the fixed targets occupying the same spatial resolution cell. A receiver and processor configuration which performs the necessary filtering to reject fixed targets (clutter) while retaining moving targets is called a moving-target-indication (MTI) or pulsed doppler system. In the pulsed doppler system, targets having specific velocities are selected by narrowband filters, while in MTI there is a broad filter responding to moving targets and a rejection band for clutter.

It should be noted, from Figure 5.3.2 and (5.3.3), that the fraction of the spectrum occupied by lines from fixed clutter is proportional to the ratio $t_r/t_o = 1/n$, where $n = f_r t_o$ is the number of hits per scan for the radar. Hence, the ability of the radar to recover the moving targets will be related directly to n. In an electronically scanned radar, where the beam steps to each position and remains fixed for n pulses, there will be no signal amplitude change during batch processing of these pulses, and MTI can form a notch with a depth that depends only on radar and clutter instabilities. These instabilities will also broaden the lines in a scanning radar, but scanning is often the dominant component in establishing the width of the line and the performance of the system.

Ambiguity Function of a Pulse Train

The waveform and spectrum shown in Figure 5.3.1 define, through (5.1.3), the characteristics of a matched filter for detection of targets at any given velocity. The response of such a filter will extend, at significant level, over a time interval several times t_o, and over a frequency interval $2/\tau = 2B$ (with sidelobes beyond this value). Near the center of this area (Figure 5.3.2), the response at the -6 dB level will consist of narrow lobes with width τ and spacing t_r in time, and width B_f and spacing f_r in frequency. The resolution area of each such lobe is $\tau B_f = 1/Bt_o$, and the effective number of lobes (considering the reduced amplitudes and broader widths in regions removed from the center) is $Bt_o = 1/B_f\tau$. Thus, the volume of the ambiguity function is preserved.

The pulse repetition interval t_r determines the *unambiguous range* of the waveform:

$$R_u = ct_r/2 = c/2f_r \tag{5.3.4}$$

While a signal cannot arrive from negative range, it often comes from $R > R_u$, and so will appear in the response at an apparent range $R' = R - iR_u$, where i is the largest integer for positive R'. When R_u exceeds the maximum target range, the waveform is called *low PRF*. Similarly,

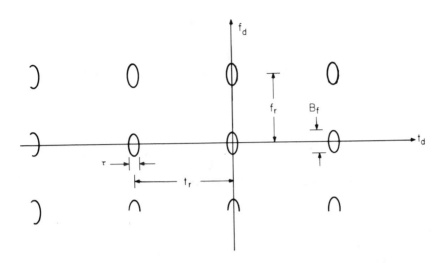

Figure 5.3.2 Central portion of ambiguity function of pulse train.

the pulse repetition frequency and transmitted wavelength determine the unambiguous velocity or *blind speed* of the waveform:

$$v_b = f_r \lambda/2 = \lambda/2t_r \tag{5.3.5}$$

When v_b is greater than the maximum velocity span of expected targets and clutter, the waveform is called *high PRF*. In some cases, ambiguity between approaching and receding targets is accepted, and the condition for high PRF operation is taken as $v_b > |v_t|_{max}$. For an airborne radar, the velocity span extends from the maximum closing velocity of targets approaching from the nose to the maximum receding velocity of targets or clutter behind the aircraft.

The area of the time-frequency plane may be divided into rectangles of area $t_r \times f_r = 1$, each centered on a response lobe of the ambiguity function. As the PRF is changed, the rectangles change shape, but not area, and the product of unambiguous range and blind speed remains constant:

$$R_u v_b = (c/2f_r)(f_r \lambda/2) = c\lambda/4 \tag{5.3.6}$$

Thus, for a given transmitted wavelength, a large unambiguous range can be obtained at the expense of a small blind speed, and *vice versa*. This immutable relationship has a profound effect on the design of all types of radar. From (5.3.6) we see that the use of a sufficiently long wavelength can make a radar unambiguous in both range and velocity for a given class of targets. For example, if aircraft up to 200 km range and 600 m/s velocity are to be observed unambiguously from a fixed radar, the minimum wavelength is

$$\lambda_{min} = 4R_u v_b/c = 1.6 \text{ m}$$

Such radar operation is unusual, and as a result most systems will be ambiguous in one or the other coordinate. A PRF chosen to produce ambiguity in both range and velocity is called *medium PRF*.

Optimum MTI Filter

When the radar receiver input interference consists of noise and clutter, the optimum filter for target detection is no longer the matched filter described by (5.1.3), but becomes [5.7]:

$$H(f) = A^*(f)/U(f) \tag{5.3.7}$$

where $U(f)$ is the voltage spectrum of the combined noise and clutter. This filter is known as the Urkowitz filter, and can be implemented as a prewhitening filter $1/U(f)$, which makes the interference spectrum uniform, followed by the conventional matched filter. The optimum MTI filter, for targets of unknown velocity, takes the form shown in Figure 5.3.3. The depth of the clutter rejection notches would be such as to reduce the residual clutter spectral density to the level of thermal noise. This filter process would be followed by envelope detection and noncoherent integration, to recover target energy over t_o.

Implementation of MTI Filters

The classic MTI filter forms a rough approximation of the Urkowitz comb filter by passing baseband video signals through a delay-line canceler (Figure 5.3.4). To produce the coherent transmission, a stable local oscillator (STALO) is used at frequency f_s, and the transmission frequency f_0 is obtained by offsetting the STALO frequency by the coherent oscillator (COHO) at intermediate frequency (IF), f_c. The transmission is at $f_0 = f_s + f_c$. The received echo signals, after down-conversion by the STALO,

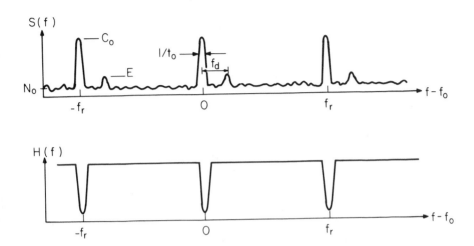

Figure 5.3.3 Interference spectrum and optimum filter.

appear at $f_c + f_d$, and are within the IF bandpass B centered at f_c. They are amplified and again down-converted to baseband, appearing at the output of the phase detector as *bipolar video* (oscillating between positive and negative amplitudes) at $f'_d = f_d - if_r$, where i is the integer that places f'_d between $-f_r/2$ and $f_r/2$. In the diagram shown, I and Q phase detectors are used to recover both components of the signal and to avoid folding of the clutter spectrum. When only a single phase detector is used, the positive and negative frequencies are folded into a band 0 to $f_r/2$. Signals that appear at the phase detector in quadrature with the COHO reference will not be passed (the blind phase phenomenon).

The delayed video, passing through the delay line of length t_r, is subtracted from the direct output of the phase detector, and the resulting bipolar output is full-wave rectified. When I and Q cancelers are used, the output video is formed as $\sqrt{I^2 + Q^2}$. The fixed echoes, which maintain the same phase over t_r, are canceled. Moving targets, with phase change over t_r, will not cancel, and will be passed with the response:

$$H_1(f) = 2|\sin(\pi f_d/f_r)| \tag{5.3.8}$$
$$H_1(v) = 2|\sin(\pi v/v_b)|$$

Targets at zero radial velocity and at integral multiples of the blind speed, $v_b = f_r \lambda/2$, also have zero response in the MTI filter.

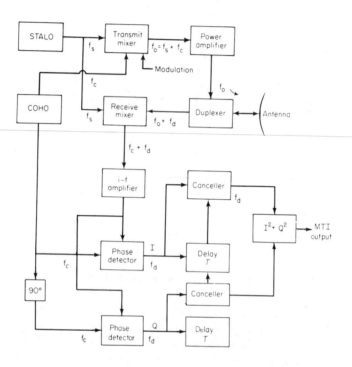

Figure 5.3.4 Coherent MTI radar system.

The single-delay canceler has an infinite null for clutter at precisely zero velocity, but cannot cancel components that result from antenna motion, clutter motion, or internal instabilities in the radar. Most MTI radars place a second canceler in cascade with the first, obtaining a response:

$$H_2(f) = 4 \sin^2(\pi f_d/f_r)$$
$$H_2(v) = 4 \sin^2(\pi v/v_b)$$

(5.3.9)

These two response curves are shown in Figure 5.3.5. Additional cancelers may be added, but the response band for targets becomes narrower and adequate detection is increasingly difficult to ensure for targets which are not at the optimum speeds $v/v_b = (2i - 1)/2$.

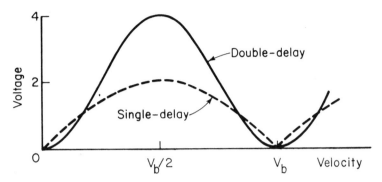

Figure 5.3.5 Velocity response of single- and double-canceler MTI.

Staggered-PRF and PRF-Diversity MTI

The low-PRF radar must detect targets having radial velocities from near zero to several times v_b. To avoid loss of detection on many targets near the blind speed and its multiples, the common practice is to introduce pulse-to-pulse stagger in the repetition interval t_r, producing the type of response shown in Figure 5.3.6. The deep null is retained at $v = 0$, but response at other blind speeds is increased. The length of the delay line is incremented by the same amount as the pulse repetition interval (PRI), to align at the cancelation point the pulses from successive transmissions. This procedure is adequate for radars in which all the clutter lies within the unambiguous range R_u. However, any clutter with range $R_c > R_u$ will arrive at the cancelation point with incorrect delay, and will not be canceled. This problem can be avoided in PRF-diversity MTI, where the PRI is held constant until echoes from the longest-range clutter are received and processed through the delay lines. The PRI is then changed on a burst-to-burst basis. After each constant-PRI burst is batch-processed in the cancelers, the output must be blanked until the first of the longest-range echoes from the next burst has passed the cancelers. Envelope-detected outputs from the successive bursts may be integrated, or separate detection decisions may be made on each.

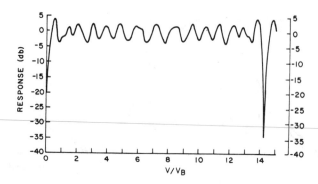

Figure 5.3.6 Response of staggered-PRF MTI [5.8]. (Copyright 1970 by McGraw-Hill Book Co. Reprinted with permission.)

MTI with Magnetron Transmitter

When the transmitter is a pulsed oscillator (e.g., a magnetron), it is impossible to ensure phase coherence between output pulses. The transmitted and received spectra are then continuous, rather than resolved into lines at the PRF as in Figure 5.3.1. It is still possible, however, to produce a coherent line spectrum at the output of the phase detector in Figure 5.3.7. It is necessary to store for each PRI the random phase by which the transmitter pulse departs from the reference sinusoid at $f_s + f_c$. One procedure, shown in the figure, is to down-convert with the STALO a sample of the transmitter pulse, producing a COHO lock pulse at f_c. The COHO is turned off just before the transmission, and as the COHO is restored to oscillation the pulse adopts the phase of the lock pulse, retaining it for the rest of the PRI. As each random-phased echo is received at the phase detector, the random phase of the transmission is subtracted, leading to a coherent bipolar output signal. This output is processed by the canceler in the same way as in the coherent system of Figure 5.3.4. This type of MTI system is called *coherent-on-receive* to distinguish it from that using a coherent transmitter and from the noncoherent MTI described below.

Magnetron radars with digital processing may use the process shown in Figure 5.3.8. The transmitted (reference) phase is measured at baseband by I and Q phase detectors, after down-conversion by continuous STALO and COHO signals. This phase is subtracted digitally from the received

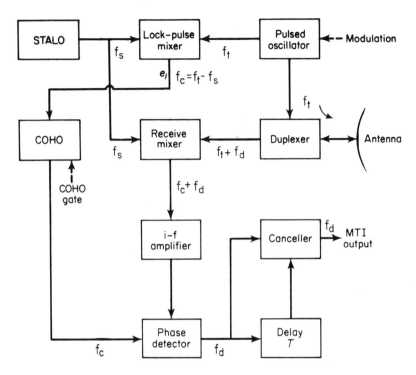

Figure 5.3.7 Coherent-on-receive MTI.

signal phase before digital cancelation. If the A/D conversion is executed with sufficient precision, the results are better than those usually available from a locked COHO, because the stability of the continuously operating COHO is better than one which is turned off and on with each transmission.

It is important to note that MTI systems using random transmitter phase and a locked COHO or stored phase data cannot cancel clutter from beyond the unambiguous range. For systems which must use staggered PRF, this poses no additional problem, because no cancelation of ambiguous clutter is available in any case. However, if ambiguous clutter is to be canceled, a coherent transmitter with PRF diversity, rather than stagger to fill blind speeds, must be used.

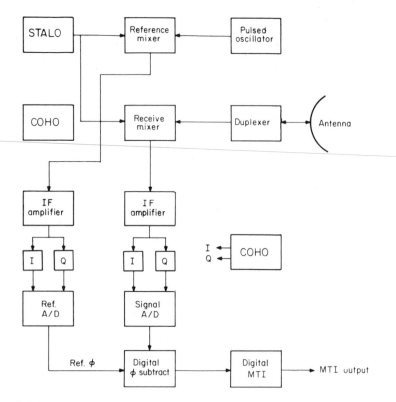

Figure 5.3.8 Digital storage of phase in coherent-on-receive MTI.

Noncoherent MTI

An alternative MTI process uses the clutter itself as the phase reference for detection of moving targets. The transmitter may be either an amplifier or a pulsed oscillator, and a STALO is used in the receiver. The IF signals are envelope detected, and no COHO is required. In a field of fixed clutter, successive video outputs will be identical, and can be canceled by the usual delay-line canceler (only one channel is needed, since there is no quadrature output). Presence of a moving target is evidenced by amplitude modulation of the $S + C$ pulse, at the frequency f_d. The modulation passes the canceler with the same response as for the coherent MTI, (5.3.8) and (5.3.9). A limitation of this process is that moving target echoes in a region not containing clutter have no modulation and are canceled. A clutter gate can be used to switch the output from normal (uncanceled) video to canceled video, when clutter is present. This switching process is difficult to control properly when the clutter appears intermittently.

A noncoherent MTI process that dispenses with the clutter gate is shown in Figure 5.3.9. Here, a phase detector is used, and its reference is derived from the hard-limited output of range cells adjacent to the signal cell. If clutter is present, it provides a zero-doppler reference, and target phase variation at f_d is detected and passed to the canceler. If only noise is present, the target output will appear with random phase and will pass the canceler.

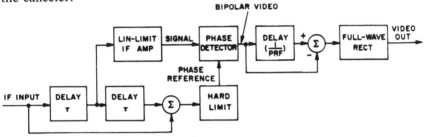

Figure 5.3.9 Noncoherent MTI using adjacent range cells for phase reference [5.8]. (Copyright 1970 by McGraw-Hill Book Co. Reprinted with permission.)

Area MTI

Rejection of fixed echoes can also be obtained by storing the normal, envelope-detected radar output in a map (implemented with a storage tube or in digital form). On each scan of the antenna, the new video is compared with the contents of the map for that resolution cell, and an output appears as the difference between the two amplitudes. The new data are stored in the map, either replacing the old data or to be put through a recursive filter process, which averages the data from several scans. The process is not sensitive to doppler shift, but rather to the motion of the signal envelope in range and angle over the map. Hence, the process is not as sensitive to small velocity components, unless very high spatial resolution is used. On the other hand, the area MTI responds to target motion other than radial, and operation over long scan periods can give good sensitivity to low velocities.

Performance of MTI

The performance measures of the MTI process are defined as follows.

Improvement Factor. The MTI improvement factor is defined as the ratio of output ratio $(S/C)_o$ to input ratio $(S/C)_i$, averaged over all target velocities:

$$I_m = \frac{(S/C)_o}{(S/C)_i}\bigg|_{\text{all } v_t} \tag{5.3.10}$$

The improvement factor is approximately equal to the *normalized clutter attenuation,* which is defined as the ratio of input $(C/N)_i$ to output $(C/N)_o$:

$$CA = \frac{(C/N)_i}{(C/N)_o} \tag{5.3.11}$$

The normalization by noise takes into account the average power gain of the canceler over all frequencies, which is $\overline{G} = 2$ for a single canceler and $\overline{G} = 6$ for a double canceler. If targets are also distributed uniformly over all doppler frequencies ($v - iv_b$ uniform over 0 to v_b), $I_m = CA$.

Subclutter Visibility. SCV is defined as the maximum input $(C/S)_i$ at which the target is still detectable with given probabilities of detection and false alarm. SCV can be expressed in terms of the improvement factor and the *clutter detectability factor* D_{xc} (the output $(S/C)_o$ required for detection):

$$SCV = I_m/D_{xc} \tag{5.3.12}$$

The relationship between improvement factor and subclutter visibility is shown in Figure 5.3.10. Note that SCV implies coincidence of target and clutter in range delay, so that effects of any receiver nonlinearity will be disclosed in measurement.

The improvement factor available in an MTI radar, for a clutter spectrum centered at zero, depends on the ratio of blind speed to clutter velocity spread. It is convenient to define the ratio $z = 2\pi\sigma_v/v_b$, where σ_v is the standard deviation of the clutter velocity spectrum. When the spectrum results from scanning of an antenna, this standard deviation is

$$z_a = 2\sqrt{\ln 2}\ \omega/f_r\theta_3 = 1.665/n \tag{5.3.13}$$

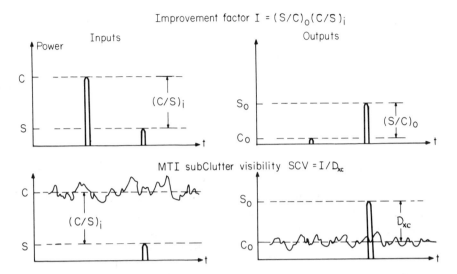

Figure 5.3.10 MTI terminology.

The improvement factors for cancelers with one, two, and three delays, preceded by linear processing, can be found as

$$I_1 = [1 - \exp(-z^2/2)]^{-1} \approx 2/z^2 = 0.72n^2 \tag{5.3.14}$$

$$I_2 = [1 - (4/3) \exp(-z^2/2) + (1/3) \exp(-2z^2)]^{-1}$$
$$\approx 2/z^4 = 0.26n^4 \tag{5.3.15}$$

$$I_3 \approx 4/3z^6 = 0.063n^6 \tag{5.3.16}$$

where the expressions involving n assume $z = z_a$ (scanning component is dominant). In general, the clutter velocity spread can be found as the rss sum of several components:

$$\sigma_v = [\sigma_{va}^2 + \sigma_{vs}^2 + \sigma_{vt}^2 + \sigma_{vi}^2 + \ldots]^{1/2} \tag{5.3.17}$$

where σ_{va} results from antenna scanning, σ_{vs} from wind shear over a clutter volume (3.6.14), σ_{vt} from internal turbulence in the clutter, and σ_{vi} from radar system instabilities. Motion of the radar platform also contributes to spectral spreading, giving a component:

$$\sigma_{vp} = (\theta_3 v_p/2.36) \sin\alpha \tag{5.3.18}$$

where α is the angle between the beam of width θ_3 and the platform velocity vector of magnitude v_p.

Internal turbulence has been measured as follows:

Wooded hills: $\sigma_{vt} = 0.01$ to 0.32 m/s
Sea echo: $\sigma_{vt} \approx v_w/8$ (at wind speed v_w)
Rain, chaff: $\sigma_{vt} = 1$ to 2 m/s

Many radars have insufficient dynamic range to pass the peaks of the clutter without saturation, or insufficient improvement factor to reduce these peaks to the noise level. To avoid false alarms in such cases, an IF limiter may be placed before the phase detectors to hold the peak clutter within the A/D range or below the level $I_m N$ (from which level the residue will be reduced to the noise level). It has been shown that this limiting action, at any level below peak clutter, spreads the clutter spectrum and degrades MTI performance (Figure 5.3.11). The Gaussian input spectrum with standard deviation $\sigma_f = 2\sigma_v/\lambda$ has a half-power bandwidth $B_3 = 2.35\sigma_v$. Figure 5.3.12 shows that the spectrum at the limiter output, while having essentially the same half-power width, spreads by a factor of five at the -30 dB level, and a factor of 35 at the -50 dB level, when the average clutter power is more than 20 dB above the limiting level.

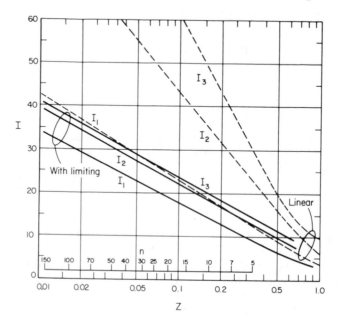

Figure 5.3.11 MTI improvement factor for linear and limiting processors.

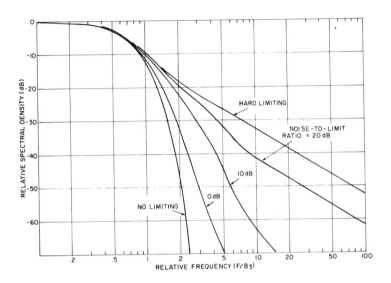

Figure 5.3.12 Spectral spreading caused by limiter [5.9].

With limiting, the two-delay and three-delay cancelers both achieve the performance predicted by (5.3.14) for the single-delay canceler under linear operating conditions. This implies that the limit level must be set above noise by $2/z^2$, if residue is to be at noise level, and that clutter above the limit level will suppress targets by the ratio of clutter to limit level. Such MTI operations, with $n = 15$ to 20 hits per scan, will achieve actual improvement factors of 20 to 25 dB, whether double or triple cancelers are used. These systems have been shown to work reasonably well in actual land clutter, however, because a large fraction of the resolution cells contain clutter of less than 20 dB above noise, and targets remain visible in these cells, even though they are suppressed in the stronger clutter cells. This mode of operation depends on an *interclutter visibility* property of radars having good spatial resolution [5.10]. In Section 7.2, the use of a clutter map to control the dynamic range of signals entering the MTI is discussed. Such maps can also provide for interclutter visibility on targets at the blind speeds.

Radar system instabilities sometimes appear as random from pulse to pulse, in which case the relationships of Table 5.1 may be used. Other limitations appear as a result of staggered PRI and quantizing of signals in an A/D converter [5.8]. The stagger limit, for a radar with n hits per scan, is

$$I_s = [2.5n/(\gamma - 1)]^2 \tag{5.3.19}$$

where γ is the PRI stagger ratio. The quantizing limit is

$$I_q = 0.75 \, (2^m - 1)^2 \approx 6m - 1 \text{ dB} \qquad (5.3.20)$$

where m is the number of bits, including the sign bit. These limits are independent of the number of delays in the canceler. The improvement factor limits may be combined to obtain the overall improvement factor:

$$I_m = [(1/I_x) + (1/I_s) + (1/I_q) + \ldots]^{-1} \qquad (5.3.21)$$

where I_x is the limit imposed by the standard deviation of the clutter spectrum for the x-delay canceler (see (5.3.14) to (5.3.16)), and individual instability limitations from Table 5.1 are included by adding their reciprocals within the brackets.

Table 5.1 Instability Limitations to Improvement Factor [5.8]

Pulse-to-pulse instability	*Limit on improvement factor*
Transmitter frequency	$I = 20 \log[1/(\pi \Delta f \tau)]$
STALO or COHO frequency	$I = 20 \log[1/(2\pi \Delta f_d)]$
Transmitter phase shift	$I = 20 \log(1/\Delta\phi)$
COHO locking	$I = 20 \log(1/\Delta\phi)$
Pulse timing*	$I = 20 \log[\tau/(\sqrt{2} \, \Delta t \, \sqrt{B\tau})]$
Pulsewidth	$I = 20 \log[\tau/(\Delta PW \, \sqrt{B\tau})]$
Pulse amplitude	$I = 20 \log(A/\Delta A)$

$$
\begin{aligned}
\text{where } \Delta f &= \text{interpulse frequency change} \\
\tau &= \text{transmitted pulse length} \\
t_d &= \text{transmission time to and from target} \\
\Delta\phi &= \text{interpulse phase change} \\
\Delta t &= \text{time jitter} \\
B\tau &= \text{time-bandwidth product of pulse compression} \\
&\quad \text{system } (B\tau = \text{unity for uncoded pulses}) \\
\Delta PW &= \text{pulsewidth jitter} \\
A &= \text{pulse amplitude, volts} \\
\Delta A &= \text{interpulse amplitude change}
\end{aligned}
$$

*For some critical timing pulses (e.g., A/D converter strobes), the improvement factor becomes $I = 20 \log [1/\sqrt{2} \, B\Delta t]$.

Performance of Noncoherent MTI

When the clutter itself serves as the reference for detection of targets, the spectrum of the clutter presented to the canceler is broadened by action of the envelope detector (or phase detector of Figure 5.3.9). The resulting improvement factor for single delay cancelers remains as in (5.3.14), but for double delay it becomes

$$I_2 = 2[1 - (4/3) \exp(-z^2) + (1/3) \exp(-4z^2)] \approx 1/z^4$$

$$= 0.13 \, n^4 \tag{5.3.22}$$

One advantage of noncoherent MTI is that it can be effective in canceling clutter with nonzero average velocity. Noncoherent MTI can also cancel clutter from an ambiguous range interval. Although the transmitter must be reasonably stable, pulse-to-pulse coherence is not required for cancelation of range-ambiguous clutter, provided that this clutter does not overlap range cells with clutter from other ambiguities.

MTI with Moving Clutter

Airborne clutter sources (rain, chaff, and birds) will have spectra centered on the mean velocity v_w of the air mass in which they are embedded, often far from zero. In addition, the radar platform may move to contribute a radar-clutter radial velocity component. The maximum improvement factor in such cases, for narrow clutter spread, will be

$$I_m = \overline{G}/|H(v_w)|^2 \tag{5.3.23}$$

For example, using a double canceler with $v_b = 50$ m/s, clutter moving at $v_w = 10$ m/s will have a response $H_2^2 = 1.9$, or 5 dB below the average gain of six. Adaptive control of the center of the rejection notch will be required to place the notch on the clutter spectrum.

Adaptive MTI processors will measure the relative clutter residue above and below the notch, and shift the notch (or the COHO) to minimize the total residue. If this adaptation is done with rapid response, ground clutter can be canceled with the notch at zero at short range, and moving rain or chaff with a displaced notch at long range. When the clutter spectrum is Gaussian around its mean velocity, proper adaptation of the notch can preserve the improvement factors given by (5.3.14) to (5.3.16). It is also possible to adapt the notch of one canceler while leaving the other fixed at zero, or to broaden the notch with controllable feedback, to cancel simultaneously both clutter components. However, the loss in target response makes it impossible to maintain high detection probabilities when these steps are taken.

Losses in MTI Systems

When MTI processing is used, there will be losses in detectability of targets, beyond those encountered in normal processing. These losses must be included in the loss factor L_x of the Blake chart (Figure 1.2.4), or applied to increase the detectability factor D_x in other forms of the radar equation.

Noise Correlation, $L_{mti(a)}$. Use of MTI does not change the average S/N of the target, but it does introduce a requirement for greater signal input, which must be included as an increase in the detectability factor D_x in the radar equation or as an equivalent loss. Passage of noise through an m-pulse MTI canceler introduces partial correlation over m successive outputs [5.11]. The effect is to reduce the number of independent noise samples available for integration in subsequent processing from n to an, where

$$a = 2/3 \text{ for } m = 2$$
$$= 18/35 \approx 1/2 \text{ for } m = 3$$
$$= 20/47 \approx 0.43 \text{ for } m = 4$$

The corresponding loss is the ratio of detectability factor for an integrations to that for n integrations:

$$L_{mti(a)} = D(an)/D(n) \tag{5.3.24}$$

If the quadrature canceler is not used, there will be further reduction factor of two in the number of independent noise samples integrated:

$$L_{mti(a)} = D(an/2)/D(n) \tag{5.3.25}$$

When batch processing is used, with an m-pulse canceler operating on n/m batches at different PRFs, the number of outputs becomes n/m, and

$$L_{mti(a)} = D(n/m)/D(n) \qquad (5.3.26)$$

Batch processing will not normally be used with the I-channel canceler only, but the further reduction to $n/2m$ effective samples will apply in this case as well.

Blind Phase Loss, $L_{mti(b)}$. Use of only the in-phase bipolar video leads to rejection of the quadrature components of both noise and signal. The effect on integration is given by (5.3.25), but there is a further effect on target fluctuation loss. The Rayleigh target, having a chi-square distribution with $2K = 2$ degrees of freedom, is converted to a Gaussian target (with one degree of freedom). If the target changes phase sufficiently to appear at full amplitude in the I channel output during t_o, the fluctuation loss is not significantly increased. However, in a batch-processed system in which there is only one output sample, the fluctuation loss (in dB) for any target with $2Kn_e$ degrees of freedom is doubled:

$$L_{mti(b)} = L_f(Kn_e/2)/L_f(Kn_e) \qquad (5.3.27)$$

Velocity Response Loss, $L_{mti(c)}$. The S/N remains the same in passage through the canceler when averaged over all target velocities, but a particular target may be canceled or significantly suppressed by the MTI velocity response. In order to achieve detection probability above 0.5, the average input signal must be increased enough to compensate for the loss in detection of targets near the blind speed. For $P_d = 0.9$, this loss is 7.6 dB for a single canceler, and about twice this for a double canceler. Because of the high value of this loss, staggered PRF or PRF diversity is almost always employed. The loss is then found by averaging detection probability over all target velocities, and increasing the input signal by the amount necessary to restore the desired probability.

Range Straddling Loss, L_{er}. When range gating or digital sampling is used at the input to the processor, the samples may not occur at the time of peak signal. The resulting loss is discussed in Section 5.4.

Resulting Detectability Factors

In calculating the detectability factor $D_x(n) = D(n)ML_pL_x$ for regions in which MTI is used, the loss L_x must include L_{mti}, the product of components (a), (b), and (c) described above, along with range straddling loss and other components. The result will be a substantial reduction in detection range, even when no clutter is actually present at the target location. In addition, for those locations where clutter is present, its residue must be included as an interference component, with appropriate adjustment for the ratio D_{xc}/D_x as noted in Section 1.6. In the calculation of D_{xc} from (2.6.1), the correlation time of the clutter residue at the MTI output should be used. To the extent that random instabilities are the cause of the residue, it will be uncorrelated from pulse to pulse, and subsequent integration will provide the full gain $G_i(n)$. If clutter peaks have been reduced by limiting, the clutter distribution loss will be low, unless high values of P_d are required. If the target is suppressed by clutter peaks with probability P_p, the detection probability will be approximately $P'_d = P_d(1 - P_p)$, and high detection probabilities will be difficult to obtain.

Detection Range with MTI Processing

Using the radar example from Figure 1.2.5, the detection range can be calculated for a particular MTI process: three-pulse digital MTI with I and Q processing and staggered PRF. The comparison of losses is shown in Table 5.2.

Table 5.2 Losses With and Without MTI (dB)

		With MTI	Without MTI
Noise correlation,	$L_{mti(a)}$	2.0	0
Blind phase,	$L_{mti(b)}$	0	0
Velocity response,	$L_{mti(c)}$	1.0	0
CFAR,	L_g	2.0	2.0
Range straddling,	L_{er}	1.0	1.0
Total misc.,	L_x	6.0	3.0
Matching,	M	0.8	0.8
Beamshape,	L_p	1.3	1.3
Total loss		8.1	5.1
Detectability factor,	$D_1(24)$	11.0	11.0
Detectability factor,	$D_x(24)$	19.1	16.1
$+ 10 \log 24$		13.8	13.8
Detectability factor	$D_x(1)$	32.9	29.9

The MTI, even given optimistic assumptions of low velocity response loss due to stagger, has introduced 3 dB of extra loss, reducing the range on 1.0 m^2 targets from 86.1 km to 72.4 km, when the target competes only with noise passed through the MTI processor.

The clutter detectability factor would be increased by approximately $G_i(24)/G_i(4) = 3.5$ dB as a result of the correlation of residue from scanning. Clutter peaks, corresponding to $\sigma_c = 10^4$ m^2, would rise some 26 dB above the average clutter, suppressing targets in a few cells, but introducing only a slight clutter distribution loss (e.g., $L_{cd} = 1$ dB). The final figure for D_{xc} (from Eq. (2.6.1)) would be about $D_x + 3.7$ dB $= 22.8$ dB. When land clutter is present in the target cell within the horizon, (1.5.18) yielded, for the example, an average $(S/C)_o = +20.9$ dB with $I_m = 25$ dB (corresponding to input $(S/C)_i = -4.1$ dB). This is the maximum value of improvement factor that can be supported by 24 hits per scan in a limiting system, Figure 5.3.11. The 1 m^2 target would thus be detectable at a level just below $P_d = 0.9$, in land clutter.

5.4 PULSED DOPPLER

Definitions

A *doppler radar* is one which utilizes the doppler effect to determine the radial component of relative radar-target velocity or to select targets having particular radial velocities, and a *pulsed doppler (PD) radar* does this with pulsed transmissions [5.12]. Thus, the PD radar processes a train of received pulses in a narrowband, coherently integrating filter, which resolves and enhances targets within a particular velocity band. Usually, a bank of such filters will process in parallel different target velocities, while rejecting clutter and other echoes from objects outside the velocity band of interest. In tracking, only one filter may be needed, once the target is acquired. The PD approach differs from MTI in that the doppler filter performs coherent integration of the selected signal within a narrow band, while MTI passes targets through a broad response band. When a bank of doppler filters is used to pass targets over a broad velocity band, omitting filters which contain clutter, the envelope of the filter response may correspond closely to the optimum MTI filter response of Figure 5.3.3.

PD radars are further described as low, medium, or high PRF, according to the definitions accompanying (5.3.4) to (5.3.6). High and medium PRF radars have been used principally in airborne applications, where the clutter spectrum extends over hundreds of m/s, and where short-range clutter is minimized by the aircraft altitude. Use of these range-ambiguous waveforms in surface-based radar is made difficult by the presence of short-range clutter, which competes with targets at long range and has a large power advantage over the targets because of the $(R/R_c)^4$ term in (1.5.19). Low-PRF PD radars have become increasingly important in surface-based surveillance roles because of their better performance in clutter rejection and the decreasing cost of modern digital processors.

The performance of PD radars is described in terms of clutter attenuation and subclutter visibility. The *clutter attenuation* CA can be defined as the ratio of the input $(C/S)_i$ ratio to the output $(C/S)_o$ ratio, for a target at a specific velocity:

$$CA(v_t) = (C/S)_i/(C/S)_o|_{v_t} \tag{5.4.1}$$

This is illustrated in Figure 5.4.1, where input clutter and signal powers can be measured on a single-pulse basis in the wideband portion of the receiver, and output levels are measured after the narrowband filter containing the signal. The output clutter residue results from sidelobe response of the target doppler filter, and from spreading of the clutter spectrum into the filter as a result of radar system instabilities. In a stable system, CA is determined by the filter response, and is equal to the sidelobe rejection level at velocity v_t. In a system with ideal filters, CA is determined by system instabilities, measured in terms of the noise sideband level at $f_d = 2v_t/\lambda$ from the carrier.

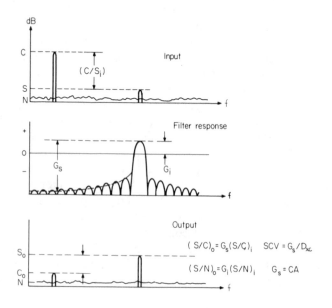

Figure 5.4.1 Pulsed doppler terminology.

The subclutter visibility is defined as with MTI (see Eq. (5.3.12)), but for targets in a specific velocity band:

$$SCV = CA(v_t)/D_{xc} \qquad (5.4.2)$$

where D_{xc} is the required $(S/C)_o$ for detection, given the subsequent processing (which may include noncoherent integration of successive filter outputs).

The other properties of PD radar which must be defined are the losses relative to the matched filter for the transmitted waveform. These losses are discussed in the final paragraphs of this section.

Low-PRF PD (MTD) Radar

The moving-target detector (MTD), developed at the Lincoln Laboratory during the 1970s, is an example of low-PRF PD radar applied to air surveillance. In its original form, the waveform consisted of two bursts, each of 10 constant-PRI pulses, transmitted by a magnetron or coherent transmitter of a 2D scanning radar. PRI diversity between bursts was used to fill blind speeds (Figure 5.4.2). Each burst was processed in two parallel channels, a zero-velocity channel and one for moving targets (Figure 5.4.3). The moving-target channel is processed by a double-delay MTI filter in cascade with an eight-point discrete Fourier transform (DFT), to produce seven filters spaced at $v_b/8$ with very low response at zero velocity. The zero-velocity channel uses a 10-point DFT matched to targets at zero and iv_b, $i = \pm 1, 2, \ldots$ This channel feeds a clutter map through a recursive filter, constituting an area MTI processor for targets in the rejection notch of the other channel.

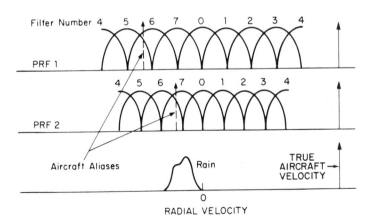

Figure 5.4.2 MTD filter responses and clutter spectrum [5.13].

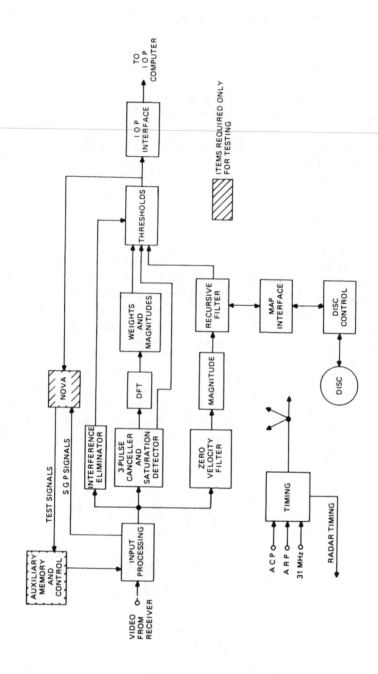

Figure 5.4.3 Block diagram of MTD processor [5.13].

The seven moving-target channels are envelope detected and applied to range cell averaging CFAR thresholds, which adapt the system response to any range-extensive clutter (rain and chaff). Thus, in Figure 5.4.2, the presence of rain just below zero velocity will cause filters 0, 6, and 7 to be desensitized by increased thresholds, leaving filters 1–5 to pass targets. The PRI diversity is chosen to minimize the chance of a target's lying in the desensitized filters in both bursts, as long as the clutter spectrum is not too wide relative to the average blind speed of the bursts.

Other versions of the MTD omit the MTI precanceler and optimize the weights on pulses in each burst to achieve low response at zero velocity. The number of pulses in a burst can vary, but the best control of the response is available with large numbers of pulses. For example, in a low-sidelobe filter operating on n_c pulses received over time t_f, the transition between the stopband (e.g., at -40 dB) and the passband (-6 dB) occupies a frequency interval $1/t_f \approx B_f$, corresponding to a velocity interval v_b/n_c (Figure 5.4.4). If the stopband is to cover a clutter spectrum of width Δ_c, the band lost to effective target detection is

$$\Delta_v = \Delta_c + 2v_b/n_c \tag{5.4.3}$$

For a 16-pulse burst, at least 1/8 of the velocities will lie below -6 dB in response, and this rises to 1/4 if the clutter spectral width is $v_b/8$.

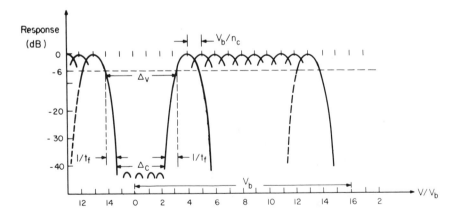

Figure 5.4.4 Clutter rejection notch width.

Another factor favoring long bursts is the need to cancel clutter from beyond the unambiguous range of the waveform. Even the low-PRF PD radar may not have a low enough PRF to avoid second-time-around clutter from rain and chaff, and land clutter can appear from very long ranges due to ducting in the atmosphere. When processing an n_c-pulse burst with low-sidelobe weighting (Figure 5.4.5), the second-time-around clutter echoes will arrive with an extra time delay t_r, the first echo will receive the second weight, and the last echo will be lost (or included with the incorrrect delay in the next processed burst). The resulting velocity response will be as shown in Figure 5.4.6, with sidelobes much higher than intended. In a PRI-diversity system, operating at constant RF, a pulse of clutter from the previous burst will produce the output level shown, and an independent output will be produced by absence of the first echo from the current burst.

    The required rejection for the longer range clutter is less than for similar clutter within the unambiguous range, but only by the ratio $[R/(R + R_u)]^2$ for volume clutter filling the beam. Hence, it will be necessary to ensure significant clutter attenuation on ambiguous clutter when it appears. This requires either that one or more *fill pulses* be transmitted prior to the n_c pulses to be processed in each burst, or that n_c be large enough that the increased sidelobe levels for ambiguous clutter are tolerable. In either case, long bursts are needed to preserve efficiency in both time and transmitter energy usage.

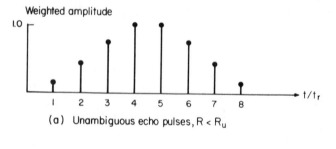

(a) Unambiguous echo pulses, $R < R_u$

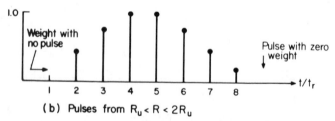

(b) Pulses from $R_u < R < 2R_u$

Figure 5.4.5 Weighting of unambiguous and ambiguous clutter.

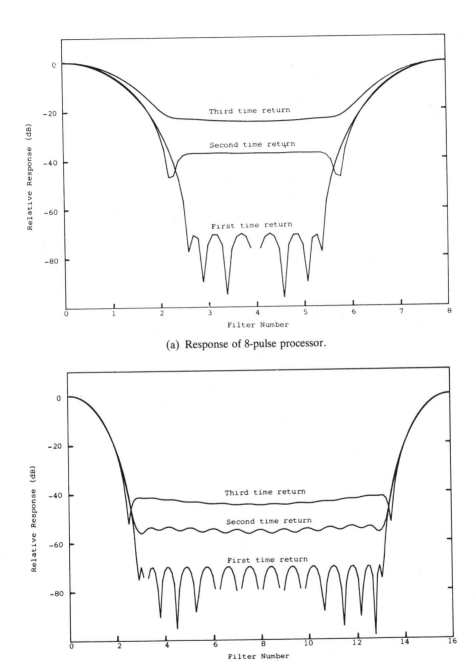

(a) Response of 8-pulse processor.

(b) Response of 16-pulse processor.

Figure 5.4.6 Response of 70-dB Dolph weighted filter to unambiguous and ambiguous clutter [5.9].

Medium-PRF PD Radar

In the presence of a broad clutter spectrum, it may be impossible to obtain a high enough blind speed to reject clutter and retain a reasonable fraction of the targets, using low-PRF radar. For example, if rain or chaff is present with significant wind shear, the clutter spectral width may exceed 30 m/s, requiring blind speeds in excess of 120 m/s to preserve target detection with reasonable PRF diversity. In airborne radar, the spreading of clutter within the main lobe of the antenna pattern (5.3.18) may lead to σ_{vp} = 5 to 10 m/s, and $2\pi\sigma_{vp}$ = 30 to 60 m/s when scanning to the side. If microwave radar is to be used, range ambiguity will almost surely occur when the PRF is set high enough to achieve rejection of clutter in these cases.

The principal problem in surface-based medium-PRF radar is to achieve the clutter attenuation required to detect long-range targets competing with short-range clutter. Equations (1.5.6) and (1.5.19) give the input C/S for volume and surface clutter occupying multiple range ambiguities. Usually, the clutter from the first range interval will govern, according to (1.5.4) and (1.5.16). With limited CA, the first several range cells after the transmitted pulse will contain strong clutter which may eclipse the target, adding to the eclipsing by the transmitter. This creates bands of width Δ_r in which the apparent target range $R - iR_u$ appears in the eclipsed region (Figure 5.4.7), running across the bands of width Δ_v caused by blind velocities. To obtain high probability of detection in each observation t_o, m bursts at different PRFs must be used, with average duration $t_f = t_o/m$. The number m depends on the fraction of clear area in range-velocity space:

$$F_a = (1 - \Delta_r/R_u)(1 - \Delta_v/v_b) \qquad (5.4.4)$$

In searching for targets which are spread widely over the ambiguity area, the detection probability P_1 per burst will be approximately $F_a P_d$, where P_d is the probability of detection in the clear region. In an observation over $t_o = mt_f$, with proper choice of diversity, the cumulative probability of detection will be approximately

$$P_c \approx 1 - (1 - P_1)^m = 1 - (1 - F_a P_d)^m \qquad (5.4.5)$$

The probability that the target is in a blind region is not actually independent over the m diversity bursts, but for typical choices of PRF, the results differ little from those predicted by (5.4.5). As F_a decreases, the

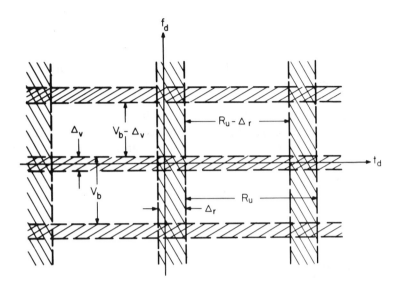

Figure 5.4.7 Blind regions in medium-PRF PD radar.

number m must be increased to obtain high P_c. In the process, however, the potential integration gain n_c and clutter attenuation are reduced, and both Δ_v and Δ_r are increased, causing further reduction in F_a. Optimization of the waveform and processing then depends on the density of the clutter and the stability of the radar system.

The duration t_f of the coherent burst depends on the allowable width of the transition band between stopband and target passband, and also on the maximum clutter time delay $t_{dc} = 2R_c/c$. Coherent processing cannot start until the first echo from maximum-range clutter has been received. The portion of the burst which contributes to the processed output then has a duration of

$$t'_f = t_f - t_{dc} \qquad (5.4.6)$$

and the filter bandwidths and transition band will depend on t'_f rather than on t_f itself. It is generally undesirable to use t_f less than about $4t_{dc}$.

In airborne medium-PRF radar, the velocity region over which targets may be detected will depend more on the antenna sidelobe levels than on the radar stability. A typical land clutter spectrum is shown in Figure 5.4.8, showing the main lobe of width $2.36\sigma_{vp}$, and sidelobes occupying the remainder of the velocity space. The width of the sidelobe spectrum is twice the aircraft velocity v_p, and the spectral density can be calculated by integrating over the areas where the range and velocity ambiguities intersect the surface below the aircraft. Each area is weighted by the two-way antenna pattern. If these sidelobe levels are low enough, a moving target can be detected, except when it lies within the clutter main lobe (i.e., when the target radial velocity is within $\pm\pi\sigma_{vp}$ of a blind speed). The fractional clear area F_a is found by using the same procedures as for surface-based radar, (5.4.4), and the requirements for PRF diversity are governed by (5.4.5). Since airborne radar can receive significant clutter from very long ranges, when the antenna points to the horizon, long bursts may be required to meet the constraint $t_f > 4t_{dc}$.

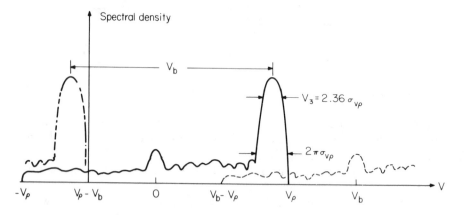

Figure 5.4.8 Clutter spectrum in airborne medium-PRF PD radar.

Tracking at medium PRF can be accomplished without long diversity bursts, if the PRF is continuously adapted to the target range and velocity. At any time, a PRF which places the target in a clear range and velocity region may be found, and only one burst at this PRF need be transmitted for each observation.

High-PRF PD Radar

When targets can be distinguished from clutter on the basis of radial velocity, and adequate clutter attenuation provided, the high-PRF wave-form offers the greatest energy and time efficiency. For example, in a surface-based radar where objects having radial velocities less than 40 m/s can be defined as clutter, a high-PRF waveform processed with a

40 m/s clutter notch can see all targets which are not in an eclipsed range cell. With sufficient CA, only the cell during the transmitted pulse need be eclipsed, giving a clear fraction:

$$F_a = 1 - \tau/t_r = 1 - D_u \tag{5.4.7}$$

The PRF for microwave radars will usually be high enough that the antenna beam does not reach the clutter surface in the shortest-range cells, and the clutter power may be almost uniform over all cells. In that case, the clutter power for a horizontal beam can be found by multiplying the clutter of a continuous-wave (CW) radar [5.14, p. 339] by the duty factor of the PD radar:

$$C/P_t = \frac{\sigma^0 \lambda w}{400 h_r^2} D_u \tag{5.4.8}$$

where w is aperture width and h_r is height of the phase center of the antenna above the clutter surface. For the constant-γ clutter reflectivity model, this becomes

$$C/P_t = \frac{\lambda^2 w \gamma}{533 h_r^2 h} D_u \tag{5.4.9}$$

where h is the aperture height.

For example, consider an X-band radar with a circular antenna ($w = h$) operating at height $h_r = 3$ m above a typical land surface ($\gamma = 0.1$), with $P_t = 10$ kW and $D_u = 0.1$ ($P_{av} = 1$ kW). From (5.4.9),

$$C/P_t = 1.9 \times 10^{-9}, \quad C = 1.9 \times 10^{-5} \text{ W}$$

The average clutter power $CD_u = 1.9 \times 10^{-6}$ W, and the clutter energy received in $t_f = 1$ ms will be 1.9×10^{-9} W-s. The noise spectral density will be $N_0 = kT_s = 4 \times 10^{-21}$ W/Hz, for $T_s = 3000$ K. At the output of a filter matched to the clutter over 1 ms, the ratio $(C/N)_o = 4.6 \times 10^{10} = +106.6$ dB. A target filter at $v_t > 40$ m/s must have zero-velocity sidelobes $G_s = 113$ dB down from its main-lobe response in order to place clutter residue at the noise level, or 103 dB down to have the output ratio $(C + N)/N = 1.25 = +1$ dB. This becomes the specification for CA to avoid 1-dB performance degradation when the antenna is pointed horizontally. The difficulty in meeting this clutter attenuation requirement has mitigated against use of land-based high-PRF PD radar.

The requirements for airborne radar are less stringent. When the radar antenna looks down at a clutter surface, at a depression angle θ greater than the antenna beamwidth, the clutter received is

$$C/P_t = \frac{D^2\gamma}{50R^2} D_u \tag{5.4.10}$$

where D is the antenna diameter. Consider an X-band radar with $P_t = 10$ kW, $D_u = 0.1$, $D = 0.7$ m above the same clutter surface. When the radar looks down for a target at $R = 10$ km, the clutter power $C = 10^{-8}$ W, and $C_{av} = 10^{-9}$ W, calculated as the duty factor times the clutter cross section within the radar beam. With the same values of t_f and T_s as are used for the land-based radar, we find $(C/N)_o = 2.5 \times 10^7 = +74$ dB, in the filter matched to clutter. For 1-dB degradation relative to the noise limited case, CA > 80 dB is required in the target filter.

Oscillator Effects on Performance of PD Radar

The major issues in PD radar performance are clutter attenuation and efficiency in target detection. The two main limits to clutter attenuation are the spreading of the clutter spectrum into the target filter main lobe, caused by radar instabilities, and the filter sidelobe levels at velocities corresponding to the main lobe of the clutter spectrum. Clutter spectral spreading caused by antenna scanning, platform motion and random variation in pulse amplitude and phase was discussed in the previous section on MTI.

For PD systems, with higher PRFs and greater CA requirements, the basic limits are set by the stability of the STALO from which both transmitter and receiving LO signals are derived. The phase noise of the oscillators primarily contributes to the output residue, since saturation in the transmitter tends to remove amplitude variations, and the usual types of receiver mixer also cancel amplitude noise. A high-quality X-band STALO, obtained by multiplication of a crystal source, will produce phase noise sidebands described by curves (a) or (b) of Figure 5.4.9. The noise sidebands are plotted as spectral density in dB below the carrier per Hz. The best performance, curve (c), is provided by a cavity-stabilized klystron oscillator operating directly at X-band at a power level measured in watts. This type of oscillator is used for CW radar, and may also be used in high-PRF land-based radar.

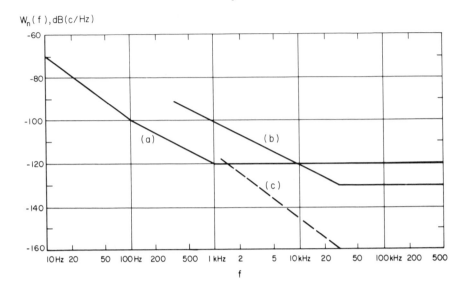

Figure 5.4.9 Noise sidebands on X-band STALO: (a) and (b) multiplied crystal sources; (c) cavity-stabilized klystron [5.15, p. 90].

Pulsed doppler radars using high- or medium-PRF waveforms will use the same STALO source for the receiver and, with an IF offset, for the transmitter. Phase noise sidebands from the STALO will then be correlated between transmission and reception, and will tend to cancel, at least for the short-range clutter, which poses the most serious problems. The correlation factor directly multiplies the phase noise density $W_n(f_d)$, and can be written

$$A(f_d, R_c) = (2 \sin\pi f_d t_d)^2 = (2 \sin2\pi f_d R_c/c)^2 = [2 \sin(R_c/2R_1)]^2 \quad (5.4.11)$$

where the range R_1 is defined as $c/4\pi f_d$. The correlation factor is plotted in Figure 5.4.10 as a function of R/R_1. Also plotted is the average value of A obtained by integration over clutter illuminated by a horizontal beam of width θ_e, from an antenna height h_r. The lower 3-dB point of this beam reaches the surface at range $R_b = 2h_r/\theta_e$, and this range is used in the abscissa of the plot. When the beam reaches the surface at $R_b > R_1$, the clutter arrives with time delays that decorrelate the receiving and transmitting noise components, doubling the effective noise sideband power.

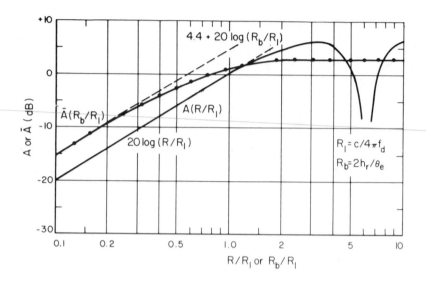

Figure 5.4.10 Correlation factor for common oscillator.

The output clutter power in a target filter at f_d, from clutter at range R_c, can be written

$$C_o(f_d, R_c)/C_i(0, R_c) = B_f \sum_{i=-B/2f_r}^{B/2f_r} W_n(f_d - if_r)A(f_d - if_r, R_c) \qquad (5.4.12)$$

where B is the signal bandwidth. In most cases, the contributions of all but one ambiguous sideband, $i = b$, are uncorrelated, leading to

$$C_o(f_d, R_c)/C_i(0, R_c) = B_f W_n(f_d - bf_r)[A(f_d - bf_r, R_c) - 1]$$
$$+ 2B_f \sum_{i=-B/2f_r}^{B/2f_r} W_n(f_d - if_r) \qquad (5.4.13)$$

where b is the integer nearest f_d/f_r.

For a low-PRF radar with low duty factor, clutter at very short range is not usually important. The contribution of the lowest sidebands can be ignored, and the clutter noise entering the target filter at f_d will be the sum of the sidebands at PRF ambiguities, unmodified by the correlation factor. For example, consider a radar with $f_r = 1000$ Hz, $B = 1$ MHz, $t_f = 0.01$ s, $n_c = 10$. There will be $B/f_r = 1000$ sidebands aliased into the filter, and for a filter other than at zero doppler, most of their power, for curve (a) of Figure 5.4.9, will come from the flat spectrum with density $W_n(f) = W_0 = 10^{-12}$ (c/Hz). The clutter residue will be

$$C_o/C_i = B_f(B/f_r)(2W_0)$$
$$= 100 \times 1000 \times 2 \times 10^{-12} = 2 \times 10^{-7} = -67 \text{ dB}$$

The limiting value for CA would be 67 dB, if there were no other contributors to residue. For curve (b), the density of the nearest sideband or two will determine the residue only for clutter at long range. At ranges within about 10 km, CA will be based on $W_0 = -130$ dB (c/Hz), and will be 77 dB.

As PRF is increased, for a given signal bandwidth, there will be fewer lines contributing to the residue, and the nearest lines (for filters not centered too near the clutter) will be farther from zero. For example, with $f_r = 10$ kHz, $B = 1$ MHz, $t_f = 0.01$ s, $n_c = 100$, there will be 100 sidebands aliased into the filter, and the residue for curve (a) will be 77 dB below the input. For curve (b), CA will be about 85 dB for clutter within the first range ambiguity.

For high PRF, for example, 100 kHz, with $B = 1$ MHz, $t_f = 0.01$ s, there will be 10 lines contributing. The strongest clutter sources will lie within the first range ambiguity, where the correlation factor will reduce the noise by a large factor. For curve (a), CA will be 87 dB, and for curve (b), it will be 97 dB. Curve (b) gives better results because of the reduced contribution of the aliased lines, since the low-frequency noise is reduced by the correlation factor.

Filter Sidelobe Effects on Performance of PD Radar

The major contributors to filter sidelobes are the weighting applied to the input pulse train, and the quantization introduced by the A/D converter at the input of the digital process. The curves of antenna efficiency and beamwidth broadening of Figure 4.2.3 are directly applicable to filters as well. The low-sidelobe weighting can be implemented either in an analog filter network or after A/D conversion. In the latter case, the number of bits in the converter becomes a critical factor in performance.

An estimate of the maximum clutter attenuation available from a system using m bits (including sign) is

$$CA = 6(m - 2) + 10 \log n_c \qquad (5.4.14)$$

The reduction by two in the effective number of bits results from two considerations: (a) the converter must pass clutter peaks, and (b) the rms residue must be at a level corresponding to two or three quanta at the low end of the converter range, in order to preserve the detection statistics.

A 12-bit converter with subsequent integration of $n_c = 10$ pulses will then support $CA = 70$ dB, adequate for most low-PRF system requirements. For medium-PRF radar, the number of pulses integrated increases along with the required CA, and processing of 12-bit data remains adequate. High-PRF airborne systems can also be implemented with A/D converters, but for ground-based systems it is necessary to use an analog clutter filter prior to the A/D converter to reduce the dynamic range.

Loss Factors in PD Radar

As with MTI systems, there are several loss factors which will be encountered in PD radar processing.

Filter Matching Loss, L_{mf}. This is the loss relative to matching to the envelope of the input pulse train, $a_0(t/t_o)$ in Figure 5.3.1. This loss is controlled by the filter bandwidth and the weighting applied to the pulses to obtain the desired filter sidelobes. In many cases, the coherent processing is carried out over a time $t_f < t_o$, and a loss from this must be included either as an increase in L_{mf} or as a noncoherent integration loss $L_i(n')$, where $n' = t_o/t_f$. In the modified Blake chart (Figure 1.2.4), the signal energy was calculated for t_f, and the detectability factor $D(n')$ includes the required integration loss.

Range Gate Matching Loss, L_m. When a range gate is used at the input to the doppler filters, the combined response of this gate and the preceding IF filter H_1 (Figure 5.1.4) should match the transmitted pulse. If a wideband IF is used with a rectangular gate, the range gate loss will be

$$L_m = \tau_g/\tau, \quad \tau_g > \tau$$
$$\quad = \tau/\tau_g, \quad \tau > \tau_g \qquad (5.4.15)$$

If a short sampling strobe is used in place of the range gate, L_m is determined by the preceding IF filter.

Range Straddling Loss, L_{er}. The range gates or strobes will be spaced by approximately τ, but a signal having a peak that arrives other than in the center of the gate or strobe will not be passed with full amplitude. The amount of straddling loss depends on the pulse shape and gate duration, as well as on the spacing. For rectangular pulses and gates, preceded by a wideband IF, the loss will be as shown in Figure 5.4.11. Overlapping gates may be used to reduce this loss, at the expense of greater processor throughput.

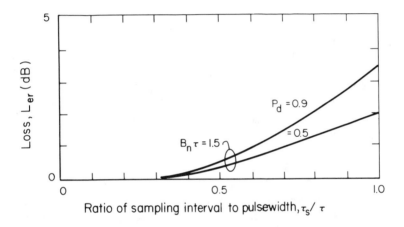

Figure 5.4.11 Range straddling loss for rectangular pulses.

Filter Straddling Loss, L_{ef}. Signals may also be centered at frequencies between the doppler filters, introducing a filter straddling loss (Figure 5.4.12). When heavily weighted filters are used, they will overlap within the -3 dB points, making this loss insignificant.

Angle Straddling Loss, L_{ea}. When the antenna scans continuously, and burst processing is used, the center of the burst may not coincide with the angle of the target. As a result, the peak of the burst envelope may straddle two processing intervals, introducing an angle straddling loss. This loss is 2.0 dB for contiguous processing intervals $t_f = t_o$, when using a 40-dB filter weighting. When the observation time is divided into two or more processing intervals, there is negligible straddling loss. Overlapping processing intervals may be used, but this is inconsistent with burst-to-burst

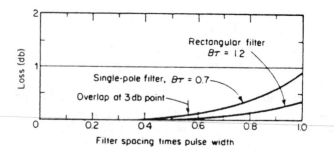

Figure 5.4.12 Filter straddling loss for unweighted bursts.

PRF or RF agility, so the usual choice is to make $t_f < t_o/2$ to minimize this straddling loss.

   Eclipsing Loss, L_{ec}. Medium- and high-PRF systems will lose some target pulses during search as a result of eclipsing by the transmitted pulse (and sometimes also by short-range clutter). On the average, the fraction of energy lost will be Δ_r/R_u, as used to establish the visible target fraction of (5.4.4). Accordingly, for search or target acquisition with probabilities near 0.5, the loss will be

$$L_{ec} = R_u/(R_u - \Delta_r) \tag{5.4.16}$$

For higher probabilities, the loss will increase, and may be found by integrating P_d over all target ranges and increasing signal power until the desired value is achieved, with the burst-to-burst diversity procedure to be employed in the system.

   Velocity Response Loss, L_{ev}. Low- and medium-PRF systems will lose some targets in the clutter rejection notch. The calculation is similar to that for eclipsing. For P_d near 0.5,

$$L_{ev} = v_b/(v_b - \Delta_v) \tag{5.4.17}$$

   Other losses, including those caused by collapsing of coordinates, noncoherent integration following doppler filtering, binary integration, *et cetera,* will be similar to those in MTI and other systems.

   Transient Gating Loss, L_{eg}. All PD systems must transmit enough pulses, at the beginning of each burst and before processing begins, to ensure that the echoes from clutter at maximum range have been received. During these transmissions, the processor must be gated off to prevent adding into the weighted output a sample which has no clutter echo (see Figure 5.4.5 and (5.4.6)). When an analog filter is used to reduce clutter prior to A/D conversion, time t_t must also be allowed for the initial transient at the output of this filter to decay. The transient gating loss is expressed in terms of the total transient gating time t_g and the total burst duration t_f:

$$L_{eg} = 1 + t_g/t_f \qquad (5.4.18)$$

where $t_g = t_{dc} + t_t$. The gating time t_g is usually about 1 ms, and this means that bursts shorter than about 5 ms will have significant gating loss.

Detection Range of Pulse Doppler Radar

The example of Figure 1.2.5 will be used to illustrate the performance of ground-based PD radar, and to compare with MTI performance described in Section 5.3. The S-band surveillance radar parameters for low-PRF (MTD) and high-PRF operation will be assumed as shown in Table 5.3.

Note that the high-PRF system has an additional 12.6 dB of integration gain, which exactly compensates for its reduced peak power. If the modified Blake chart were used to calculate range, the average power and coherent integration time t_f would be used, and identical results obtained for the two systems. The difference between the two systems lies entirely in the loss budget, the response to clutter, and the response to targets of different velocities. Comparing these results with those of the noncoherent radar of Figure 1.2.5, the integration gain for that radar was 10.6 dB, 2.5 dB lower than the low-PRF PD radar of Table 5.3. Whether that gain is translated into increased range will depend on the comparative loss budgets and response to clutter and targets.

Before estimating losses, the requirements for doppler processing must be addressed. For the MTD system, the requirements are the same as for the MTI radar discussed in Section 5.3. In that case, the input $(S/C)_i$ ratio, using clutter averaged over the cells within the horizon, was -4.1 dB, and an improvement factor of 25 dB was not quite sufficient to achieve $P_d = 0.9$ with this clutter. Clutter peaks ($\sigma_c = 10^4$ m^2) would rise another 26 dB, causing some target suppression. That clutter environment poses no particular problem to the MTD system. If the clutter attenuation is limited only by oscillator stability (curve (a) of Figure 5.4.9), the available 67 dB would be sufficient to preserve target detection in these clutter peaks.

Table 5.3 S-Band Pulsed Doppler Radar Parameters

	Low PRF	High PRF
Peak power, P_t (kW)	100	5.5
Pulsewidth, τ (μs)	1	1
Pulse repetition frequency, f_r (kHz)	1.1	20
Duty factor, D_u	0.0011	0.02
Average power, P_{av} (W)	110	110
Unambiguous range, R_u (km)	136	7.5
Unambiguous velocity, v_b (m/s)	55	1000
Time on target, t_o (s)	0.022	0.022
Hits per scan, n	24	440
Coherent integration time, t_f (s)	0.0072	0.0072
Coherent pulses integrated, n_c	8	144
Coherent integration gain, G_c (dB)	9	21.6
Noncoherent samples integrated, n'	3	3
Noncoherent integration gain, G_i (dB)	4.1	4.1
Total integration gain (dB)	13.1	25.7

For the high-PRF system, clutter attenuation is an extreme problem. Equation (5.4.9) gives the peak power of clutter as

$$C = 5500 \times 0.1^2 \times 1.5 \times 0.1 \times 0.02/533 \times 10^2 = 3.1 \times 10^{-6} \text{ W}$$

where $w/h = 1.5$, $h_r = 10$ m, and C is averaged over all range cells. Since the antenna beam grazes the surface at a range $R_b = 2h_r/\theta_e = 570$ m, which is 1/3 the unambiguous range, the clutter power will change only slightly over the 49 range cells. The output $(C/N)_o$ ratio in a filter at zero doppler would be

$$(C/N)_o = CD_u t_f/N_0 = 3.3 \times 10^{10} = +105 \text{ dB}$$

To suppress this clutter to -6 dB relative to noise in a target filter at f_d, $CA = 111$ dB is required, and from (5.4.13),

$$B_f W_n(f_d - bf_r)[A(f_d - bf_r, R_c) - 1]$$
$$+ 2B_f \sum_{i=-B/2f_r}^{B/2f_r} W_n(f_d - if_r) = 8 \times 10^{-12} = -111 \text{ dB}$$

The value of $B_f = 1/t_f = 140$ Hz $= 21.4$ dB(Hz). If the summation includes nine ambiguities at constant level W_0, that level must be at least

-145 dB(c/Hz) if the requirement is to be met. Only the cavity-stabilized klystron (curve (c) of Figure 5.4.9) will be adequate, since it was designed for a CW radar with even more stringent requirements.

The loss budget for PD systems can now be addressed, as in Table 5.4. The final values of detectability factor represent the requirement after coherent integration for $t_f = 0.0072$ s. In terms of the matched filter processor for $t_o = 0.022$ s, the total losses will be higher by the three-sample integration loss $L_i(3) = 0.7$ dB: 9.9 dB for low PRF and 7.5 dB for high PRF. The total losses, relative to the matched filter, for the noncoherent system and MTI system are 8.3 and 11.3 dB, and $D_x(1) = 29.9$ and 32.9 dB, respectively. Thus, the high-PRF PD system is 0.8 dB more efficient than the noncoherent system of Figure 1.2.5, and 3.8 dB more efficient than the MTI system, while achieving rejection of all land clutter with negligible target suppression. The low-PRF PD system is 1.6 dB less efficient than the noncoherent system and 1.4 dB more efficient than the MTI system, in return for which it cancels land clutter with very little target suppression. Perhaps the most significant difference is that the high-PRF system can also reject moving clutter (rain, chaff, birds), while

Table 5.4 Loss Budgets for PD Radar

	Low PRF	High PRF
CFAR loss, L_g	2.0	1.0
Filter matching loss, L_{mf}	2.0	2.0
Velocity response loss, L_{ev}	2.0	0
Range straddling loss, L_{er}	1.0	1.0
Filter straddling loss, L_{ef}	0.1	0.1
Angle straddling loss, L_{ea}	0	0
Eclipsing loss, L_{ec}	0	0.1
Transient gating loss, L_{eg}	0	0.5
Total misc. loss, L_x	7.1	4.7
Matching loss, L_m	0.8	0.8
Beamshape loss, L_p	1.3	1.3
Total loss	9.2	6.8
Detectability factor, $D_1(3)$	17.5	17.5
Detectability factor, $D_x(3)$	26.7	24.3
$+10\log3$	4.8	4.8
Total energy ratio, $D_x(1)$	31.5	29.1

the low-PRF system cannot. However, because of the many range ambiguities, the high-PRF system must use doppler processing in all range cells, whereas the low-PRF system may restrict this processing to the actual clutter regions.

The RGCALC printout for the high-PRF PD system is shown in Figure 5.4.13.

```
Radar and Target Parameters (inputs) --

Peak Pulse Power (kilowatts) ...................           .1
Pulse Duration (usec) .......................      7200.0000
Transmit Antenna Gain (dB) ....................         40.0
Receive Antenna Gain (dB) .....................         40.0
Frequency (MHz) ...............................       3000.0
Receiver Noise Factor (dB) ....................          5.0
Bandwidth Correction Factor (dB) ..............           .8
Antenna Ohmic Loss (dB) .......................           .2
Transmit Transmission Line Loss (dB) ..........          1.0
Receive  Transmission Line Loss (dB) ..........          1.0
Scanning-Antenna Pattern Loss (dB) ............          1.3
Miscellaneous Loss (dB) .......................          4.7
Number of Pulses Integrated ...................            3
Probability of Detection ......................         .900
False-Alarm Probability (Negative Power of Ten)          6.0
Target Cross Section (Square Meters) ..........       1.0000
Target Elevation Angle (Degrees) ..............         1.00
Average Solar and Galactic Noise Assumed
Pattern-Propagation Factors Assumed = 1


     ***********************************


Calculated Quantities (Outputs) --

Noise Temperatures, Degrees Kelvin --
          Antenna (TA) ..........................    111.3
          Receiving Transmission Line (TR) .......     75.1
          Receiver (TE) ..........................    627.1
          TE X Line-Loss Factor = TEI ............    789.4
          System (TA + TR + TEI) .................    975.8
Two-Way Attenuation Through Entire Troposphere (dB)     2.8


 Swerling      Signal-      Tropospheric   Range,     Range,
 Fluctuation   to-Noise     Attenuation,   Nautical   Kilometers
 Case          Ratio, dB    Decibels       Miles
 ---------     ---------    -----------    --------   ----------

     1          17.50         1.28          48.5        89.8
```

Figure 5.4.13 Pulsed doppler range calculation using RGCALC.

REFERENCES

[5.1] D.O. North, An Analysis of the Factors which Determine Signal/ Noise Discrimination in Pulsed Carrier Systems, *RCA Laboratories Tech. Rep. PTR-6C,* June 25, 1943; reprinted in *Proc. IEEE* **51,** No. 7, July 1963, p. 1016.

[5.2] P.M. Woodward, *Probability and Information Theory with Applications to Radar,* Artech House, 1980.

[5.3] A.W. Rihaczek, *Principles of High Resolution Radar,* McGraw-Hill, 1969.

[5.4] C.E. Cook and M. Bernfeld, *Radar Signals,* Academic Press, 1967.

[5.5] B.L. Lewis, F.F. Kretschmer, Jr., and W.W. Shelton, *Aspects of Radar Signal Processing,* Artech House, 1986.

[5.6] F.E. Nathanson, *Radar Design Principles,* McGraw-Hill, 1969.

[5.7] H. Urkowitz, Filters for detection of small signals in clutter, *J. Appl. Phys.* **24,** No. 8, August 1953, pp. 1024–1031.

[5.8] W.W. Shrader, MTI Radar, Chap. 17 in *Radar Handbook* (M.I. Skolnik, ed.), McGraw-Hill Book Company, 1970.

[5.9] H.R. Ward, The effect of bandpass limiting on noise with a Gaussian spectrum, *Proc. IEEE* **57,** No. 11, November 1969, pp. 2089–2090.

[5.10] D.K. Barton and W.W. Shrader, Interclutter visibility in MTI systems, *IEEE Eascon Record,* 1969, pp. 294–297.

[5.11] W.M. Hall and H.R. Ward, Signal-to-noise loss in moving target indicator, *Proc. IEEE* **56,** No. 2, February 1968, pp. 233–234.

[5.12] IEEE Standard Dictionary of Electrical and Electronic Terms, *Std. 100-1984,* 1984.

[5.13] L. Cartledge and R.M. O'Donnell, Description and Performance Evaluation of the Moving Target Detector, *MIT Lincoln Laboratory Project Rep. ATC-69,* March 8, 1977.

[5.14] D.K. Barton, ed., *Radars,* Vol. 7, *CW and Doppler Radar,* Artech House, 1978.

[5.15] W.P. Robins, *Phase Noise in Signal Sources,* Peter Peregrinus, 1982, p. 90.

[5.16] E.C. Farnett, T.B. Howard, and G.H. Stevens, Pulse-Compression Radar, Chapter 20 in *Radar Handbook* (M.I. Skolnik, ed.), McGraw-Hill, 1970.

Chapter 6
Radar Propagation

The general theory of wave propagation, with applications to both radar and communications systems, has been the subject of several excellent books [6.1–6.3], and will not be covered here. Instead, we shall survey the material most directly applicable to radar system design and analysis, and we shall present the results of some recent studies, which will assist the radar engineer in solving the most common types of problem.

The effects of propagation conditions on the performance of search and tracking radars may be discussed under four general topics.

(a) *Atmospheric Attenuation*: Losses in radar energy due to absorption in the propagation medium (air, clouds, precipitation, and the ionosphere). The accompanying noise temperature contribution will also be discussed.

(b) *Surface Reflection*: Modification of the free-space field resulting from reflection of the waves from the surface beneath the direct path. These reflections cause multipath lobing effects on target detection, and multipath errors in tracking and measurement.

(c) *Diffraction*: Modification of the free-space field resulting from interaction of the wave with surfaces or obstacles beneath the direct path or blocking that path. Diffraction affects both detection and measurement.

(d) *Refraction*: Bending of the rays caused by the varying refractive index of the propagation medium. Both tropospheric and ionospheric refraction will be considered. Radar detection coverage and measurement accuracy are both affected by refraction.

6.1   ATMOSPHERIC ATTENUATION

Clear Atmosphere

Attenuation of radar waves in the normal, clear atmosphere is negligible at the lowest radar frequencies, becoming significant in the microwave bands, and imposing severe limits on radar operation in the millimeter-wave bands.

At sea level, the attenuation coefficient (in dB/km for two-way radar propagation) is shown as a function of frequency in Figure 6.1.1. Below 1 GHz, the values are low enough that the beam can penetrate the entire earth's atmosphere with less than 4 dB two-way attenuation, and this rises to only 5 dB at 6 GHz (C-band). Only when operation at or above X-band is considered will clear-air attenuation have to be considered as a major issue in radar analysis.

The operating bands in the upper microwave and millimeter-wave regions are determined largely by the presence of so-called *windows* in the atmosphere, defined as minima in the attenuation coefficient. As shown in Figure 6.1.1, these windows are not entirely clear, sea level values being roughly as shown in Table 6.1.

Table 6.1 Attenuation Coefficients in Radar Windows

Radar Band	Center Frequency (GHz)	Attenuation Coefficient (dB/km)
L	1.3	0.012
S	3	0.015
C	5.5	0.017
X	10	0.024
K_u	15	0.055
K	22	0.3
K_a	35	0.14
V	60	35.0
W	95	0.8
	140	1.0
	240	15.0

Curves for attenuation as a function of range, for different beam elevation angles and frequencies, were prepared by Blake [6.4], and are replotted on logarithmic range scales in Figures 6.1.2 and 6.1.3. The initial slopes of these curves represent the attenuation coefficients, and the leveling at longer ranges results from the beam having penetrated to a region

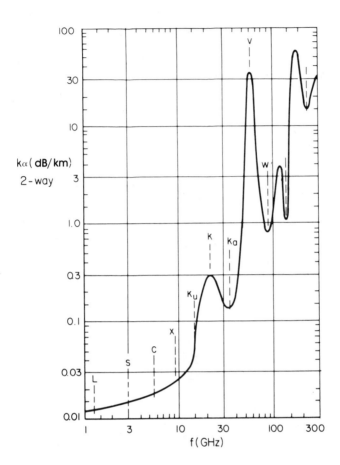

Figure 6.1.1 Attenuation coefficient *versus* frequency (clear atmosphere
at sea level).

of lower atmospheric density. For targets in space, the atmospheric loss
in penetration of the entire atmosphere is shown in Figure 6.1.4.

Weather Attenuation

Rain and wet snow have the most serious effects on radar attenuation,
while dry snow, clouds, and fog have lesser effects. Figure 6.1.5 shows the
attenuation coefficients of rain at different rates r in mm/h, as functions
of frequency. At L-band only the most intense rain can equal the clear-
atmosphere attenuation coefficient, and such rainfall cannot exist over any

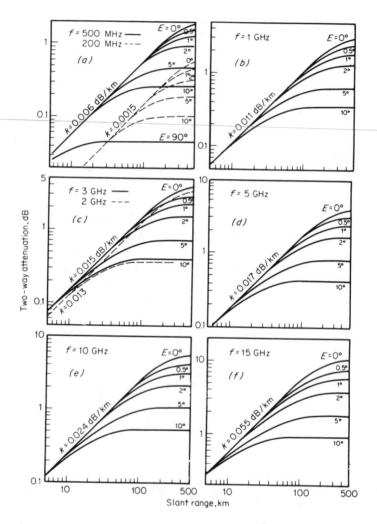

Figure 6.1.2 Atmospheric attenuation *versus* range for different elevation angles (0.2 to 15 GHz) [6.5].

substantial path length. At S-band the clear-atmosphere value is matched by rain at 10 mm/h, doubling the total attenuation in regions having this medium rain condition. Higher frequencies are much more affected by rain than by the clear air.

A rule of thumb for weather effects is that a radar should be capable of significant performance in the presence of 3 mm/h rain at altitudes below 5 km. This occurs in Washington, DC with about 1.2% probability, or 100

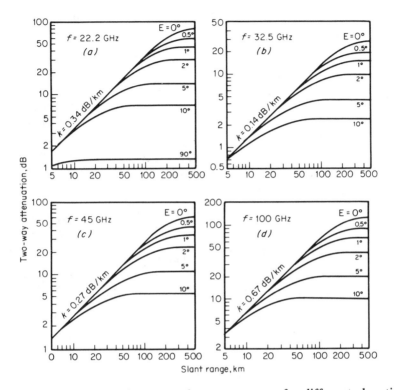

Figure 6.1.3 Atmospheric attenuation *versus* range for different elevation angles (20 to 100 GHz) [6.5].

hours per year. The attenuation coefficients for the different radar bands under these conditions are shown in Table 6.2. When the total attenuation reaches about 6 dB, the radar analysis becomes dominated by weather effects that cannot be predicted with great accuracy, and hence any predicted performance must be regarded as a coarse estimate. The ranges at which this occurs, on low-altitude paths with $r = 3$ mm/h, are shown in the table.

Dry snow has a much smaller attenuation, as shown by the dashed curve in Figure 6.1.5. The snow becomes an important contributor to radar attenuation only when it passes through a melting layer.

The smaller particles suspended as clouds or fog will have less effect on the radar wave, as shown in Figure 6.1.6. For a density of condensed water of 0.3 g/m³, the attenuation coefficients will be about half those for rain at 3 mm/h, in microwave bands. However, this attenuation continues to rise in the millimeter-wave bands, and above 100 GHz the attenuation for 0.3 g/m³ is greater than for 3 mm/h rain.

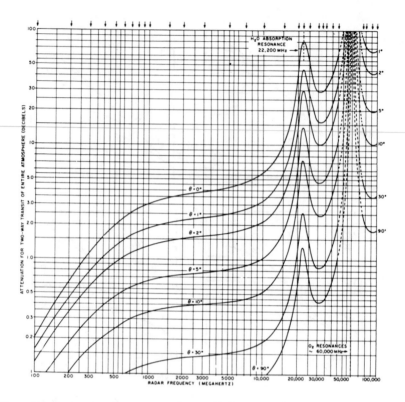

Figure 6.1.4 Attenuation for a radar beam passing through the entire atmosphere from sea level [6.4].

The attenuation from both rain and clouds is temperature dependent, but the variation caused by temperatures within the normal range is less than the inherent uncertainty in atmospheric conditions, and use of attenuations calculated at zero degrees Celsius is the standard practice.

Attenuation through a Wet Radome

When considering rain effects, it is also appropriate to include the attenuation encountered when the ray passes through a film of water on a radome. Table 6.3 shows values measured at two frequencies for different values of film thickness. Radomes are usually treated with a hydrophobic surface coating to prevent the formation of continuous water films, but as the material ages, with exposure to atmospheric contaminants, and as

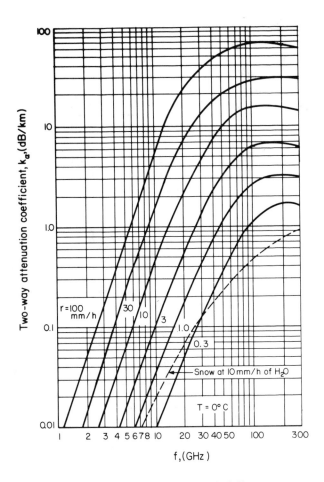

Figure 6.1.5 Attenuation coefficient *versus* rainfall rate.

erosion from sand and rain occurs (especially in airborne radomes), the ability to break the water into droplets is degraded, and films will appear in rain and even as a result of condensation on the radome. If the water flows in streaks across the surface, or the radar-radome geometry is asymmetrical, the attenuation will be unequal for vertical and horizontal polarizations, preventing the transmission of pure circular polarization to cancel rain clutter. Data are also given in the table for water films on a reflector surface. Since the electric field drops to zero at the surface of the reflector, there is little coupling to the water molecules, and the loss is small relative to that of a radome with the same film thickness.

Table 6.2 Attenuation Coefficients in 3 mm/h Rain

Radar Band	Center Frequency (GHz)	Attenuation Coefficient (dB/km)		Range for 6 dB Attenuation (km)
		Rain	Total	
L	1.3	0.001	0.013	
S	3	0.004	0.019	
C	5.5	0.023	0.04	150
X	10	0.11	0.13	45
K_u	15	0.25	0.3	20
K	22	0.7	1.0	6
K_a	35	1.7	1.8	3
V	60	4.0	39.0	0.15
W	95	6.0	6.8	0.9
	140	7.0	8.0	0.7
	240	6.5	21.5	0.28

Table 6.3 Losses in Water Films

Film Thickness (mm)	Radome Loss (dB)		Reflector Loss (dB)	
	$f = 3.65$ GHz	$f = 16$ GHz	$f = 3.65$ GHz	$f = 16$ GHz
0.05		3.2*		
0.13	1.1	5.3	≤0.01	0.03
0.25	2.6	8.7	≤0.01	0.2
0.38	4.2	10.9	≤0.01	0.9
0.5	5.6	12.3	≤0.01	2.7

NOTE: Theoretical data from [6.7]. The asterisk indicates experimental data reported in [6.8].

Ionospheric Attenuation

Attenuation in the ionosphere is a problem for radars operating in the HF and VHF bands, but seldom exceeds 1 dB at frequencies above 100 MHz. A possible problem more serious than attenuation is the rotation of polarization (Faraday rotation) as the signal penetrates the ionosphere. This effect is nonreciprocal, and can lead to serious polarization mismatch losses if a single linear polarization is used at frequencies below L-band on paths through the ionosphere.

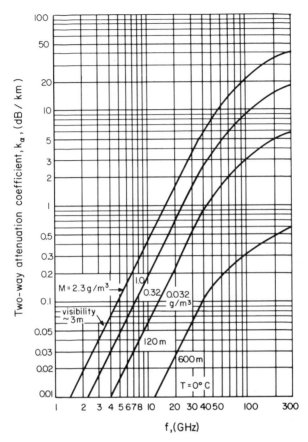

Figure 6.1.6 Attenuation coefficient of clouds and fog.

Attenuation in Chaff

It is sometimes postulated that chaff may introduce attenuation in addition to (or instead of) backscatter effects. The chaff densities required to produce significant attenuation, however, are extremely large. For example, a high-density chaff cloud may be characterized as having a volume reflectivity $\eta_v = 10^{-6}$ m²/m³. This means that 10^{-6} of the power entering a 1 m³ cube of chaff will be scattered, while the remainder penetrates as a coherent wave. If the chaff dipoles are made ideally absorbing, this same small fraction of power is converted to heat, rather than backscattered. In a 10 km path through such chaff, 1% of the incoming power will be absorbed, for a loss of only 0.04 dB, or 0.004 dB/km. There is no feasible method by which chaff can be made to have attenuation effects as large as those commonly encountered in rain or fog.

Blake's Method for Range with Attenuation

In the Blake chart, the free-space range R_0 is first calculated, and then adjusted for the propagation factor F, giving

$$R' = R_0 F \qquad\qquad (6.1.1)$$

The atmospheric attenuation for range R' at low altitude is then computed (in dB):

$$L_{\alpha 1} = R' k_\alpha \qquad\qquad (6.1.2)$$

where k_α in dB/km is the attenuation coefficient of the atmosphere at sea level. The first estimate of radar range is then

$$R_1 = R' \times 10^{-L_{\alpha 1}/40} \qquad\qquad (6.1.3)$$

From R_1, a second estimate $L_{\alpha 2}$ is made, and $\Delta L_\alpha = L_{\alpha 1} - L_{\alpha 2}$ is found. Then, the second range estimate is

$$R_2 = R_1 \times 10^{\Delta L_\alpha/40} \qquad\qquad (6.1.4)$$

It is important to determine the accuracy of this method, after the first calculation of R_1 and after the iteration leading to R_2. Figure 6.1.7 shows curves for the following, as a function of the true path attenuation $L_{\alpha m}$ to the final target range:

(a) *Attenuation $L_{\alpha 1}$ for the range R'*. This will be the decibel value of error in received power, if no correction is made for attenuation.
(b) *Difference $L_{\alpha 1} - L_{\alpha m}$*. This is the error in received power if the range is corrected for $L_{\alpha 1}$ without proceeding to the second step.
(c) *Difference $L_{\alpha 2} - L_{\alpha m}$*. This is the error remaining after the second step.

The results from Figure 6.1.7 can be summarized as follows:

(1) A range error of 1% will result from 0.17 dB attenuation if no correction is made. The range estimate will be on the high side.
(2) Attenuation up to 1.7 dB actual (1.8 dB at R') requires the first correction step if 1% range accuracy is to be achieved. The range estimate will be on the low side.
(3) Attenuation up to 3.6 dB actual (4.5 dB at R') can be corrected for 1% range error by using the second step. The range estimate will be on the low side.

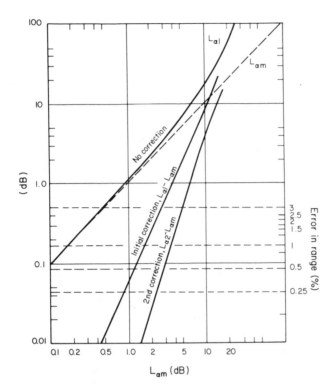

Figure 6.1.7 Accuracy of attenuation and range estimates using Blake's procedure.

(4) Range estimates to within 3% can be made when the actual attenuation is 5.3 dB (7 dB at R'), using the two steps. When the attenuation is this high or higher, uncertainty in the value of the attenuation coefficient is likely to be the major source of error, rather than lack of further iteration. (5) Corrected range estimates are more accurate than shown in Figure 6.1.7 when the target is at an altitude having lower attenuation coefficient.

Atmospheric Noise Temperature

The coupling of radar waves to the molecules of the atmosphere causes attenuation of the wave, but also leads to coupling of the thermal energy of the molecules to the radar receiver. This noise temperature component T_a' was one of the terms in the calculation of system input temperature, (1.2.15), and curves of apparent sky temperature were given in Figure 1.2.2 for a clear atmosphere. In the presence of precipitation or clouds, there will be an increase in T'_a, according to

$$T'_a = T_p(1 - 1/\sqrt{L_{at}}) \tag{6.1.5}$$

where $T_p \approx 290$ K is the physical temperature of the loss medium and $\sqrt{L_{at}}$ is the one-way attenuation through the entire atmosphere, including rain or clouds. The value $T'_a \approx 100$ K for $\theta = 1°$ in S-band, shown in Figure 1.2.2, corresponds to $L_{at} = 3.7$ dB, as in Figure 6.1.4. If 3 dB of rain attenuation were added, T'_a would increase to 156 K. This would not have a major effect on the radar range, unless an extremely low-noise receiver were being used.

Atmospheric Lens Loss

Waves leaving the radar at low elevation angles are refracted toward the earth, resulting in paths that are often described by using an effective earth's radius of 4/3 times the actual value. The increasing refraction at lower elevations causes a dilution of power in this region, resulting from the spreading of the beam to cover the propagation beyond the actual horizon. This lens loss is the same for all radar bands and is a function of elevation angle, as shown in Figure 6.1.8. This loss is not a dissipative loss, which would contribute to sky temperature, but rather is a reduction in field strength properly considered as a component of the propagation factor, or as a separate RF loss along with L_t in the denominator of the radar equation. Lens loss is significant only for targets at long range, and since the loss varies slowly with range there is no need to apply the iterative procedures developed by Blake for atmospheric attenuation.

6.2 SURFACE REFLECTION EFFECTS

Propagation Factor

The *pattern-propagation factor F* is defined as the ratio of the field strength created at a point in space by a radio transmission (but excluding attenuation effects of the medium), to that which would have been created by the same system operating in free space with the on-axis gain of the antenna. For radar, the fourth power F^4 describes the two-way power ratio. Several factors can change the free-space propagation conditions to produce greater or lesser fields. Atmospheric refraction, reflections from the underlying surface, and diffraction over this surface are the effects that must be considered. To separate the effects of the antenna from those of

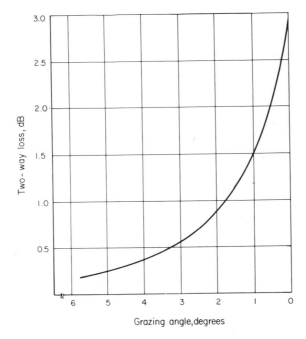

(a) Lens loss for paths into space.

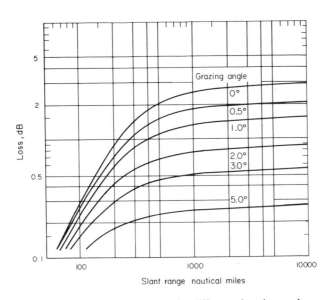

(b) Lens loss *versus* slant range for different elevation angles.

Figure 6.1.8 Lens loss in the standard atmosphere (after [6.9]).

the propagation path, it is convenient to define a *propagation factor* as the value of F which would be obtained with a broad antenna beam, such that the underlying surface of the earth is fully illuminated.

Surface Reflections

The geometry of surface reflection for a flat earth is shown in Figure 6.2.1. For a radar antenna height h_r, a target range R, and altitude h_t, the following relationships apply:

Target elevation, $\theta_t = \sin^{-1}[(h_t - h_r)/R] \approx (h_t - h_r)/R$ (6.2.1)

Grazing angle, $\psi = \sin^{-1}[(h_t + h_r)/R] \approx (h_t + h_r)/R$ (6.2.2)

Path length difference, $\delta_0 = R[(\cos\theta_t/\cos\psi) - 1] \approx 2h_r h_t/R$ (6.2.3)

Range to reflection point, $x_0 = h_r/\tan\psi \approx Rh_r/(h_r + h_t)$ (6.2.4)

Relationships for the spherical earth are more complicated, and may be found in [6.4].

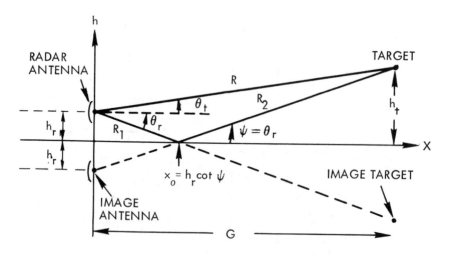

Figure 6.2.1 Geometry of specular reflection from a flat surface.

The pattern-propagation factor in the presence of surface reflections may be written

$$F = |f(\theta_t) + f(-\psi)\rho D \exp(-j\alpha)| \qquad (6.2.5)$$

Here $f(\theta)$ is the voltage pattern of the antenna, relative to on-axis gain, at elevation angle θ, ρ is the magnitude of the surface reflection coefficient, D is the spherical earth divergence factor [6.4], and α is the phase angle of the reflected wave, given by

$$\alpha = (2\pi\delta_0/\lambda) + \phi \qquad (6.2.6)$$

where ϕ is the phase angle of the reflection coefficient. This formulation for F applies only when $\delta_0 > \lambda/6$. For smaller path differences, the effects of diffraction are dominant, as described in Section 6.3. Accordingly, for surface-based radar, the conditions leading to $D \neq 1$ almost always coincide with those for diffraction, and the inclusion of D in (6.2.5) is merely to remind the analyst that the curved earth may need to be considered.

Substituting into (6.2.6) and (6.2.5) the approximation for δ_0 from (6.2.3), using the approximation for θ_t with $h_t \gg h_r$, assuming a broad antenna pattern, and writing the result for F, we find

$$F^2 = 1 + \rho^2 + 2\rho \cos(\phi + 4\pi h_r\theta_t/\lambda) \qquad (6.2.7)$$

The propagation factor F thus has a cyclic variation between $F_{max} = 1 + \rho$ and $F_{min} = 1 - \rho$, with a period equal to $\theta_n = \lambda/2h_r$ in elevation angle. If we also let $\rho \approx 1$, $\phi = \pi$, this reduces to

$$F = 2 |\sin(\pi\theta_t/\theta_n)| \qquad (6.2.8)$$

which gives coverage lobes extending to twice the free-space range, separated by nulls with zero range. These assumptions for the reflection coefficient are justified for horizontally polarized radars, as we will see below.

The surface reflection coefficient may be written as the product of three factors:

$$\rho = \rho_0 \rho_s \rho_v \qquad (6.2.9)$$

where ρ_0 is the *Fresnel reflection coefficient* for the surface material, ρ_s is the *specular scattering coefficient* for a rough surface, and ρ_v is the coefficient for vegetative absorption.

The complex Fresnel reflection coefficient is given, for horizontal polarization, by

$$\rho_0 \exp(-j\phi) = \frac{\sin\psi - \sqrt{\epsilon_c - \cos^2\psi}}{\sin\psi + \sqrt{\epsilon_c - \cos^2\psi}} \qquad (6.2.10)$$

In this equation, ϵ_c is the complex dielectric constant:

$$\epsilon_c = \epsilon_r - j\epsilon_i = \epsilon_r - j60\lambda\sigma_e \qquad (6.2.11)$$

where ϵ_r is the relative dielectric constant and σ_e is the conductivity of the surface material. For vertical polarization,

$$\rho_0 \exp(-j\phi) = \frac{\epsilon_c \sin\psi - \sqrt{\epsilon_c - \cos^2\psi}}{\epsilon_c \sin\psi + \sqrt{\epsilon_c - \cos^2\psi}} \qquad (6.2.12)$$

The magnitude ρ_0 is plotted in Figure 6.2.2. The phase shift ϕ is also indicated in the figure, and will be near π at low angles for both polarizations. For vertical polarization, there is an abrupt change in phase near the Brewster angle ψ_b, and the phase is near zero above that angle. Curves for this phase angle, with different surfaces, are given in [6.3]. Typical values of ϵ_r and σ_e are given in Table 6.4. Further data are given in [6.3, pp. 15–19].

Table 6.4 Electrical Properties of Typical Surfaces

Material	e_r	σ_e (mho/m)
Good soil (wet)	25	0.02
Average soil	15	0.005
Poor soil (dry)	3	0.001
Snow, ice	3	0.001
Fresh water ($1 = 1$ m)	81	0.7
($1 = 0.03$ m)	65	15
Salt water ($1 = 1$ m)	75	5
($1 = 0.03$ m)	60	15

Rough Surfaces

For a rough surface, with an rms deviation σ_h from a smooth surface, the specular scattering coefficient is given by

$$\rho_s^2 = \exp[-(4\pi\sigma_h/\lambda)^2 \sin^2\psi] \tag{6.2.13}$$

This coefficient is plotted in Figure 6.2.3, as a function of grazing angle in mr.

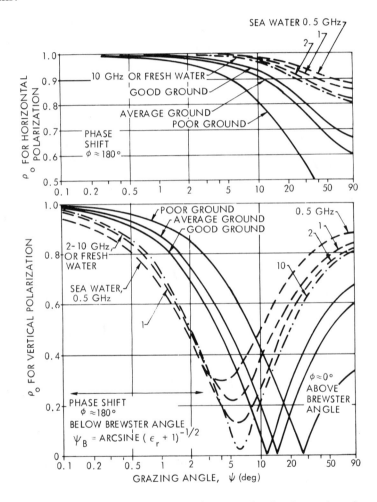

Figure 6.2.2 Fresnel reflection coefficients ρ_0 for horizontal and vertical polarizations.

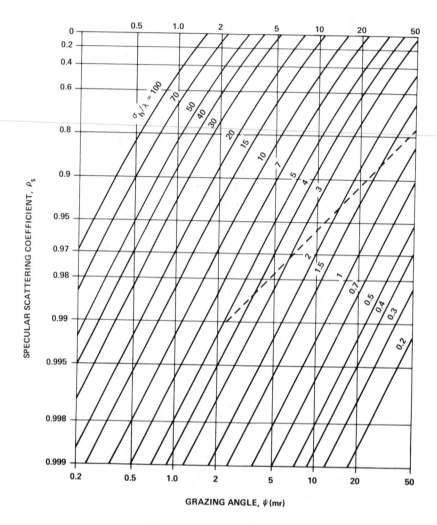

Figure 6.2.3 Specular scattering coefficient *versus* grazing angle for different values of surface roughness.

Vegetation

The third coefficient entering into ρ is the vegetation factor ρ_v. A layer of vegetation covering the surface absorbs much of the incident radar energy, and scatters the remainder in an irregular pattern. Little data are available to establish values for ρ_v, but it is typically found to lie between 0.03 and 0.3 when the vegetation layer is more than one wavelength thick, and when there are no bare areas visible to the radar and the target. For any but thin layers of vegetation, the effect of surface reflections on the propagation factor is small.

For any given surface conditions, an approximate plot of F *versus* θ_t may be made by locating the positions of nulls at $i\theta_n$, and peaks at $(i + 0.5)\theta_n$, $i = 0, 1, 2, \ldots$ The locus of the peaks is then plotted at $1 + \rho$, and of nulls at $1 - \rho$, where $\rho(\theta_t)$ is evaluated from Figures 6.2.2 and 6.2.3, with a vegetation coefficient ρ_v if appropriate. Then, if the pattern factor is to be included, the loci are modified to $f(\theta_t) + \rho f(-\psi)$ and $f(\theta_t) - \rho f(-\psi)$. The actual lobes may be drawn in, remembering that the curves cross the free-space curve at $\theta_n/6$ and $5\theta_n/6$ for $\rho f(-\psi) \approx f(\theta_t)$, where the direct and reflected field vectors form an equilateral triangle. This process is illustrated in Figure 6.2.4, for the example radar of Figure 1.2.5 operating over a smooth sea with horizontal polarization. In this case, the two-degree elevation beamwidth, with the beam center elevated to 1°, determines the loci of peaks and nulls, as $\rho = 1$ applies throughout the sector of interest.

A computer program, VCCALC, to generate and plot the coverage pattern with lobing has been published [6.20], and results for the example radar are shown in Figure 6.2.4(b). The features agree with those of the hand calculation. Above 1° the lobes of the hand calculation are more pronounced because a Gaussian beam was used, rather than the $(\sin x)/x$ beam assumed in VCCALC.

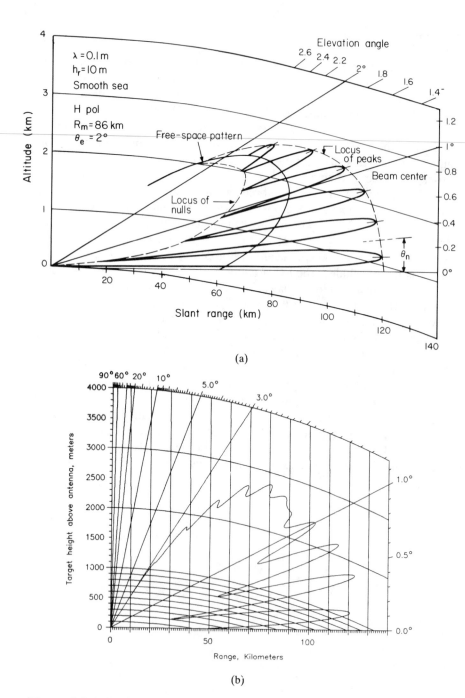

Figure 6.2.4 Coverage diagram with lobing.

6.3 DIFFRACTION

Smooth-Sphere Diffraction

Radar waves propagating near the surface of the curved earth will interact with that surface to reduce the field strength, even on paths that are not below the horizon. The transition from lobing to diffraction conditions takes place when the extra length of the reflected path is reduced to less than $\lambda/2$, corresponding to elevation angles below the first peak of the lobing pattern. Radar coverage below this angle, for a short-range radar operating over the sea, is illustrated in Figure 6.3.1, for different frequency bands. The geometrical horizon is drawn for the 4/3 earth's radius, and this determines the coverage limit as radar frequency is increased into the millimeter-wave bands.

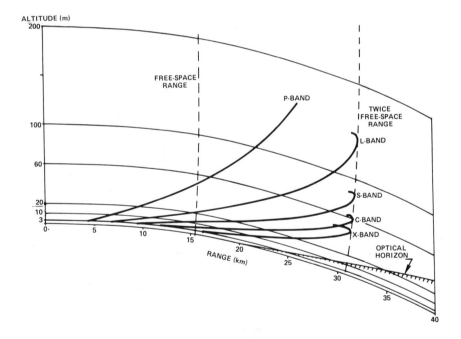

Figure 6.3.1 Radar coverage below the first lobe ($h_r = 20$ m, $R_0 \approx 15$ km).

Evaluation of the propagation factor in the diffraction region, for a smooth earth, can be done by using methods described in [6.1] and [6.4], analytic approximations given in [6.10], or exact procedures of [6.11]. The Burroughs and Atwood approximations [6.10] are convenient for computations on small computers, and are used in the TI-59 program Propagation Factor (Appendix A). An example of the results from that program is shown in Figure 6.3.2. Except when the exact procedures are used, involving many terms in the series expansion for spherical earth diffraction [6.11], the diffraction results must be merged with those for the reflection-interference region (Section 6.2) by interpolation, leading to some variability in data just below the peak of the first reflection lobe. The two curves are both provided as outputs from Propagation Factor, and the interpolation is performed manually.

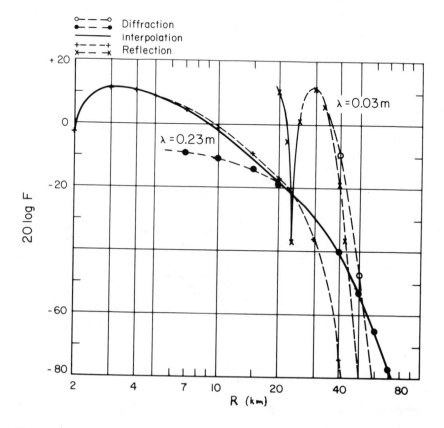

Figure 6.3.2 Example of results from Propagation Factor program for TI-59 calculator.

Knife-Edge Diffraction

Obstacles projecting above the earth's surface introduce another type of diffraction effect: knife-edge diffraction. The object need not have a sharp contour, because for radar wavelengths, almost any surface feature will produce this effect at low grazing angle. Trees, buildings, and even low ridges and raised ledges will establish a horizon at a masking angle above that of the smooth sphere. The factor F at this horizon will be 0.5, giving $F^4 = -12$ dB for two-way propagation. Below the mask angle, F drops to zero more slowly than for the smooth sphere (Figure 6.3.3), enabling some radars to detect targets in the shadow region. This capability is most pronounced for longer wavelengths when the obstacle projects well above the surface, but it has been exploited even at microwaves in communications systems using the *obstacle gain* of a mountain or ridge.

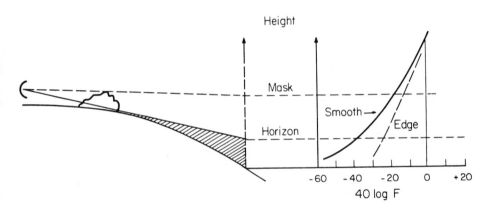

Figure 6.3.3 Comparison of diffraction effects.

The propagation factor for knife-edge diffraction depends on the vertical distance h between the top of the obstacle and the direct path between the radar and the target (Figure 6.3.4). A diffraction parameter v is defined as

$$v = \sqrt{2h\theta/\lambda} = h\sqrt{(2/\lambda)(1/d_1 + 1/d_2)} \tag{6.3.1}$$

Positive v implies a blocked path, and approximations to F can be written

$$20 \log F = -(6 + 8v), \quad -1 < v < 1 \tag{6.3.2}$$
$$20 \log F = -[6.4 + 20 \log(\sqrt{v^2 + 1} + v)], \quad v < -1$$

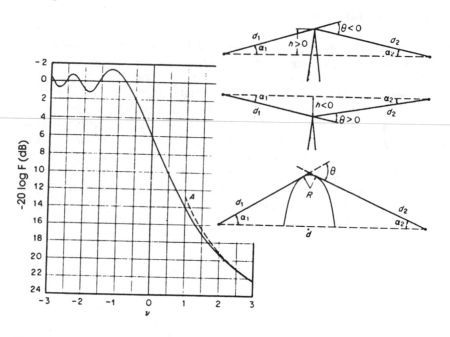

Figure 6.3.4 Knife-edge diffraction [6.12].

Obstacle gain occurs when this factor is greater than the corresponding value for smooth-sphere diffraction in the absence of the obstacle. However, it is important to establish that the top of the obstacle is in a free-space field. For example, Figure 6.3.5 shows a typical diffraction situation with an isolated obstacle. In this case, the extra path length δ_0 for reflections from the terrain between the radar and the obstacle is $0.6\,m$ (from (6.2.3)). Knife-edge diffraction should describe the fields beyond the obstacle for $\lambda < 6\delta_0 = 3.6$ m. For longer wavelengths or lower antenna or obstacle heights, the top of the obstacle would not be fully illuminated, and F would be reduced by reflection-interference or smooth-sphere diffraction from the intervening terrain.

The obstacle need not be particularly sharp to qualify as a knife-edge for diffraction purposes. For the example in Figure 6.3.5, the radius of curvature of the obstacle can be as great as 9 km without significant effect on the fields produced at the target.

Rough Surface Effects

Many sea and land surfaces follow the spherical earth contour, but with small-scale roughness such as waves and comparable undulations on

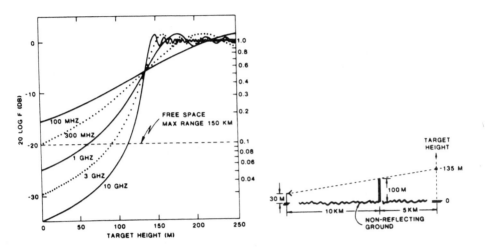

Figure 6.3.5 Example of knife-edge diffraction [6.3].

land. This roughness is often sufficient to disperse, at the higher elevation angles, the specular reflections that would otherwise produce lobing. As the elevation angle is reduced, propagation conditions will be intermediate between those for a smooth sphere and for knife-edge diffraction. When combined with uncertainties caused by nonstandard refraction or ducting, the effect is to make it impossible to predict accurately the characteristics of radar propagation at low elevations. The smooth-sphere procedure gives a pessimistic result, while assumption of ducting or a knife edge at the horizon gives optimistic results.

Some insight into the conditions that can give knife-edge diffraction may be found by considering a random rough surface with rms deviation σ_h and correlation distance such that there is only one significant ridge between the radar and the target (Figure 6.3.6). Assume that the radar antenna is located at a height h_r over ground which in turn is $2\sigma_h$ above the average surface, and that the ridge which forms the horizon also has height $2\sigma_h$ above the average terrain. The top of the ridge will receive full illumination if the radar-ridge path has first Fresnel zone clearance above the many high points near the middle of the path, $1\sigma_h$ above the average terrain. Solution of the triangle for the extra $\lambda/2$ length of paths via the high points centered between the radar and the ridge leads to the requirement:

$$\sigma_h > \sqrt{\lambda R_h}/2 = \sqrt{\lambda}\sqrt[4]{2kah_r}/2 = 32\sqrt{\lambda}\sqrt[4]{h_r} \qquad (6.3.3)$$

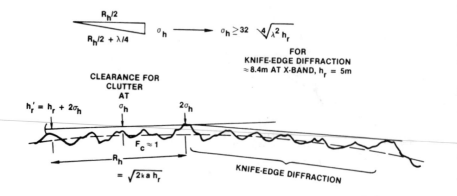

Figure 6.3.6 Conditions for Fresnel clearance on path from radar to obstacle.

where $ka = 8.5 \times 10^6$ m is the effective earth's radius. This requirement is summarized in Figure 6.3.7. For radar antenna heights between 5 and 10 m, the horizon range is between 9 and 13 km, and the required terrain roughness is $\sigma_h > 50\sqrt{\lambda}$, or about 15 m for S-band. As a result, many terrain situations will give values of F less than for knife-edge diffraction, but larger than for the smooth sphere. Over the sea, smooth-sphere diffraction conditions can always be expected.

6.4 ATMOSPHERIC REFRACTION

Radar waves passing through the earth's atmosphere are bent downward by the changing refractive index of the troposphere, and then by the ionosphere. This produces an error in elevation angle, the ray at the antenna having a somewhat larger angle than the direct geometric path to the target (Figure 6.4.1). At the same time an extra time delay is produced, giving a larger range reading than the true range. Random variations in refractive index produce smaller, random errors in measured coordinates. In the troposphere, three effects must be considered:

(a) *regular refraction,* resulting from the gradual reduction in refractive index with altitude and causing elevation and range bias errors;
(b) *tropospheric fluctuations,* resulting from random variations in local refractive index and causing slowly varying errors in all measured coordinates;
(c) *ducting,* resulting from steep gradients in refractive index (usually near the surface), creating low-loss propagation paths to low-altitude targets

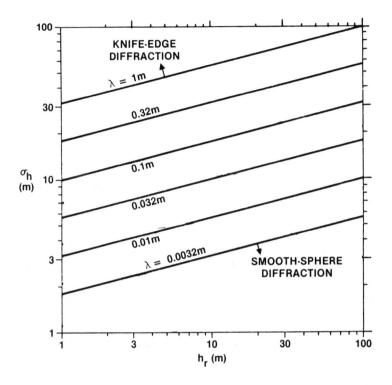

Figure 6.3.7 Criterion for surface roughness to support knife-edge diffraction.

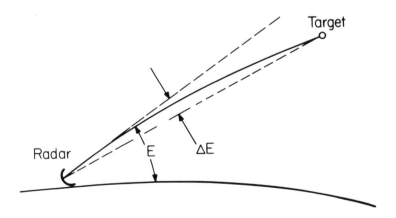

Figure 6.4.1 Ray path through the atmosphere.

beyond the normal horizon, and leaving gaps in the coverage for targets just above the duct.

Exponential Reference Atmosphere

The refractive index of the troposphere for all radar frequencies can be expressed as

$$N = (n - 1) \times 10^6 = (77.6/T)(P + 4810p/T) \qquad (6.4.1)$$

where T is temperature in Kelvins, P is pressure in millibars, p is the partial pressure of the water-vapor component, n is the refractive index, and N is a scaled-up value known as the *refractivity*. Small variations in this relationship occur near the resonances for water vapor (22 GHz) and oxygen (60 GHz), but outside of these strong absorption bands the constants in (6.4.1) can be used with negligible error to the highest millimeter-wave bands.

At sea level, N is typically between 300 and 350, with mean and extreme height profiles as shown in Figure 6.4.2. The approximate straight-line relationship of these data on the log-linear plot suggests the use of the exponential form for a mathematical model:

$$N(h) = 313 \exp(-0.1439h) \approx 313 \exp(-h/7) \qquad (6.4.2)$$

where h is in km above sea level and the sea level value $N_0 = 313$.

At low altitudes, the gradient of refractive index is nearly constant at a value which produces the *effective earth radius ka* $= (4/3)a$, often used to calculate the radar horizon and to permit radio paths to be plotted as straight lines over a surface profile. At altitudes above about 4 km, however, the actual gradient is not this steep, and radar coverage charts based on the exponential model are more accurate [6.15].

Elevation and Range Bias Errors

The errors in measured elevation and range data, caused by refraction in the exponential reference atmosphere, are shown in Figures 6.4.3 and 6.4.4 for a radar at sea level with surface refractivity $N_s = N_0 = 313$. For other radar altitudes and refractivity conditions, the errors may be scaled directly to N_s. The limiting errors, for targets well beyond the atmosphere, are given, for $E > 5°$, by

$$\Delta E = N_s \cot E \times 10^{-6} \text{ (radian)} \qquad (6.4.3)$$

$$\Delta R = 0.007 N_s \csc E \text{ (m)} \qquad (6.4.4)$$

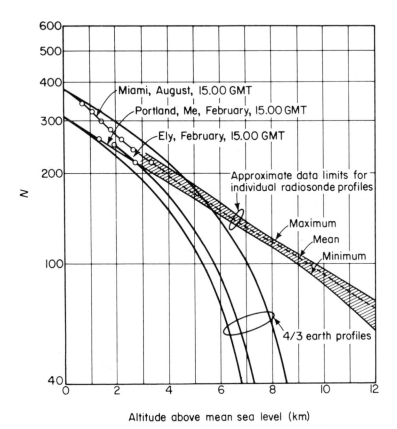

Figure 6.4.2 Refractivity *versus* altitude for mean and extreme conditions, compared to 4/3 earth radius profiles [6.14].

At lower elevations, the elevation error can be predicted more accurately by

$$\Delta E = (bN_s + a) \times 10^{-6} \text{ (radian)} \qquad (6.4.5)$$

where the coefficents b and a are plotted in Figure 6.4.5.

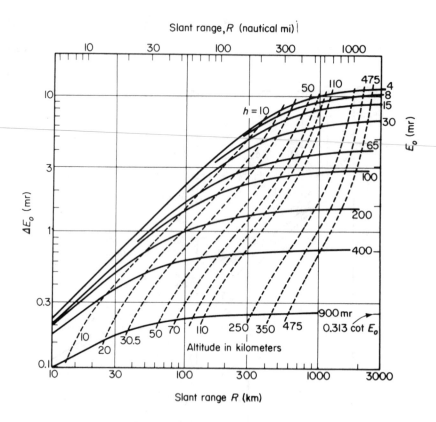

Figure 6.4.3 Elevation bias error *versus* range for exponential reference atmosphere, $N_s = 313$.

Correction of Bias Errors

The curves of Figures 6.4.3 to 6.4.5 provide a basis for correcting refraction errors at the output of the radar system. Simple polynomial approximations give

$$\Delta E = (bN_s + a)\, h/(11 + h) \approx (N_s \cot E)\, h/(11 + h) \qquad (6.4.6)$$

$$\Delta R = (0.0072\, N_s \csc E)\, (h + 0.33h^2)/(15 + h + 0.33h^2) \qquad (6.4.7)$$

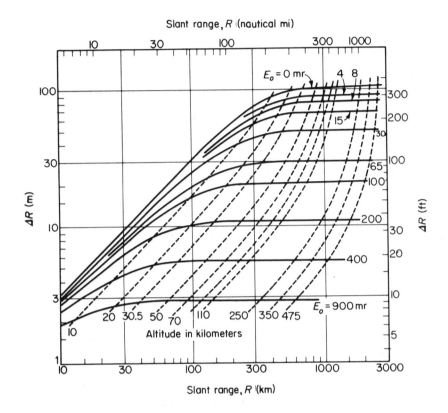

Figure 6.4.4 Range bias error *versus* range for exponential reference atmosphere, $N_s = 313$.

These approximations are accurate to about 5% of the maximum error for $E > 3°$. Since the maximum errors at 3° are about 5 mr in elevation and 40 m in range, the corrections are accurate to 0.25 mr and 2 m in range. For greater accuracy or lower elevation angles, actual refractivity profiles must be measured, and ray-tracing procedures applied, with accuracies to about 1% of the actual refraction errors.

Tropospheric Fluctuations

Irregular variations in refractivity occur in the troposphere, with scale sizes from fractions of one meter to hundreds of kilometers. A detailed model of these fluctuations is given in [6.17], and only a summary will be presented here. The model is based on the *frozen turbulence* model of

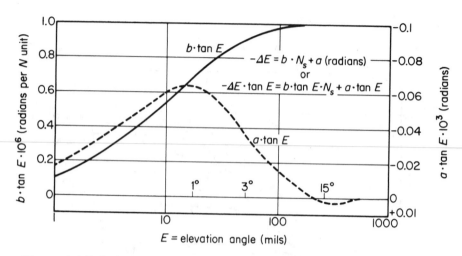

Figure 6.4.5 Refraction correction coefficients for targets well beyond atmosphere [6.16].

refractivity variations, in which the regions of differing refractivity drift across the radar beam with the relative motion of the air mass and the beam. This permits the long-term refractivity spectra, measured at various locations, to be translated into spatial spectra for estimation of phase gradients and angle errors. The resulting estimates of azimuth angle error, for antennas and interferometers of different sizes, are shown in Figure 6.4.6. A key result is that errors are reduced only as the fourth root of the baseline length, even for lengths measured in km. The azimuth errors are small, however, reaching only 0.1 mr for antennas of 1 m dimension. Elevation errors will be similar for an aperture oriented normal to the direction of arrival, but will increase for interferometers with antennas located in a horizontal plane (Figure 6.4.7). The source of the larger error is not merely the smaller projected length of the baseline, but the presence of refractivity variations in the extra path leading to the further antenna of the interferometer (Figure 6.4.7b). The resulting errors will be on the order of 0.01 cot$E$ (mr) for all baseline lengths, when low-frequency error components are included. Presumably the same considerations would apply to phased array antennas when scanned away from broadside.

Ducting

The reference atmosphere has a refractivity gradient $dN/dh = -45\,N$ units per km at the surface. When this gradient increases beyond -157 N units per km, rays leaving an antenna can be trapped in a duct and

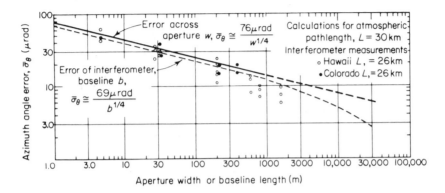

Figure 6.4.6 Azimuth error from tropospheric fluctuations, as a function of aperture width or baseline length [6.17].

propagated for great distances with low loss. The maximum ray elevation which can be trapped is given by

$$\theta_d = \sqrt{2(N_s - N_d - 0.157h_d)} \ (mr) \tag{6.4.8}$$

where N_d is the refractivity at the height h_d (m) of the top of the duct. The maximum wavelength which can be trapped is

$$\lambda_{max} = 0.0025h_d \sqrt{N_s - N_d - 0.157h_d} \ (m) \tag{6.4.9}$$

For example, a duct of $h_d = 100$ m with $N_s - N_d = 20$ N units can trap rays leaving the antenna at elevations up to 2.9 mr, at wavelengths up to 0.52 m. Data on the occurrence of ducting conditions is given in [6.18].

Ionospheric Refraction

When radar measurements are made on targets above about 100 km, the effects of the ionospheric layers must be considered. These effects are all dependent on the operating frequency of the radar, varying directly with the square of the wavelength for VHF and higher bands. Figure 6.4.8 shows typical measured day and night profiles of electron density *versus* altitude, compared to the Chapman distribution and with the simple, rectangular distributions used in an Air Force study of ionospheric effects [6.20]. These profiles may be used to calculate the refractivity of the ionosphere as a function of frequency:

$$N_i = -40N_e/f^2 = -0.5(f_c/f)^2 \times 10^6 \tag{6.4.10}$$

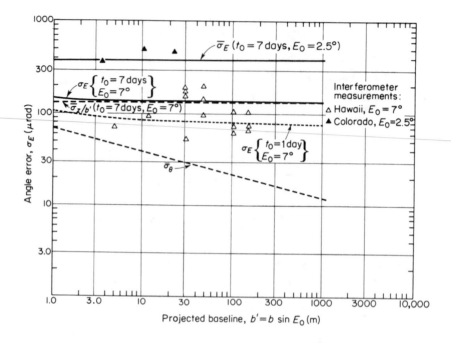

(a) Elevation error from tropospheric fluctuations.

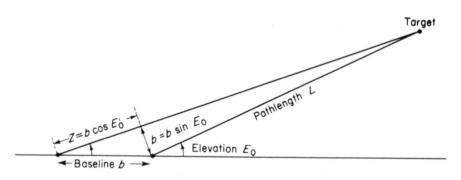

(b) Geometry of elevation measurement.

Figure 6.4.7 Elevation measurement with interferometer.

where N_e is the electron density per m³, f is in Hz, and f_c is the *critical frequency* ($f_c = 9\sqrt{N_e}$). When N_e is expressed in electrons per cm³, the same expression can be used with f and f_c in kHz. These relationships are plotted in Figure 6.4.9, which shows the operating region for accurate measurement systems, defined as having errors less than one part in 10^4.

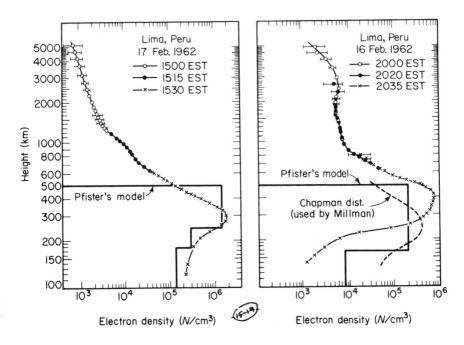

Figure 6.4.8 Typical ionospheric densities compared to models.

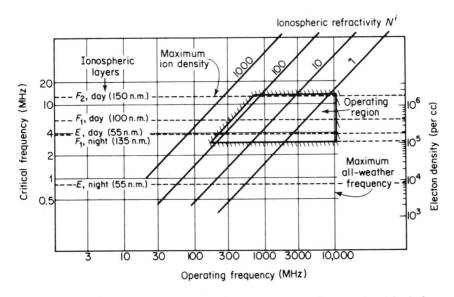

Figure 6.4.9 Ionospheric refractivity *versus* operating and critical frequencies and electron density.

Using the rectangular distributions of Figure 6.4.8, calculations were made of elevation and range errors on paths through the daytime iono- sphere to different heights. The results are shown in Figure 6.4.10. Night- time operations will have about 1/3 of this level of error, and operations during periods of extreme ionospheric disturbance will be about three times as high. In Figure 6.4.11 is plotted the expected range fluctuation error for normal and disturbed conditions.

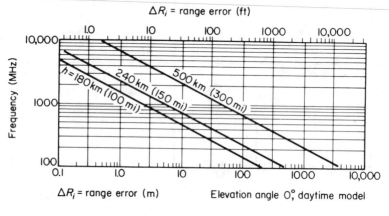

(a) Elevation angle error.

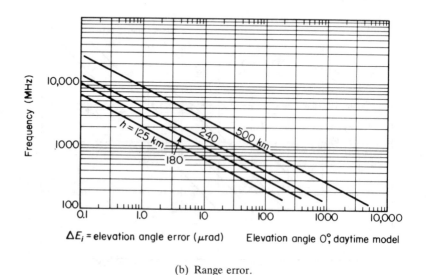

(b) Range error.

Figure 6.4.10 Ionospheric refraction error *versus* frequency.

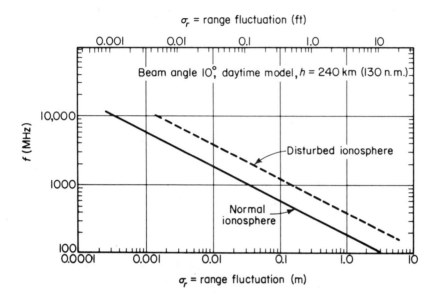

Figure 6.4.11 Ionospheric range fluctuation *versus* frequency.

REFERENCES

[6.1]   D.E. Kerr, *Propagation of Short Radio Waves*, McGraw-Hill, 1951.

[6.2]   H.R. Reed and C.M. Russell, *Ultra High Frequency Propagation*, John Wiley and Sons, 1953.

[6.3]   M.L. Meeks, *Radar Propagation at Low Altitudes*, Artech House, 1982.

[6.4]   L.V. Blake, A Guide to Basic Pulse-Radar Maximum-Range Calculation, *U.S. Naval Research Laboratory Rep. 6930*, December 23, 1969; reprinted in part in D.K. Barton, ed., *Radars*, Vol. 2, *The Radar Equation*, Artech House, 1974.

[6.5]   D.K. Barton, Radar Principles, Secs. 25.1–25.58 in D.G. Fink and D. Christiansen, eds., *Electronic Engineers' Handbook*, McGraw-Hill, 1982.

[6.6]   R.L. Olsen, D.V. Rogers, and D.B. Hodge, The aR^b relation in the calculation of rain attenuation, *IEEE Trans.* **AP-26,** No. 2, March 1978, pp. 318–329.

[6.7]   B.C. Blevis, Losses due to rain on radomes and antenna reflecting surfaces, *IEEE Trans.* **AP-13,** No. 1, January 1965, pp. 175–176.

[6.8]   J. Ruze, More on wet radomes, *IEEE Trans.* **AP-13,** No. 5, September 1965, pp. 823–824.

[6.9]   T.A. Weil, Atmospheric lens effect: another loss for the radar equation, *IEEE Trans.* **AES-9,** No. 1, January 1973, pp. 51–54; reprinted in D.K. Barton, ed., *Radars,* Vol. 1, *The Radar Equation,* Artech House, 1974.

[6.10] C.R. Burroughs and S.S. Atwood, *Radio Wave Propagation,* Academic Press, 1949.

[6.11] S. Ayasli, SEKE: A computer model for low altitude radar propagation over irregular terrain, *IEEE Trans.* **AP-34,** No. 8, August 1986, pp. 1013–1023.

[6.12] R.C. Kirby, Radio-Wave Propagation, Secs. 18.65–18.113 in D.G. Fink and D. Christiansen, eds., *Electronic Engineers' Handbook,* McGraw-Hill, 1982.

[6.13] D.K. Barton, Land clutter models for radar design and analysis, *Proc. IEEE* **73,** No. 2, February 1985, pp. 198–204.

[6.14] B.R. Bean and G.D. Thayer, CRPL Exponential Reference Atmosphere, *National Bureau of Standards Monograph No. 4,* U.S. Government Printing Office, October 29, 1959.

[6.15] L.V. Blake, Radio ray (radar) range-height-angle charts, *Microwave J.,* October 1968, pp. 49–53.

[6.16] B.R. Bean and B.A. Cahoon, The use of surface weather observations to predict total atmospheric bending of radio waves at small elevation angles, *Proc. IRE* **45,** No. 11, November 1957, pp. 1545–1546.

[6.17] D.K. Barton and H.R. Ward, *Handbook of Radar Measurement,* Artech House, 1969.

[6.18] B.R. Bean and E.J. Dutton, *Radio Meteorology, National Bureau of Standards Monograph No. 92,* U.S. Government Printing Office, March 1, 1966.

[6.19] W. Pfister and T. J. Keneshea, Ionospheric Effects on Positioning of Vehicles at High Altitudes, *Air Force Surveys in Geophysics,* No. 83, March 1956, DDC Doc. AD 98777.

[6.20] J.E. Fielding and G.D. Reynolds, *VCCALC: Vertical Coverage Diagram Software and User's Manual,* Artech House, 1988.

Chapter 7
Search Radar Systems

A *search radar* is one which is used primarily for the detection of targets in a particular volume of interest. A *surveillance radar* is used to maintain cognizance of selected traffic within a selected area, such as an airport terminal area or air route [7.1]. The difference in terminology implies that surveillance radar provides for the maintenance of track files on the selected traffic, while the search radar output may be simply a warning or one-time designation of a target for acquisition by a tracker. Thus, search radar is the more general term, and most of the discussion in this chapter will be devoted to the detection of targets within a volume scanned by the radar.

7.1 COVERAGE OF THE SEARCH VOLUME

Beam Shape and Scan Pattern

The volume of interest is defined by boundaries of maximum and minimum elevation, azimuth, range, and radial velocity. The antenna beam must generally be scanned to cover the angle sector, although it is sometimes possible to cover the sector with a number of fixed beams. Figure 7.1.1 (a–f) shows the options for elevation coverage in a radar which scans mechanically in azimuth:

(a) Horizon scan at fixed elevation, for detection of targets on the surface or at low altitudes;

(b) Fan-beam scan of a broad elevation sector, for detection of targets below a given elevation angle;

(c) Fan-beam with cosecant-squared extension, for detection of targets below a given altitude;

(d) Inverted fan-beam scan for search from an airborne radar;

(e) Elevation-scanning pencil beam, capable of shaping coverage to any of the options (a–d);

(f) Stacked elevation beams, also capable of shaping coverage to any of these options.

The azimuth scan used with options (a–f) may be continuous through 360°, or may cover only a sector of arbitrary width. In the latter case, the antenna may be implemented as a phased array, in which the azimuth and elevation scans can be executed in any pattern or in a pseudorandom way, apportioning the transmitted energy in an arbitrary manner over the search sector. The most often used scans are of the raster type, either in lines at constant elevation, Figure 7.1.1g, or in lines at constant azimuth. When 360° azimuth coverage is required in a phased array radar, several array faces must be used, usually covering 90° per face.

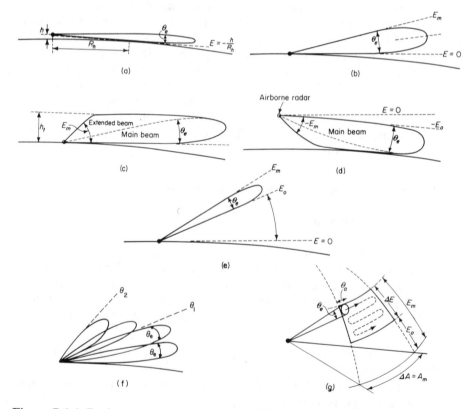

Figure 7.1.1 Basic types of search scan: (a) horizon scan; (b) fan beam; (c) cosecant-squared beam; (d) inverted cosecant-squared beam; (e) elevation-scanning pencil beam; (f) stacked elevation beams; (g) raster scan.

Search radar using options (a–d) are called *two-dimensional (2D) search radars,* while (e–g) are *3D search radars.* In some cases, a *multifunction phased array radar* operates part of the time in the *search mode* to perform the search function.

Range and Velocity Coverage

The range and radial velocity coordinates are covered in parallel channels, so that in most cases a single observation time (or *dwell time*) t_o provides for detection of all targets within the beam or beams that are provided by the antenna. This is necessary because a sequential search of the three- or four-dimensional volume will consume excessive time and lead to low efficiency. In certain cases, however, where a single waveform cannot cover the range-velocity space without gaps, this time is divided into subintervals using different waveforms to ensure that no gaps remain in the coverage. Alternatively, the waveform may be changed from scan to scan, so that gaps are filled over a period of several scans. The issues of waveform efficiency and diversity were addressed in Chapter 5.

Detection Requirements

The search radar equation (1.3.3) can be written to give the minimum requirements for power-aperture product to obtain detection at a desired probability P_d within the coverage contour:

$$P_{av}A_r = 4\pi R_m^4 \psi_s kT_s D_0 (1) L_s/t_s \sigma \qquad (7.1.1)$$

where the terms have been defined in Chapter 1. In some cases, where only warning of a target is required, a contour for cumulative probability of detection P_c is defined, from which the value of P_d on the contour can be derived [7.2]. More usually, the search radar must support formation of track files (surveillance), and the basic requirement is defined in terms of the contour for which adequate P_d (≈ 0.5 to 0.9) is achieved for this purpose.

The search loss L_s may be divided into RF losses $L_t L_\alpha$, which reduce the received peak signal and processing (or statistical) losses $L_i L_f L_p L_x$, which increase the required signal:

$$D_x (1) = D_0 (1) L_i L_f L_p L_x M = nD_x (n) \qquad (7.1.2)$$

The latter group of losses, combined with other losses that are specially applicable to clutter statistics, will also increase the required signal-to-clutter ratio, $D_{xc} (n)$. Beyond meeting the requirement of the search radar equation for the product $P_{av}A_r$, the search radar design must also provide an output $(S/C)_o = I_m (S/C) \geqslant D_{xc} (n)$, where I_m is the improvement factor produced by doppler (or MTI) processing. Military radars must also consider jamming as part of the operating environment.

The combined interference spectral density in which target detection must be obtained is then increased from $N_0 = kT_s$ to

$$I_0 = N_0 + J_0 + C_0 + \ldots \tag{7.1.3}$$

where there may be more than one source of jamming or clutter. The methods of calculating J_0 and C_0 are discussed in Sections 1.4 and 1.5. For E/I_0 to reach the required value $D_x(1)$ on the detection contour at R_m, it will be necessary to budget the allowable contributions from each source, and in general to increase $P_{av}A_r$ enough to compensate for the other interference components. Alternatively, the specification for R_m may be applied only for the benign environment, with degradation accepted when jamming or clutter is present for targets at maximum range. In the examples of this chapter, we will assume that J_0 or C_0 can be allowed to equal N_0, resulting in possible reduction in output S/I by 5 dB (32% range reduction), if all three interference components are present. On the detection contour, the requirements are then

$$E/N_0 = D_x(1), \quad E_1/N_0 = D_x(n)$$

$$E/J_0 = D_x(1), \quad E_1/J_0 = D_x(n)$$

$$E/C_0 = D_x(1), \quad E_1/C_0 = D_x(n), \quad S/C = D_{xc}(n)/I_m$$

where S/C is the input signal-to-clutter ratio and D_{xc} is the clutter detectability factor, discussed below.

For each type of search radar, the determination of performance in clutter will then involve the calculation of input S/C, clutter detectability factor D_{xc}, and available improvement factor I_m for each clutter environment. Recall, from Section 2.6, that D_{xc} depends on the number of independent samples of clutter, n_c, applied to the integrator, and on the amplitude distribution of the clutter (which determines the clutter distribution loss, L_{cd}), as well as on the processing loss factors for noise:

$$D_{xc}(n) = D_x(n) \frac{L_{cd}G_i(n)}{MG_i(n_c)} = D_0(1) \frac{L_i(n_c)\, L_f L_p L_x L_{cd}}{n_c} \tag{7.1.4}$$

The number of independent clutter samples in a system using a fixed frequency is

$$n_c = 1 + t_o/t_c = 1 + t_o(2\sqrt{2}\pi\sigma_v/\lambda) \tag{7.1.5}$$

where σ_v is the standard deviation of the velocity spectrum of the clutter. Thus, the radar wavelength and waveform parameters will enter into the calculation of E/C_0 in a very complex way, which we will try to clarify in practical examples.

Electronic and Mechanical Scan

The technological capability of phased array antennas has progressed to the point that all types of beam scanning and shaping can be supported, subject only to the basic relationships among aperture size, wavelength, beamwidth, gain, and sidelobe levels, as described in Chapter 4. However, in the search role, the key issue is not that of rapidly covering the volume, but rather of covering it slowly enough to provide for clutter rejection through doppler processing. Thus, many of the most advanced techniques in phased array antennas are inapplicable to the search radar, instead finding their place in the tracking radar. This chapter will be concerned primarily with the search capabilities of different beam configurations and scan procedures, regardless of how they are implemented in the antenna. Mechanically rotated antennas are often entirely suitable, and likely to be more practical economically. When electronic scan is used, it is often for mechanical or aerodynamic reasons, rather than to achieve beam motions that are not available in mechanical scan.

7.2 TWO-DIMENSIONAL SEARCH RADAR

The basic issues in design and evaluation of the 2D search radar are governed by the search radar equation (1.3.1), and the other equations given in Chapter 1 for detection in jamming and clutter environments:

(a) The power-aperture product of the radar must be adequate for coverage to range R_m when scanning over the search sector ψ_s in the allowable frame time t_s;
(b) In military radar, average power, tuning bandwidth, and sidelobe levels must be adequate to provide detection of quiet targets in azimuth sectors not occupied by jammers;
(c) The width of the main lobe must not be so great as to permit one jammer to mask targets in too large an azimuth sector;
(d) The combination of spatial and velocity resolution must be adequate to bring the output clutter levels below the target by the detectability factor D_{xc}.

In air and missile surveillance radar, clutter is from land, sea, weather, chaff, and birds. Clutter reduction requirements usually establish a minimum value for time-on-target, or observation time, t_o, for such radars. In airborne mapping and navigation applications, where the surface is the desired target, weather and chaff are the major clutter sources.

All of these performance issues are subject to the constraint that the output false alarms must not be so frequent that they saturate the system which receives the radar data, be it a human operator or an automatic data processor. Other performance issues have to do with the ability to resolve closely spaced targets, and the accuracy of the data in azimuth and range.

Power-Aperture Product for 2D Search Radar

A fundamental property of 2D search radar is that the height h of the aperture must not be greater than required to produce an elevation beamwidth θ_e matched to the required vertical coverage sector:

$$h = K_\theta \lambda/\theta_e \leq K_\theta \lambda/\theta_m \tag{7.2.1}$$

Smaller values of h, giving $\theta_e > \theta_m$ may be used for mechanical reasons, but at the expense of wastefully spreading energy above the required coverage sector. The width w is limited by the requirement for a given observation time t_o while scanning:

$$w = K_\theta \lambda/\theta_a = K_\theta \lambda t_s/A_m t_o \tag{7.2.2}$$

Equations (7.1.1) and (7.1.2) combine to constrain the effective aperture area that can be used in a given application:

$$A_r = (\lambda^2/A_m \theta_m)(t_s/t_o)/L_n \tag{7.2.3}$$

where $L_n = 1/K_\theta^2 \eta_a$ (≈ 1.5 for reflector systems) is the loss constant in the relationship between beamwidth and gain of the antenna, from (1.3.1), and the aperture size is assumed to be limited by performance requirements, rather than by mechanical constraints. This relationship introduces into the search radar equation a direct dependence of R^4 on λ^2 and t_s^2, and an inverse dependence on $(A_m \theta_m)^2$:

$$P_{av} = \frac{4\pi R_m^4 A_m^2 \theta_m^2 k T_s D_0(1) L_s L_n^2 t_o}{\lambda^2 t_s^2 \sigma F^4} \tag{7.2.4}$$

The appearance of λ^2 in the denominator reflects the fact that minimum values of θ_e and θ_a have been established by (7.2.1) and (7.2.2) for the given elevation coverage and observation time.

For air search applications over broad elevation sectors, the result very strongly favors the lower radar frequencies in 2D radar. Higher microwave frequencies are often used for surface search, where the target is a sea or land vehicle, a fixed structure, or the surface itself (for navigation). Microwave operation is feasible because, for surface-based radars, the required elevation sector is very small, while for airborne radars the elevation sector is small enough to match the beamwidth of the available aperture height.

Pattern-Propagation Factor

The minimum elevation beamwidth θ_e has been determined for 2D radar by the required vertical coverage sector θ_m, as in (7.2.1). Additional constraints on this beamshape and the wavelength are imposed by the pattern-propagation factor F in the radar equation. To avoid strong multipath lobing (Figure 6.2.4), the pattern $f(-\theta)$ for angles below the horizon should drop rapidly relative to $f(\theta)$. At the same time, the appearance of a low-elevation gap in the coverage is undesirable, so the high slope in $f(\theta)$ cannot be obtained by too much upward tilting of the antenna beam. The usual compromise directs the peak of the beam to an angle $\theta_b \approx \theta_e/3$ above the horizon, reducing the horizon gain by about 1.3 dB (2.6 dB two-way gain reduction, or 14% range reduction). The enhancement in detection coverage in the lobes more than compensates for this reduced gain, but serious nulls appear in the coverage.

When continuity of coverage on air targets is important, as in support of track files on detected targets, the antenna may have to be enlarged to narrow the basic elevation beamwidth, producing the sharp cut-off on the low side of the beam. The required vertical coverage as then obtained by adding feeds for different elevation angles, and combining these feeds in a microwave network between the antenna and the duplexer. The antenna shown in Figure 7.2.1 generates a basic beamwidth of 2.8°, and by using 12 horns in the feed it produces uniform coverage up to $\theta_1 = 6°$, with csc^2 extension to $\theta_2 = 30°$. The low-elevation cut-off below 0.5° is controlled by the basic beamwidth, while gain is controlled by θ_1 and the csc^2 extension. Vertical polarization is used to minimize the reflection lobing, and circular polarization is provided as an option for rain reduction.

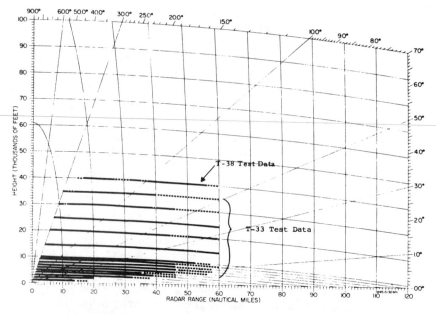

Figure 7.2.1 AN/TPN-24 2D radar antenna, with csc^2 coverage provided by multiple feed horns [7.3].

Coverage at very low altitudes is controlled not by beamwidth, but rather by wavelength and height of the antenna phase center above the surface, from (6.2.8). The first lobe is at an angle:

$$\theta_n/2 = \lambda/4h_r$$

above the horizon. In the null below this lobe, for $\theta < \theta_n/6$, the propagation factor is controlled by diffraction over the surface, as shown in Figures 6.3.1 and 6.3.2. For surface search and detection of low-altitude flying targets from surface-based radars, only the use of the higher microwave frequencies can provide the required low coverage. Attempts to fill this gap with longer wavelength transmission of higher power give large signals from birds, which fly above the gap, as well as from rain and chaff.

Clutter Considerations for 2D Air Search Radar

Clutter input levels and processing improvement factors are dominant in selecting wavelength and waveform for a search radar. The elevation beamwidth and effective aperture height h have been established by the coverage contour, through (7.2.1). Azimuth beamwidth, unless determined by target resolution or accuracy requirements, is selected on the basis of observation time t_o in (7.2.2):

$$\theta_a = A_m t_o/t_s \tag{7.2.5}$$

The observation time must support signal processing for clutter rejection, and the first step in establishing this requirement is to evaluate the input clutter.

For rain and chaff that fill the spatial resolution cell of the radar at range R_c, the clutter cross section is

$$\sigma_c = V_c \eta_v = (R_c \theta_a / L_p)(R_c \theta_e / L_p)(\tau_n c/2)\eta_v \tag{7.2.6}$$

For a 2D radar with $\csc^2$ pattern, the effective elevation beamwidth θ_m replaces θ_e in this equation. At short range, this entire volume is filled with clutter, but as range increases the clutter ceiling begins to fall below the top of the beam, and ultimately below the horizon. This reduction in clutter volume can best be expressed as a reduction in the pattern-propagation factor F_c. However, in the region of strong multipath lobing, below an elevation $\theta_e/2$, from (6.2.8):

$$\overline{F_c^4} = 16\,\overline{\sin^4(\pi\theta/\theta_n)} \approx 6 \tag{7.2.7}$$

where $0 < \theta < \theta_e/2$ is the elevation angle of the clutter. There can be an 8 dB enhancement in this volume clutter over a reflecting surface, so σ_c from (7.2.6) may be exceeded when clutter fills the beam, and may be accurate when the clutter ceiling is at an elevation near $\theta_e/6$. Beyond the range at which that occurs, F falls below unity.

The volume reflectivity η_v is strongly dependent on wavelength for rain (Figure 3.6.6), and less so for chaff (Eq. (3.6.14)). For many applications involving widespread chaff clouds, a constant $\eta_v = 10^{-8}$ m²/m³ for the chaff can be assumed as a waveform design parameter, from L-band to X-band. When rain or chaff appears beyond the unambiguous range R_u, it is convenient to express its effective cross section σ_c', which is the cross section producing the same echo power from the target range R:

$$\sigma_{ci}' = \sigma_{ci}\{R/[R + (i - 1)R_u]\}^4$$
$$= \sigma_c\{R/[R + (i - 1)R_u]\}^4 \, (\overline{F_{ci}}/\overline{F_c})^4 \tag{7.2.8}$$

where the clutter is in the ith ambiguity. The ambiguous clutter values σ_{ci} and F_{ci} are calculated by using $R + iR_u$ in (7.2.6) and the equations for F, while σ_c and F_c are calculated by using R. The ambiguity index i is increased until the clutter source is no longer present, or until F_{ci} falls to zero.

Surface clutter has a cross section:

$$\sigma_c = A_c\sigma^0 = (R_c\theta_a/L_p)(\tau_n c/2)\sigma^0 \tag{7.2.9}$$

The pattern-propagation factor $\overline{F_c^4}$, averaged over regions near the target, also determines the clutter power interfering with target detection, and the effect of the elevation pattern of the antenna appears in this factor. Surface clutter following the constant-γ model has a cross section which is constant with range:

$$\sigma_c = A_c\gamma \sin\psi = (h_r\theta_a/L_p)(\tau_n c/2)\gamma \tag{7.2.10}$$

As range increases, the average pattern-propagation factor $\overline{F_c^4}$ decreases, but for land clutter the peak values tend to remain constant, almost out to the horizon range. When elevated patches of clutter are visible (e.g., hill tops and urban structures), there will be a concentration of large F_c^4 values in those regions, giving a local $\sigma_c\overline{F_c^4}$ well above the average over all azimuths. The statistics of F_c will largely determine the amplitude distribution of the clutter, and the resulting clutter distributions loss L_{cd}.

The remaining clutter characteristics that are needed to establish the signal processing performance are the spectral (or temporal correlation) properties:

$$t_c = \lambda/2 \sqrt{2\pi} \, \sigma_v \qquad (7.2.11)$$

where t_c, defined as the reciprocal of the noise bandwidth of the clutter frequency spectrum, is the correlation time, and σ_v is the standard deviation of the clutter velocity spectrum. This spread σ_v, normalized to the blind speed of the waveform in the parameter $z = 2\pi\sigma_v/v_b$, determines the improvement factor of the MTI, and the width of the rejection notch in PD processing. The correlation time t_c also determines the number of independent clutter samples available for integration, through (7.1.5), giving an integration gain:

$$G_i(n_c) = n_c/L_i(n_c) \qquad (7.2.12)$$

Having estimated the input S/C power, σ/σ_c, using clutter cross section (7.2.6) or (7.2.9) and knowledge of target F^4 and clutter F_c^4, in a given radar band, the required improvement factor from doppler processing can be calculated as a basis for waveform design:

$$I_{req} = (C/S)D_{xc}(n) = (\sigma_c \overline{F_c^4}/\sigma F^4)D_x(n)G_i(n_c)L_{cd}M_c/G_i(n)M \qquad (7.2.13)$$

where M_c is the matching factor for clutter and M is for noise. The input ratio, C/S, is the required subclutter visibility of the radar.

It is evident that the selection of a wavelength and waveform for search radar is not a straightforward procedure of solving for unknowns depending in complex ways on λ, τ_n, and f_r. The normal procedure in waveform design is to start with a set of parameters of an existing radar which has proven successful in a similar environment, and to adjust these parameters until the performance is satisfactory in the specified environment.

Wavelength and Waveforms for Air Search Radar

The use of low radar frequencies for air and missile search is strongly favored by considerations of clutter rejection, as discussed in Chapter 5, and by power-aperture requirements as well. The reasons for this are:

(a) The reflectivity of rain is greatly reduced in the lower radar frequency bands, with η_v varying as f^4;

(b) The ability to reject rain and other clutter with MTI and MTD processing is much better at the lower frequencies, with I_m varying directly as some power of the blind speed $v_b = \lambda/2t_r$.

The strong frequency dependence of clutter and the doppler processing improvement are such that they overcome the potential benefits of improved spatial resolution for the higher frequencies, insofar as target detection is concerned.

In 2D radar, the input clutter spectrum covers the entire velocity interval of rain or chaff from all altitudes within the beam. Figure 7.2.2 shows the measured spectrum of rain observed by a 2D S-band radar. The width of the spectrum is about $\Delta_c = 30$ m/s. This is an extreme value for rain, corresponding to a wind shear coefficient $k_{sh} = 4$ (m/s)/km, but chaff clouds can have this spread even with smaller k_{sh}. Although this measured spectrum is hardly Gaussian in shape, a rough estimate of the required blind speed for a given level of MTI improvement factor can be obtained from (5.3.15), by assuming $2\pi\sigma_v \approx \Delta_c = 30$ m/s, $z = 30/v_b$. Then, for example, if $I_2 = 30$ dB were required, the wavelength and waveform would have to provide a blind speed to meet the required $2/z^4 = 1000$, $z = 0.21$, or $v_b = 30/0.21 = 142$ m/s. Approximately the same blind speed requirement can be derived for a PRF-diversity MTD system, in which bursts at two PRFs are used to fill the higher blind speed regions and $\Delta_v/v_b \approx 0.25$ is allowed. Figure 7.2.3 shows the envelopes of response for bursts of $n_c = 16$ pulses, at PRFs giving blind speeds of 120 and 160 m/s, and rejection notch widths of 20 m/s at -40 dB. Using (5.4.3), the band lost to target detection (response < -6 dB) is

$$\Delta_v = \Delta_c + 2v_b/n_c = 20 + 280/16 = 38 \text{ m/s}$$

The blind notches near 140 m/s are just covered by the diversity channels, but at 480 m/s the blind notches coincide at the two PRFs. In general, for dual diversity, the highest velocity that can be covered by at least one of the channels is

$$v_{max} = (\bar{v}_b^2/\Delta_v) - \bar{v}_b/2 \tag{7.2.14}$$

Larger values of v_{max} can be obtained with triple or quadruple diversity, at the expense of longer dwell times t_o.

Thus, both MTI and MTD systems require, for high clutter rejection, that the average blind speed be four to five times the width of the clutter spectrum, or about 140 m/s for rain extending upward to 6 or 7 km in an atmosphere with a wind shear coefficient of 4 (m/s)/km. Equation (5.3.6)

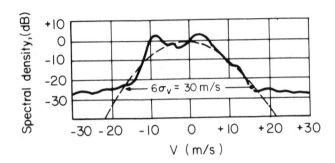

Figure 7.2.2 Spectrum of rain observed by 2D radar [5.13].

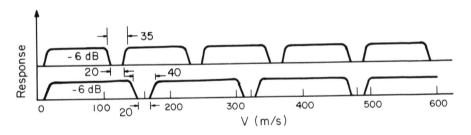

Figure 7.2.3 Response of dual-diversity MTD system for high clutter attenuation on rain spectrum.

now gives a minimum wavelength for given unambiguous range:

$$\lambda_{\min} = 4R_u v_b/c = R_u \text{ (km)}/540$$

with λ in m. An L-band radar, $\lambda = 0.23$ m, is adequate for 120 km search range, and an S-band radar, $\lambda = 0.1$ m, for 50 km. When waveforms having unambiguous range of less than 300 km are used, rain can be expected at ambiguous ranges. If this long-range clutter is intense enough to require clutter attenuation, the use of staggered PRF is ruled out as a means of filling blind regions. Burst-to-burst stagger (PRF diversity) with a coherent transmitter must be used if MTI or MTD is relied upon for rain rejection. Because these limits are not generally acceptable, one or more of the following restrictions are placed on system performance specifications for 2D radar:

(a) Circular polarization is employed to reduce the effects of rain, and detection is limited to regions in which the rain is not too intense;
(b) Doppler processing is used to reduce the effects of rain, and detection is limited to regions with moderate wind shear (and negligible ambiguous clutter, for magnetron transmitters and staggered-PRF systems);
(c) Detection in chaff is not required, or is restricted to clouds of limited altitude extent to reduce the spectral width, and limited range extent to avoid ambiguous clutter.

Example of 2D Air Search Radar

Consider the example used in Section 1.3, but with the elevation coverage increased from 2° to 10° (Figure 7.2.4), and with a requirement for $n = 24$ hits per scan, to achieve clutter rejection. Equation (7.2.4) gives, for $R_m = 86.4$ km and $F = 1$, the requirement:

$$P_{\text{av}} = 27/\lambda^2 \text{ (W)} = 2700 \text{ W at } \lambda = 0.1 \text{ m}^2$$

As a result of the increased vertical coverage, the required power-aperture product has increased by a factor of five, and the usable aperture height is reduced by a factor of five, increasing the power requirements by a factor of 25. In this case, the resulting power level is not too great, and well within the capabilities of tube or solid-state transmitters at S-band. The antenna, which was originally 5 × 3.5 m, becomes 5 × 0.66 m for a 10° fan beam.

If the requirement were expanded to include coverage of targets up to 11 km altitude, at ranges 25 to 75 km, a csc^2 type of antenna pattern would be used (Figure 7.2.3). From (1.3.5), with $\theta_1 = 10° = 0.175$ radian and $\theta_2 = 30° = 0.52$ radian,

$$\theta_m = 0.175(2 - 0.175 \cot 0.52) = 0.30$$

The effective coverage sector has increased by a factor of $0.30/0.175 = 1.69$, relative to the 10° fan beam, the effective aperture has decreased by this factor, and the power requirement increases by $1.69^2 = 2.87$ to $P_{av} = 7750$ W. The physical height of the reflector would typically be increased to about 1 m, with the lowest third shaped to produce the coverage from 10° to 30° elevation. Alternatively, the reflector size could be left at 0.66 m, and additional feed horns would then be used to produce the csc^2 coverage (Figure 7.2.1). The latter design approach leads to greater blockage from the larger feed structure, but reduces the total size of the antenna. For the particular design shown in Figure 7.2.1, the entire allowable antenna height was needed to produce a sharp cut-off at the bottom of the main lobe, and this compelled the use of the stacked horn feed.

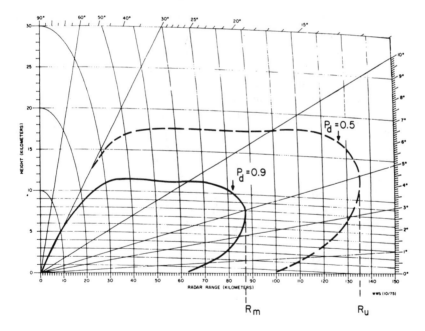

Figure 7.2.4 Coverage of 2D radar with csc^2 antenna.

Waveforms and Processing for Example Radar

Waveform selection for the example 2D radar depends on the required unambiguous range, assumed in this case to be R_u = 135 km. This dictates a PRF of 1100 Hz, to provide n = 24 hits per scan. A typical tube transmitter with peak P_t = 500 kW would then operate at a duty factor of

$$D_u = P_{av}/P_t = 7750/500,000 = 0.0155$$

The pulsewidth would be

$$\tau = D_u/f_r = 14 \ \mu s$$

A reasonable range resolution (for separation of multiple targets and to reduce the input clutter) would be provided by pulse compression, and a logical choice would be the 13-chip Barker code:

$$\tau_n = \tau/13 = 1.1 \ \mu s$$

Use of three-pulse MTI with stagger and velocity adaptation is initially assumed as a clutter reduction measure. Clutter calculations for radar parameters listed in Table 7.1 and environmental parameters listed in Table 7.2 are computed in Table 7.3. In calculating MTI performance, a detectability factor must be found for clutter residue at the MTI output, based on the broadened (but not completely whitened) spectrum of the residue and the MTI losses. For rain and chaff, with broad input spectra, we assume that t_c will be cut in half in passage through the MTI. For land clutter, t_c will be cut to 1/4 its original value. The additional integration gain for the spectrum of clutter residue can be subtracted from the required detectability factor, while MTI losses must be added.

The three-pulse MTI, preceded by a limiter, cannot quite provide the needed I_m, since Figure 5.3.11 gives only I_m = +25 dB for n = 24. With an adaptive notch, three-pulse MTI must be combined with a circularly polarized (CP) antenna (I_{cp} = 15 dB) to meet the rain rejection requirement. Since the rain from the second ambiguity can be reduced adequately by CP, the MTI may use staggered PRF to reduce velocity

Table 7.1 Parameters of Example 2D Air Search Radar

Wavelength, λ (m)	0.1
Peak power, P_t (kW)	500
Pulsewidth, τ (μs)	14
Repetition frequency, f_r (Hz)	1100
Average power, P_{av} (kW)	7.75
Processed pulsewidth, τ_n (μs)	1.1
Antenna $w \times h$ (m)	5×1
Beamwidth $\theta_a \times \theta_e$ (deg)	1.3×10
	$\csc^2$ to 30
Search time, t_s (s)	6
Time-on-target, t_o (s)	0.0217
Hits per scan, n	24
Unambiguous range, R_u (km)	135
Blind speed, v_b (m/s)	55
System temperature, T_s (K)	967
RF losses, $L_t L_\alpha L_n$ (dB)	4
Processing losses, $L_p L_x M$ (dB)	5
$L_i L_f$ (dB)	11
Detectability factor, $D_x(n)$ (dB)	16

Table 7.2 Environmental Parameters for Example 2D Air Search Radar

Rainfall rate, r (mm/h)	3
Rain reflectivity, η_v (m^2/m^3)	3×10^{-9}
Maximum rain altitude, h_{max} (km)	7
Chaff reflectivity, η_v (m^2/m^3)	10^{-8}
Maximum chaff altitude, h_{max} (km)	15
Wind shear coefficient, (m/s)/km	2
Land surface reflectivity, γ	0.1
Land height deviation, σ_h (m)	10
Antenna height, h_r (m)	10
Effective antenna height, h_r' (m)	30
Horizon range, R_h (km)	13

Table 7.3 Clutter Calculations for Example 2D Air Search Radar

	1	2	3
Range ambiguity, i	1	2	3
Range, R_{ci} (km)	86	221	356
Range ratio $(R_{ci}/R_{c1})^2$	0	+8	+12
Altitude at horizon (km)	0.4	2.9	7.5
Rain clutter:			
Max. elevation (degree)	5	1	0
$\overline{F_c^4}$ (dB)	0	−10	−∞
Effective cross section, σ_c' (dBm^2)	+10	−8	−∞
Velocity spread, σ_v (m/s)	4.5	2	−
Correlation time, t_c (s)	0.004	0.01	−
(after MTI)	0.002	0.005	−
No. of independent samples, n_c	6	3	−
(after MTI)	12	5	−
Detectability factor, $D_{xc}(n)$ (dB)	+19	+22	−
(with MTI)	+19	+21	−
I_{req} (dB)	+29	+14	−
(with MTI)	+29	+13	−
Available MTI I_m (dB)	+16	+29	0
Chaff:			
Max. elevation (degree)	10	3	1
$\overline{F_c^4}$ (dB)	+3	0	−10
Effective cross section, σ_c' (dBm^2)	+18	+7	−7
Velocity spread, σ_v (m/s)	9	7	4
Correlation time, t_c (s)	0.002	0.003	0.005
(after MTI)	0.001	0.0014	0.0025
No. of independent samples, n_c	11	8	5
(after MTI)	23	16	10
Detectability factor, $D_{xc}(n)$ (dB)	+17	+18	+20
(with MTI)	+16	+17	+19
I_{req} (dB)	+35	+25	+13
(with (MTI)	+34	+24	+12
Available MTI I_m (dB)	+6	+9	+17

Table 7.3 (cont'd)

Range ambiguity, i	1	2	3
Land clutter (for $R < R_h = 13$ km):			
Cross section, σ_c (dBm^2)	+9		
Velocity spread, σ_v (m/s)	0.6		
Correlation time, t_c (s)	0.033		
(after MTI)	0.008		
No. of independent samples, n_c	1.7		
(after MTI)	4		
Clutter distribution loss, L_{cd} (dB)	4		
(after limiter)	0		
Detectability factor $D_{xc}(n)$ (dB)	+27		
(with MTI)	+20		
I_{req} (dB)	+37		
(with limiting MTI)	+29		
Available MTI I_m (dB)	+25		

response loss. The chaff requirement, which is not reduced by use of CP, remains far beyond the capabilities of MTI with a blind speed so low.

A further calculation is needed for cases in which rain forces the MTI processor and CP antenna to be used near R_m. The $P_{av}A_r$ product was calculated on the basis of a loss budget of 20 dB, without allowance for either MTI loss ($L_{mti(a)} + L_{mti(b)} = 3$ dB with staggered PRF) or CP loss ($L_{cp} = 3$ dB). The new detection range, when using these processes, and without considering the addition of clutter residue to the output, becomes

$$R'_m = R_m/\sqrt[4]{\Delta L} = 86.4/\sqrt[4]{4} = 61 \text{ km}$$

The MTI processor may be bypassed in sectors that are clear of clutter, under control of a range-azimuth sector MTI gate which is established and updated by the operator, but unless the antenna polarization is also switched rapidly by this control, there will remain a CP loss of 3 dB, giving

$$R'_m = 86.3/\sqrt[4]{2} = 73 \text{ km}$$

Options for Improvement of Clutter Rejection

The options for improvement in clutter reduction are as follows, for a radar of this type.

(a) *Moving Target Detector (MTD).* An MTD process using two 12-pulse or three 8-pulse bursts would provide considerable improvement in performance against ground clutter, as compared to MTI, if linear processing can be ensured. This implies that the range-azimuth cells exceeding the linear range of the A/D converter must be blanked at the processor output. Performance against rain will be somewhat better than with MTI, with the advantage of more optimal adaptation to the clutter spectrum. In addition, its efficiency in operation of MTD against targets in the cluttered regions is slightly better than the MTI system (see Table 5.4). The use of triple diversity with an average blind speed near 55 m/s can provide essentially complete coverage for targets up to 600 m/s, if the velocity notch at -6 dB is no wider than 20 m/s (see Figure 7.2.5). Such a notch will provide about 20 dB attenuation of the rain clutter spectrum, and also some improvement against clutter beyond the ambiguous range.

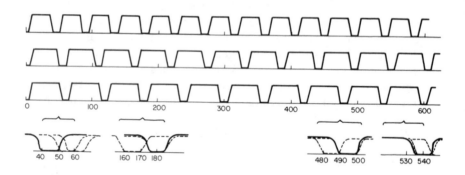

Figure 7.2.5 Velocity coverage of triple-diversity MTD.

(b) *Pulse Compression.* The compressed pulsewidth can be made considerably narrower than the 0.9 µs provided by the Barker code. In searching for aircraft or missile targets, the practical limit is set by the radial extent L_r of the target itself, and this is typically 10 to 15 m. Assuming the 10 m value, a pulse length of 0.067 µs can be used without loss in target cross section. Narrower pulses will resolve the target into two or more scattering sources, without improvement in S/C ratio. The reduction to $\tau_n c/2 = 10$ m will reduce rain, chaff, and average land clutter by 11 dB, relative to the earlier calculations for 1 µs, giving

Rain clutter $\sigma_c = -1$ dBm^2

Chaff $\sigma_c = +7$ dBm^2

Land clutter peaks, however, will not be reduced because they originate from discrete objects. The combination of pulse compression and CP will almost solve the rain clutter problem, giving an input cross section of -19 dBm^2 as compared to target $\sigma = -3$ dBm^2, for $S/C = +16$ dB (compared to the required $D_{xc} = +19$ dB). Pulse compression will reduce the required performance of the MTI or MTD systems, but will not eliminate the need to use doppler processing for land and for chaff, and will increase the required throughput of the processor by the pulse compression ratio.

(c) *Frequency Diversity*. None of the calculations so far have assumed the use of RF diversity, which has the ability to decrease target fluctuation loss and to increase the number of independent clutter samples available for integration. The diversity may be implemented as parallel channel diversity or RF diversity from burst to burst in MTI or MTD operation. In achieving $P_d = 0.9$, the diversity gain on the target is, from (2.4.11),

$$G_d(n_e) = 8.4(1 - 1/n_e) \text{ dB}$$

RF diversity between the three MTD bursts can thus reduce both D_x and D_{xc} by 5.6 dB, extending the detection range to 119 km against the noise background (or providing a performance margin at 86.4 km range). Burst-to-burst diversity with MTI could provide eight 3-pulse bursts, for a diversity gain of 7.4 dB and a net gain of 4.1 dB in noise (considering $L_{mti(a)}$, Eq. (5.3.26)). The gain in a clutter background may be more than 7.4 dB, when correlation of the clutter has previously reduced the number of integrated samples below eight. Table 7.4 shows the calculations with eightfold diversity. The combination of frequency diversity, CP, and pulse compression can provide detection in rain, and with MTI can provide detection in land clutter when a limiter is used for suppression of clutter peaks. In chaff, the diversity gain falls 24 dB short of solving the problem, and remains 12 dB short, even when combined with pulse compression to 0.067 μs.

(d) *High-PRF PD*. None of the above techniques is able to counter the chaff threat for the example S-band 2D search radar. Referring to Table 5.3, however, the high-PRF PD radar is seen to have a blind speed of 1000 m/s, and this permits the rejection of all types of clutter, including chaff with large velocity spread. For chaff with density $\eta_v = 10^{-8}$ m^2/m^3 enveloping the radar, the clutter in the first range gate (150–300 m) is 107 dB above noise level, approximately the same as land clutter. This requires extremely high stability in the oscillator, and in the remainder of the transmitter and receiver processing. In addition, the PRF diversity must fill the eclipsed range regions, which occupy about two of the 50 range cells.

Table 7.4 Clutter Calculations with Frequency Diversity

Range ambiguity, i	1	2	3
Range, R_{ci} (km)	86	221	356
Rain clutter:			
Effective cross section, σ'_c (dBm^2)	+10	−8	−∞
No. of independent samples, n_c	8	8	−
Detectability factor, $D_{xc}(n)$ (dB)	+12	+12	−
I_{req} (dB)	+20	+4	−
Available MTI I_m (dB)	+16	+29	0
Chaff:			
Effective cross section, σ'_c (dBm^2)	+18	+7	−7
No. of independent samples, n_c	8	8	8
Detectability factor, $D_{xc}(n)$ (dB)	+12	+12	+12
I_{req} (dB)	+30	+19	+5
Available MTI I_m (dB)	+6	+9	+17
Land clutter (for $R < R_h = 13$ km):			
Cross section, σ_c (dBm^2)	+9		
No. of independent samples, n_c	8		
Clutter distribution loss, L_{cd} (dB)	4		
(after limiter)	0		
Detectability factor D_{xc} (n) (dB)	+16		
(after MTI)	+12		
I_{req} (dB)	+25		
(with limiting MTI)	+25		
Available MTI I_m (dB)	+25		

(e) *Lower Frequency.* The broad spectrum of rain and chaff clutter can best be overcome at frequencies below S-band, in search operations beyond about 50 km. If the example radar were implemented at L-band, the average blind speed would increase to $v_b = f_r\lambda/2 = 126$ m/s. The aperture height could be increased from 1 m to 2.3 m, reducing the average power requirement to 2750 W. The width of the aperture could also be increased to 11.5 m, reducing the power further, or, if azimuth resolution considerations permitted, the width could remain at 5 m, and $n = 55$ hits per scan would be received with the beamwidth of 0.053 radian = 3.4°.

Triple diversity with $n_c = 16$ would now be possible, and all clutter would lie deep in the rejection notch allowed with this blind speed. The rain clutter reflectivity would also be reduced by about 15 dB at the longer wavelength. Normally, this solution would be adopted, rather than re-sorting to high-PRF operation at S-band, if operation in chaff or heavy rain were required, and if the required azimuth resolution were consistent with mechanical limitations on antenna width.

Example of Horizon Search Radar

The example radar from Section 1.3 can serve as a basis for analysis of a radar to detect the sea-skimmer antiship missile, flying at 3-m altitude. The required elevation coverage is small enough that the beamwidth will be limited by mechanical limitations on aperture height, rather than by the coverage requirement. The detection performance will be determined primarily by the low-altitude propagation factor F and the ability of the radar to reject clutter. While sea clutter will be present, the major problem is moving clutter, such as birds, rain, and chaff, at altitudes above the target.

Figure 6.3.1 shows that the optical horizon for $h_r = 20$ m, $h_t = 3$m is at $R_h = 25$ km. Microwave radar, however, cannot propagate to that point in a standard atmosphere because of diffraction over the surface of the spherical earth. At X-band and higher frequencies, coverage for $h_t = 3$ m is above the free-space value at $R = 15$ km, for antennas at $h_r = 20$ m. This fact, and elevation beamwidth considerations, suggest scaling the example radar from S-band to C-band, X-band, or higher. Consider an aperture limited in total area A, and hence in effective area A_r. The radar equation, in the form originally derived (1.2.26), can be written with gains G_t and G_r expressed in terms of A_r:

$$P_{av} = 4\pi R_m^4 k T_s D_0(1) L_s \lambda^2 / t_o A_r^2 \sigma F^4 \qquad (7.2.15)$$

The time-on-target t_o will be established to support a clutter-rejection waveform. The pattern-propagation factor F can be estimated by using the flat-earth approximation of (6.2.8), provided that h_t and h_r are reduced to account for curvature of the surface over the range R_m:

$$F = 2|\sin (\pi\theta_t/\theta_n)| \approx 2\pi\theta_t/\theta_n = 4\pi h_t' h_r'/R\lambda \qquad (7.2.16)$$

where the primes denote heights adjusted for earth's curvature. When substituted into (7.2.15), this leads to

$$P_{av} = R_m^8 \lambda^6 k T_s D_0(1) L_s / (4\pi)^3 t_o \sigma A_r^2 h_t'^4 h_r'^4 \qquad (7.2.17)$$

The λ^6 dependence strongly favors short wavelengths, but limits are set by inclusion of atmospheric attenuation L_α in L_s, and by requirements for doppler processing.

The broad velocity spectrum of rain and chaff has been described (e.g., Figure 7.2.2). The problem of birds must also be addressed when low-RCS or low-altitude targets are to be detected, because birds in the enhanced reflection lobes of Figure 6.3.1 will have effective cross sections of

$$\sigma_c' = \sigma_c F_c^4 = 16\sigma_c \qquad (7.2.18)$$

A sea bird with $\sigma_c = 0.01$ m^2 appears as 0.16 m^2, not much less than a missile target and larger than some. Birds' velocities are ± 15 m/s with respect to the local wind, requiring an azimuth-adaptive rejection notch width $\Delta_c = 30$ m/s (at about the -20 dB level). The width of the notch at which target detection is inadequate, from (5.4.3), will be near $\Delta_v = 45$ m/s for most systems. Now (7.2.14) can be solved to find the required average blind speed, in a dual-diversity system, as a function of Δ_v and the maximum target velocity to be covered:

$$v_b \approx \Delta_v \left(\sqrt{v_m/\Delta_v} + 1/4 \right) \qquad (7.2.19)$$

The input requirements for our sea-skimmer example are listed in Table 7.5. For dual-PRF diversity,

$$v_b = 45 \left(\sqrt{600/45} + 1/4 \right) = 175 \text{ m/s}$$

Triple diversity would permit $v_b = 115$ m/s, at the expense of longer dwells and lower waveform efficiency. From (5.3.6), we obtain

$$\lambda_{min} = 4R_u v_b/c = 0.047 \text{ m (C-band, dual diversity)}$$

or

$$\lambda_{min} = 0.031 \text{ m (X-band, triple diversity)}$$

The use of K_u-band and higher frequencies is ruled out by the doppler notch requirement, unless range-ambiguous operation is adopted. The λ^6 dependence in (7.2.17) favors the X-band system over C-band, in spite of reduced efficiency, higher dwell time, and greater rain attenuation. By combining PRF diversity with RF diversity, the target fluctuation loss can be reduced to recover some of the lost efficiency. The resulting detectability factor for each of the three bursts will be

$$D_x(3) = D_{xc}(3) = +22 \text{ dB}$$

Table 7.5 Requirements for Sea-Skimmer Detection

Input parameters:

R_m	= 15 km	t_s	= 2 s
h_t	= 3 m	A_m	= 2π
h_r	= 20 m	R_u	= 20 km
σ	= 0.3 m^2	$D_0(1)$	= 20
v_m	= 600 m/s	L_s	= 100 = 20 dB
T_s	= 1000 K	r	= 3 mm/h
A_r	= 1 m^2		

Derived requirements at X-band:

λ	= 0.032 m	A_r	= 0.6 m^2
f_r	= 7500 Hz	τ_n	= 1 μs
v_b	= 120 m/s	V_c	= 2.7 $\times$ 10^7 m^3
$\theta_a\theta_e$	= 0.0014 sr	t_o	= 0.0087 s

The coherent process, in order to minimize the notch width Δ_v, must use at least $n_c = 16$, or

$$t_f = n_c t_r = 16 \times 0.000133 = 0.0021 \text{ s}$$

Rain clutter, however, has $\sigma_c = 8 \text{ m}^2 = +9 \text{ dBm}^2$ at R_m. When enhanced by an average propagation factor of 8 dB, the effective $\sigma_c\overline{F_c^4} = +17$ dB, giving $(C/S)(\overline{F_c^4}/F^4) = +25$ dB. This requires $I_m = +47$ dB to place residue at the noise level, and doppler rejection will continue to be required to 150 km and beyond. Using six fill pulses at the start of each coherent burst, it is possible to achieve full processing improvement on rain out to the seventh range ambiguity ($R_c = 120$ to 140 km). In the eighth ambiguity ($R_c = 140$ to 160 km) we have

$$\sigma_c' = \sigma_{c_1} [R_m/(R_m + 7R_u)]^2$$

$$= 8(15/155)^2 = 0.075 \text{ m}^2 = -11 \text{ dBm}^2$$

$\sigma_c'F_c^4 = -4$ dB, which gives $S/C = +4$ dB. The degraded improvement (Figure 5.4.6b) on rain in this and further ambiguities should be sufficient, without adding more fill pulses. Then, for triple diversity operation,

$$t_o = 3(t_f + 6t_r) = 3(0.0021 + 6 \times 0.000133) = 0.0087 \text{ s}$$
$$\theta_a = A_m t_o/t_s = 2\pi(0.0087)/2 = 0.027 \text{ radian} = 1.6°$$
$$w = K_\theta\lambda/\theta_a = 1.15 \times 0.032/0.027 = 1.4 \text{ m}$$
$$h = A/w = 0.7 \text{ m}$$
$$\theta_e = K_\theta\lambda/h = 1.15 \times 0.032/0.7 = 0.052 \text{ radian} = 3°$$

In order to apply (7.2.17), the corrected antenna and target heights must first be determined, and the nature of the curve for propagation factor near R_m must be inspected. Figure 7.2.6 shows the results of the program described in Appendix A. The signal level is plotted relative to its free-space level at $R = 15$ km. The interpolated propagation factor at 15 km is slightly below unity. The program also yields values $h_r' = 11.2$ m and $h_t' = 2.5$ m at $R = 15$ km. Equation (7.2.17) can now be evaluated, giving
$$P_{av} = 66 \text{ W}$$

As a check on the validity of the approximation, (7.2.15) gives $F^4 = -5.4$ dB at $R = 15$ km, where the interpolated value is -4 dB. Hence, the calculated power requirement is slightly high. A reasonable waveform would be

$$P_t = 9 \text{ kW}, \quad f_r = 7500 \text{ Hz}, \quad \tau = 1 \text{ μs}$$

To validate the calculation, the program RGCALC was run for this radar, using the pulsed doppler method with $P_{av} = 66$ W, $t_f = 0.0029$ s, $n = 3$, and assuming RF diversity as well as PRF diversity between bursts. The results are shown in Figure 7.2.7. Of the 20-dB loss budget applied to (7.2.17), 9 dB appears as specifically listed losses in RGCALC, or is included in the computation of $D_2(3) = +13.68$ dB. The propagation factor of -4 dB must also be included as a loss, so the miscellaneous loss in the RGCALC worksheet is set to 15 dB. The resulting $R_m = 14.3$ km is 0.8 dB less optimistic than (7.2.17), but can be considered an adequate validation of the method.

The power requirements are quite low, but the allowance for waveform inefficiencies, processing losses, and other losses should be reviewed to ensure an adequate design margin. The fill pulses introduce a loss given by (5.4.18):

$$L_{eg} = 1 + t_g/t_f = 1 + 6/16 = 1.38 = 1.4 \text{ dB}$$

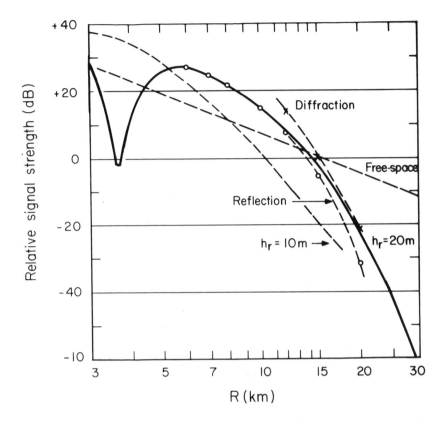

Figure 7.2.6 Signal power *versus* range for low-altitude target (h_t = 3 m, h_r = 20 m, λ = 0.032 m).

The blind velocity notch causes a velocity response loss given by (5.4.17):

$$L_{ev} = v_b/(v_b - \Delta v) = 120/(120 - 45) = 1.6 = 2 \text{ dB}$$

This loss does not apply to cases in which no targets are present within Δ_v, as in high-PRF radars intended for detection of targets having radial velocities in excess of Δ_v. In rain at 3 mm/h, Table 6.2 gives k_α = 0.13 dB/km, and

$$L_\alpha = R_m k_\alpha = 15 \times 0.13 = 2 \text{ dB}$$

When these losses are added to the usual losses for filter weighting, beam-shape, straddling, fluctuation, and RF components, it will be difficult to

```
Radar Name or Description -- Sea Skimmer Detection Radar


     Radar and Target Parameters (inputs) --

     Peak Pulse Power (kilowatts) ...................        .1
     Pulse Duration (usec) ......................... 2900.0000
     Transmit Antenna Gain (dB) ....................      38.6
     Receive Antenna Gain (dB) .....................      38.6
     Frequency (MHz) ...............................    9375.0
     Receiver Noise Factor (dB) ....................       5.0
     Bandwidth Correction Factor (dB) ..............        .8
     Antenna Ohmic Loss (dB) .......................        .2
     Transmit Transmission Line Loss (dB) ..........       1.0
     Receive  Transmission Line Loss (dB) ..........       1.0
     Scanning-Antenna Pattern Loss (dB) ............       1.3
     Miscellaneous Loss (dB) .......................      15.0
     Number of Pulses Integrated ...................         3
     Probability of Detection ......................      .900
     False-Alarm Probability (Negative Power of Ten)      6.0
     Target Cross Section (Square Meters) ..........     .3000
     Target Elevation Angle (Degrees) ..............       .00
     Average Solar and Galactic Noise Assumed
     Pattern-Propagation Factors Assumed = 1


          **********************************

     Calculated Quantities (Outputs) --

     Noise Temperatures, Degrees Kelvin --
               Antenna (TA) .........................    170.9
               Receiving Transmission Line (TR) .......    75.1
               Receiver (TE) ........................    627.1
               TE X Line-Loss Factor = TEI ............   789.4
               System (TA + TR + TEI) ...............   1035.4
     Two-Way Attenuation Through Entire Troposphere (dB)    6.5


     Swerling     Signal-     Tropospheric   Range,      Range,
     Fluctuation  to-Noise    Attenuation,   Nautical    Kilometers
     Case         Ratio, dB   Decibels       Miles
     ---------    ---------   -----------    --------    ----------

        2          15.64         .39           7.7         14.3
```

Figure 7.2.7 RGCALC results for horizon search radar.

stay within the 20 dB loss budget listed in Table 7.5. Increase in transmitter power to compensate for larger losses is not difficult, however, and even if the antenna height were reduced to $h_r = 10$ m, an increase to 660 W average, 90 kW peak, could preserve at detection range of 13.5 km (at which point the propagation factor is just 10 dB less than for the 20-m antenna at $R = 15$ km, as shown in Figure 7.2.6). It appears entirely possible to implement a low-PRF radar with $R_u = 20$ km to solve the sea-skimmer problem from either antenna height.

Advanced Techniques for 2D Search Radar

To overcome clutter from different sources, with minimum loss in detection efficiency, several new techniques were developed during the 1960s and 1970s for application to 2D search radar [7.4–7.6]. These techniques are intended for adapting the antenna pattern and processing to the specifics of the local environment, permitting the system to detect most targets with the full efficiency of the normal integration process.

(a) *Dual-Beam Antenna.* The performance of 2D radar at short ranges is often degraded by the presence of land clutter for which the input C/N exceeds the improvement factor I_m of the MTI. Use of a limiter to suppress such clutter peaks was discussed in Section 5.3. The introduction of the limiter, however, greatly reduces the improvement factor, and suppresses targets when they lie above the strong clutter. The dual-beam antenna permits the receiving gain at the horizon to be reduced, without suppressing targets at the higher elevations (Figure 7.2.8). A second feed horn is added to the conventional reflector system, and the receiver is switched to this horn for the first portion of the interpulse period. The transmitter remains connected to the low-beam horn, so high-power switching is not involved. In effect, the antenna creates an electronically scanned lower edge to the 2D beam, which can be raised to avoid the regions of strong clutter. The transition between the two receiving beams can be made by degrees (Figure 7.2.9), and only in the specific range-azimuth regions containing strong clutter. This preserves detection of the lowest targets as long as they do not lie directly over the strong clutter regions.

(b) *Adaptive Clutter Attenuator.* If the dual-beam operation proves unable to maintain all clutter within the dynamic range of the MTI, limiting is still unnecessary to avoid false alarms. The location of the cells containing excessive clutter can be stored in a clutter map [(d), below] and that information may be used to set an attenuator early in the receiver. Since the attenuator is a linear device, no spreading of the clutter spectrum is produced, and the MTI may operate with its full improvement factor, determined by the number of hits per scan and the system stability [7.4]. Targets entering the attenuated cells are also attenuated, of course, so this step is used only when necessary and in the smallest cells which can be conveniently implemented in the map. Careful attention will ensure that the transient generated when the attenuator is switched does not appear at the output of the receiver. In a system using digital processing, this consideration requires switch action to be synchronized with the A/D sampling time, and the receiver bandwidth following the switch must not be so small as to produce excessive stretching of the transient.

(c) *Adaptive Processing Paths.* The losses inherent in MTI and PD processing were discussed in Sections 5.3 and 5.4. It is desirable to restrict

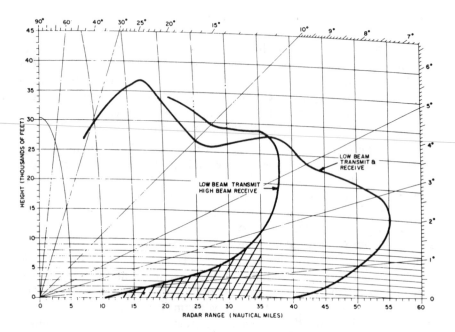

Figure 7.2.8 Coverage pattern obtained with dual-beam antenna [7.4].

these losses to the regions in which clutter must actually be suppressed by doppler processing. In conventional MTI, an *MTI gate* is used to select normal video at ranges beyond that of significant clutter. However, land clutter is seldom uniform in azimuth, and the worst azimuth may require MTI processing at much longer ranges than the average clutter. Sector control has been applied to reduce the regions of loss. The detailed clutter map provides a finer grain control of this switching process. The cell-by-cell map of the MTD [7.6] makes possible the detection of targets which would be in the rejection notch of the MTI or MTD processor, as long as they do not actually lie in strong clutter cells. There are several other optional processing paths and modes, which can be adaptively controlled to maximize target detection, such as selection of logarithmic processing to control false alarms from moving rain [7.5].

(d) *Clutter Map Control.* The key to application of the techniques discussed above is the clutter map, which maintains information over many scans and permits the antenna, receiver, and processor to be switched to the configuration best equipped to detect targets while avoiding false alarms. The simplest form of the clutter map stores the locations of large cells in which MTI is needed. Refinements include storage of actual clutter

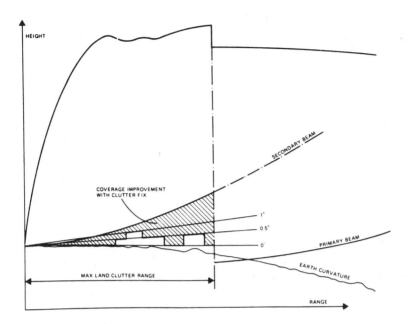

Figure 7.2.9 Adaptation of low-beam coverage to local clutter [7.5].

amplitudes, dual-beam antenna control settings, attenuator settings, processing paths, and mean clutter velocities for adaptive MTI or MTD operation, with cell sizes approaching the resolution of the radar. Because the environment does not remain fixed, the map must be updated at intervals of minutes to an hour, and means of performing the updates without interfering with the output data flow are a critical issue, especially in air traffic control radars. As the cost of digital storage and processing is progressively reduced, more refinements can be expected in new 2D radars and modifications of the existing equipment.

7.3   THREE-DIMENSIONAL SEARCH RADAR

The 3D radar has a number of significant advantages over 2D radar:

(a) The aperture height is no longer limited by the required elevation coverage;

(b) Increased aperture height provides better control of the elevation coverage pattern, with sharp cut-off at the horizon and more uniform coverage above the horizon;

(c) If the full receiving aperture is used for upper beams, the increase in effective search sector for csc² coverage is less than for 2D radar;

(d) Surface clutter is eliminated or greatly reduced in all but the lowest beam position;

(e) Rain clutter and chaff are received at reduced ranges in the upper beams, and with reduced spectral spread in all beams;

(f) Long-range jammers (SOJs) are received only in the sidelobes of the upper beams;

(g) Jammer strobes from adjacent 3D radars, giving azimuth and elevation of the jamming sources, can be combined to give range by triangulation, with greatly reduced ambiguity (ghosting);

(h) The correct ground range (and, hence, x and y coordinates) can be derived for targets at high altitude;

(i) Height (elevation) data can be obtained on each target as it is detected.

The order of importance of these advantages is approximately that of their listing. Note that height data, often believed to be the critical reason for 3D radar, are given last place on the list. Referring to the basic issues listed for 2D radar, Section 7.2, we will consider the differences in required power-aperture product, jamming vulnerability, and clutter processing requirements for 3D search radar.

Power-Aperture Product for 3D Search Radar

The basic search radar equation applies to 3D radar, without a coverage constraint on aperture height. The aperture width w remains constrained by (7.2.2), to the extent that time-on-target t_o must exceed a certain value. This is largely a function of the clutter environment. For example, if the maximum range of significant clutter in the low beam is $R_{c(max)} = 450$ km, as a result of high-altitude chaff or ducted propagation of land clutter from beyond the horizon, the time delay with which that clutter reaches the receiver is

$$t_{dc} = 2R_{c(max)}/c = 3 \text{ ms} \tag{7.3.1}$$

If the radar prf is $f_r \leq 333$ Hz, this does not present problems, since no clutter will arrive from beyond the interpulse period $t_r = 1/f_r$. However, for coherent processing at higher PRF, there will be a transient gating loss, from (5.4.18), and an increase in the required dwell time to allow the long-range clutter to enter the receiver before processing begins. The coherent processing time t_f' will then be related to the coherent transmission time t_f by (5.4.6):

$$t_f' = t_f - t_{dc} = t_f - 3 \text{ ms}$$

To each burst of a PRF-diversity waveform, an additional 3 ms must be added. For a system using dual diversity, 6 ms of signal must be discarded, leading to a requirement for $t_o \geq 20$ ms, if the transient gating loss is to be less than 1.5 dB. Even more important than loss of energy (which can be overcome with more average power) is the efficiency of time usage, which is only 70% for $t_o = 20$ ms.

In a 3D radar scanning 360° azimuth with stacked beams, substitution of $t_o = 0.02$ s, $A_m = 2\pi$ in (7.2.2) and (7.2.5), for maximum usable aperture width, gives

$$\theta_a = A_m t_o/t_s = 0.04\pi/t_s = 1/8t_s \text{ radians} = 7.2/t_s \text{ degrees}$$

$$w = K_\theta \lambda/\theta_a = K_\theta \lambda t_s/A_m t_o = 8K_\theta \lambda \approx 10\lambda t_s$$

For a radar with $t_s = 4$ s, $\theta_a > 0.03$ radian, or 1.7°, and at S-band $w \leq 6$ m. When the requirement for azimuth resolution dictates a smaller beamwidth, the radar energy and time efficiency will be reduced. In a 3D radar using a scanning pencil beam, time must be allowed for coverage of the upper beams. Even though the 3-ms penalty is not encountered at high elevation, the azimuth beamwidth must be broadened to provide the time for sequential scanning of elevation beam positions. This mode of operation is generally unsuitable for use in cluttered environments.

Elevation Coverage of Cosecant-Squared Pattern

The effective vertical angle of a $\csc^2$ antenna pattern is given by (1.3.5):

$$\theta_m = \theta_1[2 - \theta_1 \cot\theta_2]$$

However, when the full aperture area is used in receiving from the beam positions in the upper coverage ($\theta > \theta_1$), the transmitting pattern will follow $\csc^4\theta$ for which the effective sector is given instead by (1.3.6):

$$\sin\theta_m = (4/3) \sin\theta_1[1 - (\sin\theta_1/\sin\theta_2)^3/4]$$

For small θ_1 and large θ_2, the $\csc^2$ antenna has an effective angle $2\theta_1$, while the 3D with full receiving aperture has $(4/3)\theta_1$. The ratio is about 1.5, or about 1.7 dB in favor of the 3D radar. In many cases, the highest receiving beams are broadened to reduce the number of receiver channels, requiring a slight increase in transmitting gain above the $\csc^4$ pattern and reducing this advantage to about 1 dB.

The need for better control of the elevation coverage was illustrated in Figure 7.2.4, where the 2D coverage pattern for an elevation beamwidth of 10°, with csc² extension to 30°, is plotted. In this case, the beam is assumed to be pointed at 5°, leaving a region of reduced range near the horizon. It is possible, of course, to synthesize a 2D pattern having more uniform coverage by using the multiple-horn feed technique illustrated in Figure 7.2.1, at some expense in peak gain. In 3D radar antennas, however, the envelope of the coverage pattern can be made almost uniform over the sector from the horizon to θ_1, as shown in Figure 7.3.1. The dashed lines on this figure indicate the composite coverage when the detection in overlapping channels is considered.

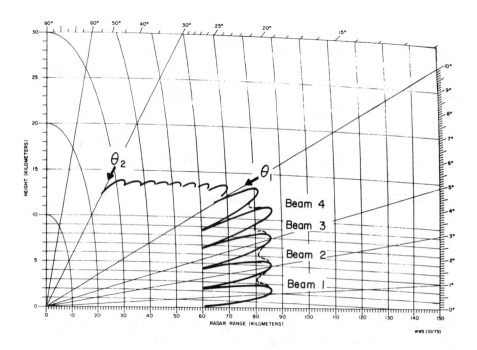

Figure 7.3.1 Coverage of stacked-beam 3D radar.

Time Budget for 3D Radars

There are three basic approaches to implementation of 3D radars:

(a) A single pencil beam (Figure 7.1.1e) may be scanned in elevation rapidly enough to cover the vertical sector during the time $t_s\theta_a/A_m$ for azimuth motion through one beamwidth;

(b) The entire vertical coverage may be obtained by stacked beams (Figure 7.1.1f), with every beam having the full time-on-target $t_o = t_s\theta_a/A_m$;

(c) A set of n_b beams may be used to cover in parallel a portion of the vertical sector, the set scanning over the entire vertical coverage during the time $t_s\theta_a/A_m$.

The need for adequate dwell times, to support doppler processing and diversity gain on the target, has been discussed. The maximum time, for a given search frame time t_s and aperture width w, is provided by the stacked-beam approach. Whether the other methods can be used success-fully depends largely on the clutter environment. If no clutter at all is present, the time required to scan over a contour bounded by maximum range R_m and maximum altitude h_m can be found by applying (1.3.7). The angle θ_1 is calculated (ignoring earth's curvature) from

$$\sin\theta_1 = h_m/R_m$$

The time to scan the coverage (Figure 1.3.1) is then equivalent to that required to scan, with constant range R_m, the sector:

$$\theta'_m = \sin\theta_1(1 + \ln\sin\theta_2 - \ln\sin\theta_1)$$

For example, with $\theta_1 = 10°$ and $\theta_2 = 30°$, the effective elevation sector θ'_m is $2.06\theta_1 = 20.6° = 0.36$ radian, some 20% greater than for a 2D radar with $\csc^2$ antenna. The effective solid angle is then $\psi'_s = A_m \sin\theta'_m = 2.2$ steradians $= 7250$ degrees-squared. For a 1×1-degree beam placing a single pulse in each beam position, with $R_{u(max)} = R_m = 150$ km, $t_r = 1$ ms, the frame time would be 7.25 s, if the PRI were precisely matched to the coverage contour at the center of each beam.

The result, for a 3D radar using a scanning pencil beam, is that (7.2.3) and (7.2.4) apply, in establishing the aperture area and average power, using the value of t_o corresponding to R_m, and with θ'_m replacing θ_m in (7.2.3) and $\theta_m\theta'_m$ replacing θ_m^2 in (7.2.4). For the 3D radar, however, the value of t_o may vary over the elevation sector in accordance with the clutter environment. For example, assume that three-pulse MTI is to be used in all beam positions of a scanning system to produce the coverage envelope of Figure 7.3.1, with $R_u = 150$ km in the lower beams, scaled to 23 km altitude for beams above 10°. The search frame time is $t_s = 6$ s, and for three pulses $t_o = 0.003$ s in the lower beams. Equation (7.2.3) gives

$$A_r = (0.1^2/2.2)(6/0.003)/1.5 = 6 \text{ m}^2$$

This is almost as large as the aperture for the original horizon-scanning beam of the radar example in Chapter 1, and is large enough to produce a 2 × 2-degree beam at S-band. The required average power, from (7.2.4), is

$$P_{av} = 1350 \text{ W}$$

The time allowance for three-pulse MTI, at $f_r = 1000$ Hz, can be met with a scanning-beam 3D radar, but its performance in rain, chaff, or long-range (ducted) land clutter would be inadequate, as noted below.

When a set of n_b beams is used, the available t_o is increased by n_b, increasing the allowable A_r in (7.2.3) and R_m^4 in (7.2.4) by n_b. Now, however, the worst clutter environment of the n_b beams must be considered in establishing the waveform and required t_o.

Clutter in 3D Radars

The lowest three beams in a typical scanning 3D radar have coverage as shown in Figure 7.3.2. The elevation beamwidth is taken as 2°. Only the first beam will have land clutter within its main lobe. For the scanning radar that transmits and receives with the same narrow beamwidth, each beam will have a response about 24 dB down at the center position of the adjacent beams. Rain clutter, visible to 300 km in beam 1, thus will be visible out to 225 km in beam 2 and 120 km in beam 3, at a level −24 dB from peak response. Chaff will be visible to 450 km in beam 1, 360 km in beam 2, and 225 km in beam 3, at −24 dB relative level. If the 24 dB reduction is not sufficient at these ranges, doppler processing for clutter reduction will be needed in all three beams to the ranges indicated, rather than just to the target detection envelope. The higher beams may require doppler processing to lesser ranges.

In a stacked-beam 3D radar, the transmitter pattern usually approximates the csc^4 shape, broad enough to cover all receiving beams. In such cases, only the one-way receiving beam pattern applies, and the −6 and −24 dB levels shown in Figure 7.3.2 become −3 and −12 dB. Clutter cross section and spectral density may be calculated by using an effective elevation beamwidth $\sqrt{2}\ \theta_e$. On the other hand, the same waveform and processing time used for the first beam are available in all beams, and much more clutter attenuation can be provided.

The narrow elevation beamwidth of 3D radar is usually assumed to reduce the spread of the rain and chaff clutter spectra, according to

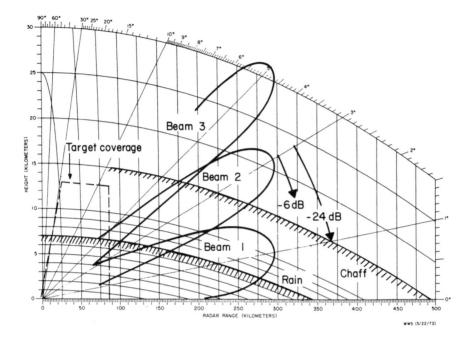

Figure 7.3.2 Rain and chaff in low beams of 3D radar.

(3.6.15). This is true for waveforms which are unambiguous in range to the maximum range of clutter. However, in the lower beams of many pencil-beam radars, having $R_u < 300$ km, rain from altitudes of 5 to 7 km will appear in the second or higher range ambiguity. This ambiguous rain comes from the highest altitudes where wind velocity differs from that in the first range ambiguity. The resulting spectrum appears as in Figure 7.3.3, with contributions from each ambiguity that is occupied by clutter. When chaff is present, the maximum range, altitude, and velocity are all greater than for rain.

Thus, a major 3D advantage in clutter, reduced spectral spread, is applicable only to beams that clear the highest clutter altitude within the unambiguous range of the waveform. The lowest beams will encounter the same spectral spread which applied to 2D radar, and the same wavelength and waveform restrictions will apply. Because these restrictions are not compatible with many modern 3D designs, it is necessary to omit, from the environmental specifications in which these radars operate, rain and chaff clouds that occupy large altitude and range intervals.

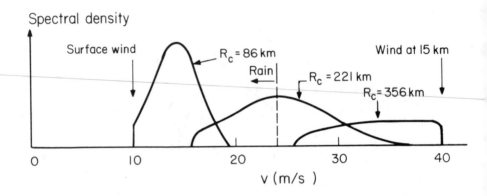

Figure 7.3.3 Spectrum of rain from first and second range intervals in pencil beam at low elevation.

Jamming in 3D Radar

There remain major advantages in 3D radar operation in jamming. The ability to maintain high-elevation coverage in the presence of long-range SOJs can be important in that it denies free access for high-altitude penetrations by quiet aircraft. The SOJ screens only the elevations of the beam in which the jammer can operate. From Figure 7.3.2, if the SOJ can fly at 15 km in altitude and 300 km in range, it can effectively jam beams 1 and 2. Jamming of the higher beams depends on their sidelobe levels.

Since penetrating jammers (SSJs and ESJs) can almost always deny range data to the radar, the use of triangulation is the only reliable means of locating these targets in range. The 2D radar network, in the presence of multiple jammers, produces large numbers of false locations (ghosts), where azimuth strobes cross. For n_j jammer sources, there are n_j^2 strobe crossings, of which $n_j(n_j - 1)$ are ghosts. The same problem would apply to 3D radar if all jammer sources were at very low altitude, where elevation data would be inaccurate. However, in most cases, the penetrators will have measurable elevation angles, and the probability of false correlations among multiple strobes will be significantly reduced.

Elevation Data from 3D Radars

The final advantage of 3D radar is its ability to provide elevation (and, hence, height) data. The two types of requirements for these data correspond to quite different levels of accuracy. The first type of requirement is to designate to a tracking radar, or to obtain correct ground range on targets, so that the x and y data may be exchanged among different radars with good correlation of tracks. For example, if a target at 20° elevation is detected by a 2D radar, the ground range will be in error by $(1 - \cos20°)R = 0.06R$, or 3 km at $R = 50$ km. A second 2D radar at a different location would see the same target, but could not correlate its track with that of the first radar because of a potential 6-km discrepancy in the ground position data. To avoid this problem, for example, maintaining 1% accuracy in ground range, targets above 8° elevation must be measured with an accuracy of

$$\Delta\theta = 0.01 \cot\theta \text{ (radian)}$$

or to about 1° at $\theta = 30°$. A similar level of accuracy will suffice for designation to tracking radars that can readily acquire targets for which angles are known to within two or three degrees. This class of requirement is readily met by any 3D technique.

The second type of requirement is for target height to assist in interception of the target by a manned aircraft or a missile, or to avoid collision in air traffic control. Here, the allowable error is measured in terms of target altitude, for example, 1-km error at maximum target range. For systems operating to 100 km in range, this implies elevation data to 10 mr, or about 0.6°. This may be the 3σ allowance, placing a significant burden on the 3D measurement technique, especially in long-range systems. The accuracy with which these measurements can be made is a direct function of the elevation beamwidth. In stacked-beam systems, the monopulse approach is used to interpolate between adjacent beams, and accuracies of a few percent of the beamwidth can be obtained. With scanning beams, the amplitudes in adjacent beams at different times (and sometimes at different frequencies) are compared, and larger errors must be expected. A major source of error in the lowest beam will be the effect of multipath reflections. At elevation angles below $\theta_e/2$, the error will normally be at least $\sigma_e = \theta_e/20$. The maximum elevation beamwidth of a 3D radar providing accurate height data is often determined by accuracy considerations. Detailed discussion of measurement errors is contained in Chapter 11.

Example of Stacked-Beam Air Search Radar

Consider the coverage requirements of the example in Section 7.2, with the added requirement for 3D operation. The coverage for $\theta_e = 2°$ would be as shown in Figure 7.3.1: $\theta_1 = 10° = 0.175$ radian, $\theta_2 = 30° = 0.52$ radian. The antenna size would be that of the horizon-scanning radar used as an example in Chapter 1: $w = 5$ m, $h = 3.5$ m. For a system using the full receiving aperture on each beam, and transmitting with csc^4 pattern, the effective solid angle of coverage, from (1.3.6), is

$$\sin\theta_m \approx \theta_m = (4/3)(0.175)[1 - (0.175/0.5)^3/4] = 0.23$$

The required average power, scaled from the example in Chapter 1, is

$$P_{av} = 110(0.23/0.035) = 723 \text{ W}$$

This is 10 dB lower than for the 2D radar with csc^2 antenna. However, if the use of 15 stacked beams were regarded as uneconomical, the beams above beam 5 could be progressively broadened, and coverage achieved with 8 to 10 beams at small increase in transmitter power (e.g., 900 W).

The azimuth beamwidth and time-on-target remain as in the previous example, the same PRF will be assumed, and the pulse length is adjusted to obtain the required average power. For peak power $P_t = 1$ MW, $\tau = 0.7$ μs would be transmitted, without pulse compression. In each beam, $n = 24$ hits per scan are available for processing.

Clutter Inputs for Example

The clutter inputs in beam 1 are as shown in Table 7.6. MTI with CP can solve the rain problem. The performance against ground clutter is almost adequate, as with the 2D radar. In chaff, the required improvement has been reduced, and the available improvement in the first range interval has been increased to the point that the requirement there is almost met. However, because of the different mean velocities of ambiguous clutter, the chaff rejection problem cannot be solved with doppler processing at S-band, with unambiguous range $R_u = 135$ km. As will be shown below, the use of combined MTI, pulse compression, and frequency diversity offers a solution.

Table 7.6 Clutter Calculations for Example 3D Air Search Radar

	1	2	3
Range ambiguity, i	1	2	3
Range, R_{ci} (km)	86	221	356
Rain clutter:			
Max. elevation (degrees)	5	1	0
$\overline{F_c^4}$ (dB)	+3	0	–
Effective cross section, σ_c' (dBm^2)	+6	−5	−∞
Mean velocity (m/s)	14	21	–
Velocity spread, σ_v (m/s)	2.5	2	–
Correlation time, t_c (s)	0.008	0.01	–
(after MTI)	0.004	0.005	–
No. of independent samples, n_c	4	3	–
(after MTI)	6	5	–
Detectability factor, $D_{xc}(n)$ (dB)	+22	+23	–
(with MTI)	+20	+21	–
I_{req} (dB)	+28	+18	–
(with MTI)	+26	+16	–
Available MTI I_m (dB)	+25	+9	0
Chaff:			
Max. elevation (degrees)	10	3	1
$\overline{F_c^4}$ (dB)	+3	+3	0
Effective cross section, σ_c' (dBm^2)	+9	+1	−6
Mean velocity (m/s)	14	25	34
Velocity spread, σ_v (m/s)	2.5	6.5	5
Correlation time, t_c (s)	0.008	0.003	0.004
(after MTI)	0.004	0.0015	0.002
No. of independent samples, n_c	4	8	6
(after MTI)	6	15	12
Detectability factor, $D_{xc}(n)$ (dB)	+22	+20	+20
(with MTI)	+20	+17	+18
I_{req} (dB)	+31	+21	+14
(with MTI)	+29	+18	+12
Available MTI I_m (dB)	+25	+4	−6
Land clutter (for $R < R_h = 13$ km):			
I_{req} (with limiting MTI) (dB)	+25		
Available MTI I_m (dB)	+25		
(*See* Table 7.4)			

Options for Improvement

The same options are available to improve clutter rejection of the 3D radar as for the 2D radar example in Section 7.2, but the results and difficulties differ in some cases.

(a) *Pulse Compression.* As with the 2D example, a reduction in rain and chaff by 11 dB is possible, with pulse compression matched to a 10-m target length. In combination with CP, pulse compression solves the rain problem without doppler processing. When used with MTI, pulse compression still remains 7 dB away from solving the chaff problem in the third ambiguity, however.

(b) *Frequency Diversity.* As with the 2D example, frequency diversity can reduce requirements by 5 to 7 dB in noise and clutter (Table 7.7). Doppler processing is still required for land clutter. When combined with pulse compression, frequency diversity provides 6 dB further improvement of the 7 dB required to solve the chaff problem. This can be regarded as adequate, given the uncertainties in the calculations. When frequency diversity is used, it also increases the detection range in the benign environment. The RGCALC worksheet for the example radar is shown in Figure 7.3.4, with the miscellaneous loss increased to allow for pulse compression and MTI loss. The transmitting antenna gain is calculated as

$$G_t = 4\pi/\theta_a\theta_m L_n = 4\pi/0.023 \times 0.28 \times 1.25 = 1560 = +32 \text{ dB}$$

(corresponding to a csc^3 pattern), while the receiving gain is

$$G_r = 4\pi/0.023 \times 0.035/1.25 = 12500 = +41 \text{ dB}$$

The calculation is based on noncoherent integration of eight pulses, with adequate diversity to obtain Case 2 target statistics. For the assumed target length $L_r = 10$ m, the correlation frequency is 15 MHz, and a total signal bandwidth of at least 120 MHz is required to achieve the needed diversity.

(c) *High-PRF PD.* This technique is applicable to the 3D radar as well as to the 2D radar, but becomes much more expensive in a 3D radar, where multiple receivers are required. High-PRF pulsed doppler remains more difficult than shifting operation to a lower frequency band, where low-PRF doppler techniques can be used with chaff.

(d) *Lower RF Frequency.* For the clutter spectral spread encountered, operation at L-band or below offers the best chance of solving the

```
Radar Name or Description -- Stacked Beam Example


   Radar and Target Parameters (inputs) --

   Peak Pulse Power (kilowatts) ...................      1000.0
   Pulse Duration (usec) .........................       .8000
   Transmit Antenna Gain (dB) ....................        32.0
   Receive Antenna Gain (dB) .....................        41.0
   Frequency (MHz) ...............................      3000.0
   Receiver Noise Factor (dB) ....................         5.0
   Bandwidth Correction Factor (dB) ..............          .8
   Antenna Ohmic Loss (dB) .......................          .2
   Transmit Transmission Line Loss (dB) ..........         1.0
   Receive  Transmission Line Loss (dB) ..........         1.0
   Scanning-Antenna Pattern Loss (dB) ............         2.6
   Miscellaneous Loss (dB) .......................         6.0
   Number of Pulses Integrated ...................           8
   Probability of Detection ......................        .900
   False-Alarm Probability (Negative Power of Ten)        6.0
   Target Cross Section (Square Meters) ..........      1.0000
   Target Elevation Angle (Degrees) ..............        1.00
   Average Solar and Galactic Noise Assumed
   Pattern-Propagation Factors Assumed = 1


      **********************************

   Calculated Quantities (Outputs) --

   Noise Temperatures, Degrees Kelvin --
         Antenna (TA) .............................       111.3
         Receiving Transmission Line (TR) .......         75.1
         Receiver (TE) ............................       627.1
         TE X Line-Loss Factor = TEI .............        789.4
         System (TA + TR + TEI) ................         975.8
   Two-Way Attenuation Through Entire Troposphere (dB)      2.8
```

Swerling Fluctuation Case	Signal-to-Noise Ratio, dB	Tropospheric Attenuation, Decibels	Range, Nautical Miles	Range, Kilometers
2	8.16	1.27	47.9	88.7

Figure 7.3.4 RGCALC results for 3D radar with MTI, pulse compression, and frequency diversity.

problems, if chaff density increases above $\eta_v = 10^{-8}$, or if the wind shear coefficient increases above 2 (m/s)/km. However, in 3D radars, the resulting antenna sizes may become inconveniently large (e.g., 7 m for $\theta_e = 2°$).

Table 7.7 Clutter Calculations for 3D with MTI, Frequency Diversity, and Pulse Compression

Range ambiguity, i	1	2	3
Range, R_{ci} (km)	86	221	356
Rain clutter:			
Effective cross section, σ_c' (dBm^2)	-5	-11	$-\infty$
No. of independent samples, n_c	8	8	$-$
Detectability factor, $D_{xc}(n)$ (dB)	$+12$	$+12$	$-$
I_{req} (dB)	$+7$	$+1$	$-$
Available MTI I_m (dB)	$+25$	$+9$	0
Chaff:			
Effective cross section, σ_c' (dBm^2)	-2	-10	-17
No. of independent samples, n_c	8	8	8
Detectability factor, $D_{xc}(n)$ (dB)	$+12$	$+12$	$+12$
I_{req} (dB)	$+10$	$+2$	-5
Available MTI I_m (dB)	$+25$	$+4$	-6
Land clutter (for $R < R_h = 13$ km):			
Cross section, σ_c (dBm^2)	$+9$		
No. of independent samples, n_c	8		
Clutter distribution loss, L_{cd} (dB)	4		
(after limiter)	0		
Detectability factor $D_{xc}(n)$ (dB)	$+16$		
(after MTI)	$+12$		
I_{req} (dB)	$+25$		
(with limiting MTI)	$+25$		
Available MTI I_m (dB)	$+25$		

7.4 SURFACE SEARCH RADAR

Search of the earth's surface for fixed or moving targets (e.g., ships or land vehicles), or for purposes of navigation or mapping of the terrain, is done with different types of radar from those applied to search for air targets. The basic search radar equation remains applicable, however, and can be modified to apply specifically to systems using synthetic apertures as well as to those that use real apertures. The real aperture is an antenna that forms a beam instantaneously, with characteristics as discussed in Chapter 4. The synthetic aperture radar uses a small antenna to sample the received field sequentially, over an extended distance traversed by the radar platform (e.g., aircraft or satellite). The received signal samples are then combined coherently to form a high resolution image of the observed surface. If the scene viewed by the radar remained static during the observations, the result would reproduce that which could be obtained by scanning with a very large antenna.

Airborne Real Aperture Radar for Surface Search

The antenna pattern of Figure 7.1.1d is used for surface search from an elevated platform. When the aperture size is determined by the electrical requirements, in accordance with (7.2.1) to (7.2.3), the 2D search equation (7.2.4) can be applied directly. The wavelength is often increased until the beamwidth requirements are met with the small available aperture, and in this case the original form of the search radar equation is equally useful, with ψ_s determined by the desired elevation coverage θ_m.

In a mapping application, the cross section can be expressed as

$$\sigma = \Delta_r \Delta_x \gamma \sin\psi \tag{7.4.1}$$

where Δ_r is the range resolution depth, Δ_x is the cross-range resolution width at maximum range, γ is the minimum reflectivity parameter for the detectable surface elements, and ψ is the grazing angle to the surface at maximum range. For ranges not too near the horizon range, the effect of earth's curvature may be neglected, giving

$$\sin\psi = h_r/R_m$$

Then, since $\Delta_x = R_m\theta_a$, we can write

$$\sigma = \Delta_r h_r \theta_a \gamma \tag{7.4.2}$$

Substituting $\theta_a = A_m t_0/t_s$, we obtain

$$\sigma = \Delta_r h_r \gamma A_m t_0/t_s \tag{7.4.3}$$

The average power can be expressed in terms of the system requirements, from (7.2.4), as follows:

$$P_{\text{av}} = \frac{4\pi R_m^4 A_m \theta_m^2 kT_s D(1)L_s L_n}{\lambda^2 t_s \Delta_r \gamma h_r} \tag{7.4.4}$$

The width of the aperture is no longer important in determining range, except in that a limit on width may require use of a shorter wavelength to achieve the required resolution.

Example of Real Aperture Surface Search Radar

Consider a requirement for mapping out to 50 km, with resolution 10×100 m, over land with minimum $\gamma = 0.01$. The angular resolution is 2 mr, requiring the use of K_a-band equipment. The atmospheric loss will be about 5 dB in clear air. The following parameters will be assumed:

$$h_r = 3000 \text{ m} \qquad A_m = \pi/2$$

$$\theta_m = 0.09 \text{ radian} \qquad \lambda = 0.0086$$

$$\Delta_r = 10 \text{ m} \qquad T_s = 1000 \text{ K}$$

$$D(1) = 10 \qquad L_s = 32 = 15 \text{ dB}$$

$$L_n = 1.25 \qquad t_s = 2.5 \text{ s}$$

With these parameters, the average power requirement is

$$P_{\text{av}} = 1220 \text{ W}$$

The selected waveform might be

$$P_t = 30 \text{ kW}, \quad f_r = 3000 \text{ Hz}, \quad \tau = 13 \text{ } \mu\text{s}, \quad B = 20 \text{ MHz}$$

The antenna dimensions would be

$$w = K_\theta \lambda/\theta_a = 1.15 \times 0.0086/0.002 = 5 \text{ m}$$

$$h = K_\theta \lambda/\theta_m = 1.15 \times 0.0086/0.09 = 0.1 \text{ m}$$

If operation in clouds were required, an additional 5 dB loss would be encountered, requiring $P_t = 100$ kW. The difficulty in rotating the 5-m antenna would require use of an electronically scanned array, oriented along the aircraft fuselage and scanning $\pm 45°$ from broadside.

Synthetic Aperture Radar

The nonscanning synthetic aperture radar moves at velocity v_p along a straight path, illuminating a region to one side of the airborne platform (Figure 7.4.1). The angular resolution Δ_a, when coherent processing is applied to the signal samples over a time t_o, covering a synthetic aperture of length $L = t_o v_p$, is

$$\Delta_a = K'_\theta \lambda / L \sin\alpha = K'_\theta \lambda / v_p t_o \sin\alpha \qquad (7.4.5)$$

where $K'_\theta \approx 0.5$ is the beamwidth constant for two-way transmission of each sample, and α is the angle from broadside to the velocity vector. The corresponding cross-range resolution is

$$\Delta_x = K'_\theta \lambda R_m / v_p t_o \sin\alpha \qquad (7.4.6)$$

When coherent processing proceeds over the entire time during which the actual antenna beam illuminates the surface, $t_o = R_m \theta_a / v_p$, the synthetic aperture length is $L = R_m \theta_a$, and the cross-range resolution becomes

$$\Delta_x = K'_\theta \lambda / \theta_a \sin\alpha = w K'_\theta / K_\theta \sin\alpha \qquad (7.4.7)$$

where $\theta_a = K_\theta \lambda / w$ is the beamwidth of the physical antenna of width w. Coherent processing over this long interval requires application of a focusing adjustment to the phase data, as a function of the range. When this is done, for $K'_\theta = 0.5$, $K_\theta = 1$, and $\alpha = 0$, the cross-range resolution cell width is $\Delta_x = w/2$ (one-half the width of the physical antenna).

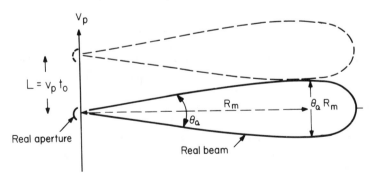

Figure 7.4.1 Synthetic aperture geometry.

The radar then forms a map of an azimuth sector $A_m = \theta_a$ in the time $t_s = t_o$. The search radar equation can be applied to this case, and is derived from (7.2.4) with the following substitutions:

$$\sigma = \Delta_r \Delta_x \sigma^0 = \Delta_r \Delta_x \gamma h_r / R_m$$

$$t_s = t_o = R_m \theta_a / v_p$$

$$A_m = \theta_a = K'_\theta \lambda / \Delta_x \sin\alpha$$

The equation gives the power required to form a map showing surface regions of reflectivity γ:

$$P_{av} = \frac{4\pi R_m^4 \theta_m^2 v_p K'_\theta k T_s D(1) L_s L_n^2}{\lambda \Delta_r \Delta_x^2 \gamma h_r \sin\alpha} \tag{7.4.8}$$

The equivalent expression for average power derived in [7.7] is (using our symbols):

$$P_{av} = \frac{(4\pi)^2 R_m^3 k T_s D_0(1) L_s}{G^2 \lambda^2 \sigma^0 \Delta_r} (2v_p/\lambda) \sin\alpha \tag{7.4.9}$$

The two equations can be shown as identical for

$$G = (4\pi)/\theta_a \theta_e L_n = 4\pi/A_m \theta_m L_n$$

$$L_n = 1, \quad K'_\theta = 0.5$$

In other literature of synthetic aperture radar, a processing factor $m = 1/K'_\theta \approx 2$ is defined as a beamwidth constant. Use of K'_θ in a way analogous to K_θ for the real aperture clarifies the relationship between the width Δ_a of the synthetic beam and the weighting applied to samples during processing to reduce sidelobes in the azimuth response.

It is usually desirable to perform coherent processing over a time $t_f = t_o/n$, and to combine the resulting n maps by addition after envelope detection (noncoherent integration of map data). The cross-range resolution cell Δ_x is broadened by the factor n, but the characteristic random fluctuations of amplitudes (following the Rayleigh distribution) in any single map cell are averaged to obtain a better image. Oscillator and inertial platform stability requirements are also reduced as t_f is decreased. Now, (7.4.6) becomes

$$\Delta_x = K'_\theta \lambda R_m / v_p t_f \sin\alpha = n K'_\theta \lambda R_m / v_p t_o \sin\alpha \qquad (7.4.10)$$

and (7.4.7) becomes

$$\Delta_x = n w K'_\theta / K_\theta \sin\alpha \qquad (7.4.11)$$

In other cases, the real beam is steered (electronically or mechanically) to maintain observation on the same region of the ground for longer periods than permitted by its physical beamwidth. This generates higher resolution, if permitted by system stability, or alternatively more noncoherent integration, or both in the resulting *spotlight map*.

Detailed discussions of SAR operation, signal processing, and stability requirements can be found in [7.7].

Example of Synthetic Aperture Radar

Consider the previous mapping problem, but with a requirement for higher resolution, $\Delta_x = \Delta_r = 3$ m at $R_m = 50$ km. Assume a platform velocity $v_p = 200$ m/s, with mapping carried out broadside to the aircraft (a strip map of 50 km in length will require 250 s of flight). There is no longer any need for very short wavelength, since the resolution (7.4.7) is independent of λ. With a low-sidelobe weighting of samples ($K'_\theta = 0.8$) and allowance for noncoherent integration of $n = 4$ maps, (7.4.11) gives the aperture width:

$$w = (\Delta_x K_\theta \sin\alpha)/n K'_\theta = (3 \times 1.15 \times 1)/94 \times 0.8 = 1 \text{ m}$$

The azimuth beamwidth can be found from this width, and the vertical aperture height is derived from the required elevation coverage angle:

$$\theta_a = K_\theta \lambda / w = 1.15 \times 0.03/1 = 0.035 \text{ radian} = 2°$$

$$h = K_\theta \lambda / \theta_e = 1.15 \times 0.03/0.09 = 0.4 \text{ m}$$

The observation time is

$$t_o = R_m \theta_a / v_p = 50 \times 35/200 = 9 \text{ s}$$

and the coherent processing time is 1/4 of this result, or 2.25 s. Equation (7.4.8), with $n K'_\theta = 2.8$ replacing K'_θ now gives the required average power:

$$P_{av} = 30 \text{ W}$$

A reasonable chirp waveform would be

$$P_t = 10 \text{ kW}, \quad f_r = 3000 \text{ Hz}, \quad \tau = 1 \text{ μs}, \quad B = 60 \text{ MHz}$$

If the 0.4-m aperture height were inconvenient, it might need to be reduced with a corresponding increase in the required transmitter power.

Shipboard Navigation Radar

The shipboard navigation radar is a special case of surface surveillance, in which the elevation beamwidth is controlled by the need to accommodate platform pitch and roll, the azimuth beamwidth by the resolution requirement, and wavelength by a compromise between low-angle propagation and sensitivity to rain clutter. Doppler processing is not used because targets and clutter are both near zero velocity. Maximum required range is set by the need for navigational data on coastal features, but need seldom exceed the horizon range for $h_r = h_t = 20$ m:

$$R_h = \sqrt{2ka} \left(\sqrt{h_t} + \sqrt{h_r} \right) = 37 \text{ km}$$

The applicable radar equation is derived simply from (7.2.4) by setting $A_m = 2\pi$ and including the factor F^4:

$$P_{av} = \frac{8\pi^2 R_m^4 \theta_a \theta_e \, kT_s \, D_0 \,(1)\, L_s L_n}{\lambda^2 \, t_s \, \sigma F^4} \tag{7.4.12}$$

The search loss budget includes the atmospheric attenuation L_α, which is wavelength dependent. Hence, the wavelength for greatest S/N at given P_{av} will be that which maximizes the quantity:

$$X = \lambda^2 F^4 / L_\alpha$$

The flat-surface approximation for F, (7.2.16), is useful in evaluating the performance on low targets, such as buoys or small boats, at short range. When this approximation is used, the quantity to be maximized becomes

$$X' = 1/\lambda^2 L_\alpha$$

Table 7.8 shows values of this quantity for three rainfall rates and three frequency bands. For light rain, K_u-band would outperform the lower bands, but its performance drops catastrophically for medium and heavy rain. X-band is always at least as good as S-band, and is much better in light and medium rain. Rain clutter comparisons favor S-band, with X-band seriously affected and K_u-band almost useless in medium rain. The general choice is for X-band, with S-band also used on ships large enough to mount a second radar.

Table 7.8 Weather Optimization for Shipboard Navigation Radar ($R = 8$ km, low target)

Radar band	S	X	K_u
Wavelength, λ (m)	0.1	0.032	0.02
λ^2 (m^2)	0.01	0.001	0.0004
Attenuation, L_α (dB)			
$r = 3$ mm/h	0.15	1.04	2.4
$r = 10$ mm/h	0.24	3.4	10.0
$r = 30$ mm/h	0.4	10.4	32.0
Index of performance, $1/\lambda^2 L_\alpha$			
$r = 3$ mm/h	100	790	1400
$r = 10$ mm/h	100	460	250
$r = 30$ mm/h	90	90	2

Example of Navigation Radar

The following input parameters will be assumed:

R_m = 30 km λ = 0.03 m

θ_a = 0.009 radian = 0.5° L_n = 1.15

θ_e = 0.25 radian = 14° t_s = 4 s

T_s = 3000 K σF^4 = 10 m^2

$D_0 (1)$ = 20 L_s = 32 = 15 dB

The antenna size is found as

$$w = K_\theta \lambda / \theta_a = 1.15 \times 0.03/0.009 = 3.8 \text{ m}$$

$$h = K_\theta \lambda / \theta_e = 1.15 \times 0.03/0.25 = 0.14 \text{ m}$$

Equation (7.4.12) gives the power requirement:

$$P_{av} = 40 \text{ W}$$

A reasonable waveform for long-range use would be

$$P_t = 32 \text{ kW}, \quad f_r = 2500 \text{ Hz}, \quad \tau = 0.5 \text{ }\mu\text{s}$$

Higher resolution at short range would be obtained by reducing the pulse-width. When mounted 10 m above the sea, viewing a radar reflector mounted on a buoy at $h_t = 3$ m, Figure 7.2.6 indicates $F = 1$ at $R = 8$ km.

7.5 SEARCH RADAR ECM AND ECCM

A stand-off jammer (SOJ), using barrage noise jamming, produces a screened sector in the search radar coverage diagram. The width of this sector depends on the jammer parameters and the azimuth beamwidth of the radar. Within the screened sector there may be some residual capability to detect targets within a *burnthrough range,* but this is seldom operationally useful. The detection range is most easily found by adding to the normal system input temperature T_s the jamming temperature given in (1.4.2):

$$T_j = \frac{P_j \, G_j \, G_r \, \lambda^2 \, F_j^2}{(4\pi)^2 \, kB_j R_j^2 \, L_{\alpha j}} \tag{7.5.1}$$

In the example of Section 1.4, a jammer with ERP of 100 kW screened the penetrating 1-m² target to a range of 4 km.

Adjacent to the screened sector is a broader region of reduced radar performance, in which the skirts of the main lobe and the principal side-lobes still receive large jamming power, but the radar maintains significant range by virtue of the reduced gain toward the jammer: $F_j^2 \ll 1$ in (7.5.1). Beyond that sector, in the region of low sidelobes, the radar coverage is expected to approach its normal range. In the example, for sidelobes of -50 dB, the range reduction was from the original 86 km to 61 km. The resulting radar coverage would appear as in Figure 7.5.1. The assumptions for this figure are:

(a) main-lobe azimuth beamwidth $\theta_a = 1.3°$;
(b) principal sidelobes from -25 dB to -40 dB over the sector from 1.5° to 10° from the main-lobe peak;
(c) sidelobes -50 dB beyond 15°.

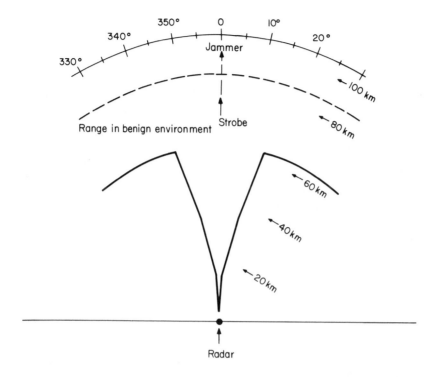

Figure 7.5.1 Azimuth coverage of 2D search radar with one SOJ.

In the region of far sidelobes, the average value of $F_j^2 G_r$ must be less than about -3 dBi (dB with respect to isotropic) for any antenna, and in a low-sidelobe design the value would be less than -15 dBi. The levels used in Figure 7.5.1 correspond to $+15$ dBi for principal sidelobes, 0 dBi at 10°, and -10 dBi beyond 15°. At the azimuth of the jammer, a properly designed military search radar will produce a single *jammer strobe*.

As the noise jammer penetrates the search volume (and becomes an escort or self-screening jammer, ESJ or SSJ), it produces greater noise density in the radar receiver, and also can increase its elevation angle. The problems caused in the radar are similar in nature to those for the SOJ, but detection range on quiet targets will be further reduced. The jamming system, however, faces the problem of increased angle coverage, especially if more than one radar site is to be jammed. Where the SOJ can obtain large ERP with directive antennas, the ESJ and SSJ are restricted in practice to low-gain antennas. They are also more likely to be limited in power, as the large types of aircraft used in the SOJ role are not expected to enter defended airspace. The attempt to maintain high ERP has led to the development of *smart jammers,* which adapt their directive beams to the angles of arrival of the radar illuminations.

Deception jamming is also used against search radar, the purpose being to cause multiple false detections and saturation of the radar data network. The jammer emission will be an attempt to replicate the radar echo signal, but with offsets in range delay, doppler shift, and angle of the jammer platform. This type of jamming, having the duty factor of the radar transmission, may be generated with much higher peak power than can continuous noise jamming. Deception jamming is most effective when it produces false detections at angles far removed from the true jammer azimuth, since large numbers of detections at one azimuth may be no more effective than generation of a noise jamming strobe.

Chaff is used both as a screening and deception technique. Its effectiveness in screening is that of artificial rain, and this has been discussed as an inherent radar environmental problem. As a deception technique, it can best be considered as an artificial point clutter source.

Search Radar ECCM

The most important approach to the ECM problem is netting of search radars, which constitutes spatial diversity. The noise jammer can be located by triangulation on the azimuth strobes from two radars. Penetrating, quiet targets masked by the jammer from one radar should be detectable by a second radar, if low sidelobes and narrow azimuth beams are used. However, there are other ECCM approaches which become increasingly important when more than one SOJ is present.

(a) *Effects of Lobing.* The reflection lobes shown in Figure 6.2.4 apply to jammers as well as targets. For a given jammer elevation angle, high enough in elevation, there will be one or more frequencies at which $F_j^2 \to 0$, even when the azimuth beam is directed at the jammer (Figure 7.5.2). High siting of the radar, above a surface of high reflection coefficient, combined with the use of horizontal polarization and tuning ability over a wide bandwidth, will ensure that the radar can tune to a hole in the frequency spectrum of each jammer. If there is only one significant jammer at each azimuth, and if radar frequency can be changed from one azimuth to another (e.g., for each t_o or t_o/m diversity burst), low F_j^2 can be maintained on a number of jammers. The SOJ will normally fly high enough to obtain good propagation at the lowest search radar band in use, so band diversity among sites will force the SOJ to be high enough to ensure that multiple lobes occur in the upper bands. Note that pulse-to-pulse frequency agility is not required, making this technique compatible with MTI and other doppler processes.

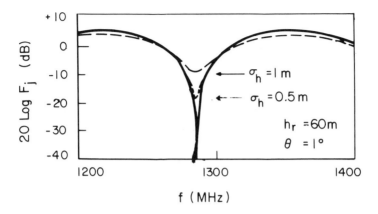

Figure 7.5.2 Jamming spectrum received over path with surface reflections.

(b) *Sidelobe Cancelation (SLC)*. Against one or a few significant
jammers, the coherent adaptive sidelobe canceler (Figure 7.5.3) can be
used. At least one loop per jammer is required. The generic circuit shown
uses an auxiliary antenna with a broad pattern which covers the envelope
of the main antenna sidelobes. The correlator at the output (a multiplier
followed by a narrowband filter) looks for a component of the main channel
output, which correlates with the output of the auxiliary channel. The
correlator output is used to change the weight of the coupling from the
auxiliary to the main channel, until there is no longer any correlated output.
The combined pattern of the antennas then has a null at the jammer
location, and the null moves automatically as the main antenna scans its
sidelobes across the jammer position. Multiple loops can form multiple
nulls. Limitations of the SLC technique are as given below.

(1) It is effective primarily against jamming with high duty factor.
If gain and correlator bandwidth are made high enough to form a null
on jamming of low duty factor, the loop may tend to null the desired signal
as well, unless the auxiliary antenna pattern has a null aligned with the
main lobe of the main antenna.

(2) When multipath lobing is present, the auxiliary antenna may lie
in a propagation null, and the loop gain will then be insufficient to form
the SLC null. Use of multiple auxiliary antennas and loops can overcome
this problem.

(3) The adding of the auxiliary pattern to the main pattern can increase the general sidelobe level while it forms the null at a specific location.
If the main antenna has low sidelobes which permit most jammers to be

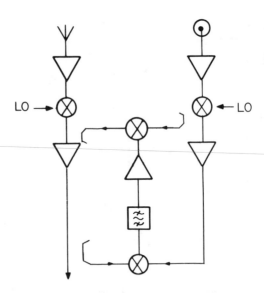

Figure 7.5.3 Block diagram of sidelobe canceler system.

ignored, the implementation of an SLC loop on one strong jammer may expose the radar to the other jammers, requiring more SLC loops to cancel the sidelobes.

(c) *Sidelobe Blanking (SLB)*. The sidelobe blanker, for jamming of low duty factor, performs the same function as the SLC for high duty factor jamming, and often uses the same auxiliary antenna and receiver preamplifier. A block diagram of a combined SLC-SLB system is shown in Figure 7.5.4. A signal from the main lobe will exceed that from a sidelobe by a large factor. Although a strong jammer in the sidelobe can appear larger than a main-lobe echo signal, the jammer will also produce a large signal in the auxiliary channel. By comparing the ratio of auxiliary to main channels, a blanking control signal may be generated. The main receiver loses only signals arriving during the blanked interval, which is assumed to have low duty factor. For a coherent burst, the entire burst must be blanked, in the affected range-azimuth cell, to prevent clutter from being spread into the doppler passband. A continuous jammer, strong enough to meet the blanking criterion, can continuously turn off the receiver. A threshold must be included to prevent this occurrence. Sidelobe blankers are necessary, even in civil radars, to eliminate accidental interference.

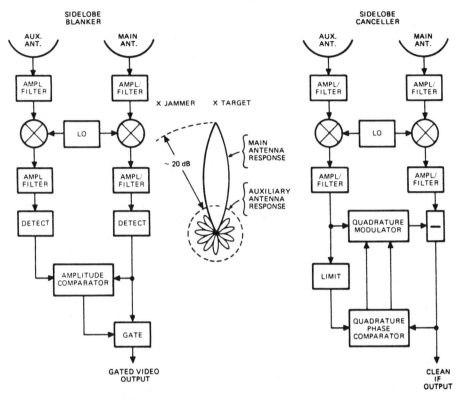

Figure 7.5.4 Combined sidelobe canceler and blanker.

(d) *Frequency Diversity.* The noise jammer attempts to increase its density J_0 by intercepting each radar and concentrating its power over the frequencies actually in use (*spot jamming*). Frequency diversity (with simultaneous transmissions in each pulse, or diversity from burst to burst) makes spot jamming difficult, especially in a netted radar system. Spot jamming cannot normally use $B_j < 10$ MHz, and when the number of 10-MHz channels reaches the point that 25% of the radar tuning band must be jammed, there is little benefit in performing the look-through intercept function (which requires cessation of jamming during the receiving interval).

(e) *Frequency Agility.* The use of random pulse-to-pulse agility has been advocated as an ECCM technique to prevent spot jamming. It is certainly effective against active jammers of most kinds (excluding barrage noise jamming), but can only be used when doppler processing is not needed for clutter rejection. Widespread use of chaff, as well as natural clutter environments, rule out the use of frequency agility in most cases. Burst-to-burst frequency diversity is used instead, but the length of the coherent MTI or doppler burst is such that a fast set-on jammer can screen the final pulses of the burst, which are needed for processing. Interleaved bursts at different frequencies can appear as random frequency agility, but these actually represent parallel diversity channels, in the same way as pulse trains in which each pulse has subpulses at different frequencies.

(f) *Deceptive Transmissions.* If the smart jammer relies on his intercept receiver to control the jamming emissions, a smart radar can assume control of the jamming. During the look-through interval (or continuously, if necessary), radar frequencies can be transmitted which are not needed nor used in radar processing. In a low-sidelobe radar, these transmissions can be through a broad auxiliary beam, or one with higher sidelobes than those of the radar antenna, reducing the power requirements on the deceptive transmissions. If the intercept and control logic of the jammer are known, these deceptive transmissions can lead most of the jammer power to portions of the spectrum not received by the radar, and can possibly hold the real radar channel open for signals. Faced with this prospect, the jammer must resort to barrage jamming, which in a smart jammer is even less efficient than in a brute-force barrage jammer. Note that low transmitting sidelobes are required if the deceptive approach is to be efficient, and the main-lobe transmissions are impossible to mask unless they are deliberately reduced in power. A multifrequency transmission with many low-power channels forces jamming of all channels, in case they are the ones being processed by the radar.

(g) *Special Receiver Fixes.* There are many circuits used in receivers to reduce vulnerability to specialized jammers. These are discussed in [7.8–7.10]. When properly implemented, their effect is to force the jammer into reliance either on broadband random noise, against which the radar processing has performance bounded by the matched filter, or on replicas of the radar signal.

3D Search Radar ECCM

There are ECCM features that are unique to 3D radars, and others which are less effective for 3D than for 2D radars. For example, exploitation of lobing is applicable to 3D radars only in the lowest beam, and then only when the antenna phase center height is much greater than the vertical dimension of the aperture $h_r \gg h$. For $h_r \approx h$, there will be only one null in the pattern, near the center of the lowest beam. Very large changes in frequency are needed to move this null appreciably, and jammers at most elevation angles cannot be nulled at any point in the radar tuning band. The techniques of sidelobe cancelation and blanking can be applied to 3D radars, but in the stacked-beam systems this leads to great complexity. The sidelobe blanker can be simply applied if a single auxiliary antenna pattern envelops the entire vertical coverage. Parallel frequency diversity (as opposed to burst-to-burst methods) also leads to greater complexity in stacked-beam systems. These are examples of the inherent conflict between clutter rejection performance (which favors stacked beams) and ECCM against active jamming.

The inherent advantage of 3D radar in a benign environment is its reliance on larger aperture A_r to reduce the required transmitter power. This is not effective in the jamming environment. The jamming noise temperature, (7.5.1), varies directly as

$$G_j F_j^2 = C A_r F_j^2$$

Unless large A_r is accompanied by lower relative sidelobes (constant with respect to isotropic gain, or in dBi), there will actually be increased T_j and J_0. Sidelobes produced by random errors in the antenna (reflector or phased array) will remain at constant level in terms of dBi, making the larger aperture irrelevant to sidelobe jamming level. However, if P_{av} has been decreased in consequence of the larger A_r, the jamming vulnerability is increased relative to a 2D radar with the same performance in a benign environment. Use of larger P_{av} in 3D radars gives a potentially large R_m in the benign environment, and this creates a strong interest in transmission of low PRFs, compromising the clutter rejection capability. Military 3D radars should include ECCM modes having higher PRFs and reduced unambiguous ranges, matched to those realistically expected in the ECM and clutter environments.

The unique ECCM capabilities of 3D radars include the following.

(a) *Elevation Resolution.* The presence of several elevation beams reduces the solid angle which is subject to main-lobe jamming. Depending on the jammer ERP, an elevation sector of about $2\theta_e$ can be screened at a level within 12 dB of the maximum J_0. Elevation sidelobes are more difficult to control than those in azimuth, and sidelobe jamming may be only 20 to 30 dB below this maximum J_0.

(b) *Elevation Null Steering.* The multiplicity of beams in the elevation coverage provides combinations which can preserve much of the elevation coverage, even at the azimuth of a jammer. In scanning beam systems,

SLC can be applied in the elevation plane. With stacked beams, adaptive nulling that uses adjacent high-gain beams can place deep nulls, even within the elevation beamwidth, limiting the screened sector to a fraction of the beamwidth. The beam-forming network must provide low dispersion between the antenna aperture and the point in the receiving system at which the beams are combined to form the null, to preserve it over the signal bandwidth.

(c) *Pseudorandom Scan.* The use of randomized scan patterns can decrease vulnerability to certain deceptive jamming techniques which take advantage of predictable scan behavior.

Summary of Search Radar ECCM

A properly designed military search radar will use antenna, wavelength, waveform, and signal processing steps which simultaneously reject clutter and minimize active ECM, while retaining targets and holding the false alarm rate within the constraints set by the using data system. In the absence of design defects, robust ECM techniques must be used against this radar: barrage noise jamming and chaff. The performance of a search radar against the combination of these two ECM techniques is necessarily degraded, relative to its capability in the benign environment, and only the use of netted radars, having considerable overlap in the benign environment, can preserve search coverage in the presence of a concerted ECM attack.

In evaluating the ECCM performance of a specific search radar, the objective of that radar in the military system is important to keep in mind. For example, if only warning of an attack is required of the radar, the presence of ECM will ensure success, even when the radar cannot detect any targets. If identification of the attack azimuth is required, the presence of jam strobes or a chaff corridor within an azimuth sector may be sufficient, if the other azimuths are found to be clear of both ECM and hostile targets. If the search radar is required to designate identifiably hostile targets to an associated tracking radar, the presence of one or a few jam strobes may be sufficient, if they correspond to targets within engagement range. Hence, to be successful, the attacking force must create jamming strobes and chaff over a wide sector, such that the actual attack corridor remains unknown to the defensive system, and to the extent that the limited number of jammers cannot be successfully engaged. The use of stand-off jammers and stand-off chaff clouds, which can screen quiet penetrating targets, is most effective. Jamming is also most likely to be successful when the radar is required to produce reliable track files on a large number of targets, without cooperation from other radars in a network.

REFERENCES

[7.1]   IEEE Standard Dictionary of Electrical and Electronics Terms, *ANSI/IEEE Std. 100-1984,* 1984.

[7.2]   E. Brookner, *Radar Technology,* Chapter 3, Cumulative Probability of Detection, Artech House, 1977.

[7.3]   H.R. Ward, C.A. Fowler, and H.I. Lipson, GCA radars: Their history and state of development, *Proc. IEEE* **62**, No. 6, June 1974, pp. 705–716.

[7.4]   W.W. Shrader, Radar technology applied to air traffic control, *IEEE Trans.* **COM-21**, No. 5, May 1973, pp. 591–605.

[7.5]   E. Giaccari and G. Nucci, A family of air traffic control radars, *IEEE Trans.* **AES-15**, No. 3, May 1979, pp. 378–396.

[7.6]   L. Cartledge and R.M. O'Donnell, Description and Performance Evaluation of the Moving Target Detector, *MIT Lincoln Laboratory Project Rep. ATC-69,* March 8, 1977.

[7.7]   S.A. Hovanessian, *Introduction to Synthetic Array and Imaging Radars,* Artech House, 1980.

[7.8]   D.C. Schleher, *Introduction to Electronic Warfare,* Artech House, 1986.

[7.9]   S.L. Johnston (ed.), *Radar Electronic Counter-Countermeasures,* Artech House, 1979.

[7.10]   L. Van Brunt, *Applied ECM,* Vol. 2, EW Communication, 1982.

Chapter 8

Angle Measurement and Tracking

A *tracking radar* is one in which the primary function is the automatic tracking of targets. This radar is distinguished from surveillance or track-while-scan radar in that the tracking radar beam responds to presence of the target and its motion, following it as part of the automatic tracking loop. Often there will also be range gates and doppler filters that move with the target, but the ease with which multiple gates and filters are produced by digital technology makes possible tracking in these coordinates on the outputs of the signal processor having fixed gates and filters, in much the same way as is done in angle with surveillance radar. This chapter will be concerned with antennas and angle measurement techniques, primarily for tracking radar, but also used in surveillance radar systems.

8.1 MEASUREMENT FUNDAMENTALS

Before a target is measured or tracked, it must be resolved from its surroundings, detected, and its signal selected as the subject for measurement. Some degree of measurement is implicit in detection, since coarse position data are provided by the identity of the resolution cell or cells that have given the alarm. The measurement process, however, consists not only of indicating the cell containing the target, but also interpolating among adjacent cells to improve the precision of the data. The basic process by which this is done is the comparison of signals in adjacent cells, and estimation of a target position based on relative amplitude (or phase) of these signals. Figure 8.1.1 shows the process in an arbitrary coordinate z, which may represent angle, time delay, or frequency. For example, two symmetrical responses may be offset by a distance $\pm z_k$ from a measurement axis z_0:

377

$$f_1(z) = f(z_1) = f(z_0 + z_k)$$
$$f_2(z) = f(z_2) = f(z_0 - z_k)$$

$$(8.1.1)$$

The two responses may be generated sequentially or simultaneously. At the axis, $f_1(z_0) - f_2(z_0)$, and this equality of response serves as an indication that the target signal lies on the measurement axis.

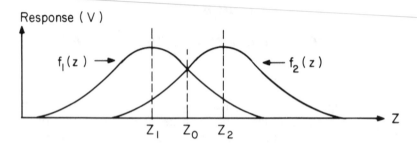

Figure 8.1.1 Basic measurement process.

To obtain a signal indicating an error in target position relative to the axis, it is convenient to form the difference between the two responses:

$$\Delta(z) = f_1(z) - f_2(z)$$

$$(8.1.2)$$

This response, shown in Figure 8.1.2a, has an S-shape passing through zero on the axis. For offsets z_k that are not too large, we may write

$$\Delta(z) \approx 2z_k f'(z_0) = 2z_k \, df/dz$$

$$(8.1.3)$$

Positive Δ indicates a target $z < z_0$, and negative $z > z_0$. However, since the magnitude of Δ depends on the strength of the target signal as well as on its location, Δ cannot show a target location except on the axis. Even there, $\Delta = 0$ is ambiguous, indicating either $z = z_0$, or $z - z_0 \ll -z_k$, or $z - z_0 \gg z_k$. To obtain an actual interpolated location, we must form another response (Figure 8.1.2b):

$$\Sigma(z) = f_1(z) + f_2(z)$$

$$(8.1.4)$$

and then observe the *normalized difference*, or error signal response:

$$\Delta/\Sigma = \frac{f_1(z) - f_2(z)}{f_1(z) + f_2(z)} \approx \frac{z_k \, df/dz}{f(z_0)}$$

$$(8.1.5)$$

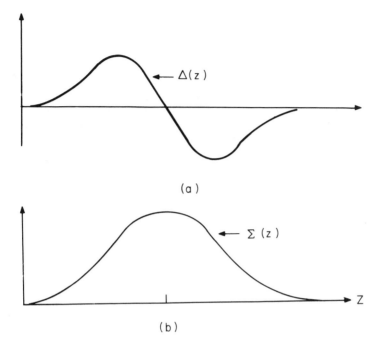

(a)

(b)

Figure 8.1.2 Response of Δ and Σ channels.

The response $f(z)$ can be controlled to make this error signal linear near $z = z_0$, and to avoid ambiguities as long as Σ is above a given detection threshold.

Measurement Sensitivity

After a further normalization of z by z_3, the -3-dB width of $\Sigma(z)$, we can obtain a dimensionless measurement slope with a magnitude of

$$k_z = -\left.\frac{d(\Delta/\Sigma)}{d(z/z_3)}\right|_{z=z_0} \tag{8.1.6}$$

For small z_k, (8.1.3) can be used to express this slope as

$$k_z = -\left.\frac{2z_k z_3}{\Sigma(0)}\frac{d^2 f}{dz^2}\right|_{z=z_0} \tag{8.1.7}$$

The sensitivity of the system to target deviations from the axis is therefore proportional to the negative curvature at the peak of the original channel response $f(z)$, which also gives the curvature of the Σ response. It has been shown [8.1] that all of the usual response shapes (antenna patterns and waveform ambiguity functions) have the property that z_3 is reciprocally related to the central curvature, restricting the normalized slope k_z to values between about 1 and 2 in practical measurement systems.

In the presence of interference, with a relatively strong signal on the axis, spurious error signals Δ_i are generated (Figure 8.1.3). The apparent error z_i corresponding to this error signal is given by

$$z_i/z_3 = -\Delta_i/\Sigma k_z \tag{8.1.8}$$

A tracking system will shift the axis to place the target at $-z_i$, generating a target error signal $\Delta_t = k_z \Sigma z_i/z_3$ and restoring the null in error signal. Where Δ_i is a randomly varying component, the error signal will vary symmetrically about the axis with an rms error corresponding to a position error:

$$\sigma_{z_1} = \frac{z_3}{k_z \sqrt{2S/I_\Delta}} \tag{8.1.9}$$

Here σ_{z_1} applies to a single interference sample appearing in the Δ channel with power Δ_i, and S is the signal power in the Σ channel. The factor $\sqrt{2}$ arises because the total interference power consists of equal in-phase and quadrature components. For $S/I \gg 1$, where I is the interference in the Σ channel, only the in-phase component affects the detected error signal (if a phase-sensing detector is used, only the quadrature component is effective, but never both). The appearance of I_Δ in this expression dictates low-sidelobe difference responses in most applications where high precision is needed.

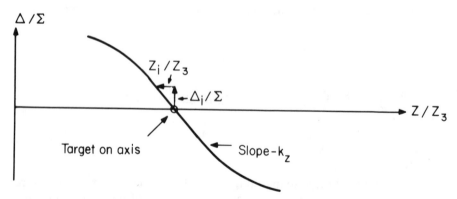

Figure 8.1.3 Response of error detector to interference.

When the errors are averaged over a period of time in which n_e independent samples of interference are received, the output error becomes

$$\sigma_z = \frac{z_3}{k_z \sqrt{2(S/I_\Delta)n_e}} \qquad (8.1.10)$$

Equation (8.1.10) is a fundamental relationship which will be applied to errors in each radar coordinate, from all types of interfering sources having varying amplitude or phase. This expression must be modified to include a second-order term as S/I becomes near or below unity.

Optimum Estimators for the Noise Environment

The theory of the optimum estimator has been extensively discussed [8.1–8.3]. In each radar coordinate, the response is given by the Fourier transform of the energy distribution in a transform coordinate, as shown in Table 8.1.

Table 8.1 Basic Measurement Relationships

Measured Coordinate	Equation for rms Error	Resolution Element	Radar Parameter in Transform Coordinate	rms Spread in Transform Coordinate
Angle, θ(rad)	$\sigma_\theta = \dfrac{\lambda}{\mathcal{L}\sqrt{\mathcal{R}}}$	Beamwidth, θ_3	Aperture width, w/λ (wavelengths)	rms aperture width, $\mathcal{L}/\lambda$
Delay time, t_d(s)	$\sigma_t = \dfrac{1}{\beta\sqrt{\mathcal{R}}}$	Effective pulse duration, τ_3	Bandwidth of signal, B_{3a}	rms bandwidth of signal, β
Frequency, f (Hz)				
Coherent case	$\sigma_f = \dfrac{1}{\alpha\sqrt{\mathcal{R}}}$	Spectral line width, B_{3a}	Observation time, t_o	rms signal duration, α
Noncoherent case	$\sigma_f = \dfrac{1}{\alpha_1\sqrt{\mathcal{R}}}$	Bandwidth of signal, B_{3a1}	Pulse duration, τ_3	rms pulse duration, α_1

$\mathcal{R} = 2E/N_0$

The curvature (second derivative) of the response peak is given by the second moment of the energy distribution in the transform coordinate, properly normalized. Thus, for example, in measurement of time delay, the curvature of the matched filter output is given by

$$\beta^2 = \frac{\int_{-\infty}^{\infty} (2\pi f)^2 |A(f)|^2 \, df}{\int_{-\infty}^{\infty} |A(f)|^2} \qquad (8.1.11)$$

where $A(f)$ is the voltage spectrum of the signal and $\int (2\pi f)^2 |A(f)|^2 \, df$ is the second moment of the energy spectrum. The parameter β has the dimension of bandwidth, and is known as the *rms bandwidth* of the signal.

For frequency measurement, the rms time spread of the signal is similarly defined, with the waveform $a(t)$ replacing $A(f)$, and t replacing f in the integration. For angle measurement, an rms aperture width $\mathcal{L}$ is defined in the aperture coordinate x, and $\mathcal{L}/\lambda$ determines the curvature of the beam pattern $f(\theta)$. With a given limit band B on $A(f)$, time limit τ on $a(t)$, and physical dimension limit w on aperture width, the corresponding optimum measurement slopes can be determined. The optimum precision of measurement is then given in time delay by

$$(\sigma_t)_{min} = 1/\beta_0 \sqrt{2E/N_0} = \sqrt{3}/\pi B \sqrt{2E/N_0}$$
$$= 1/1.81B \sqrt{2E/N_0} = \tau_3/1.61 \sqrt{2E/N_0} \qquad (8.1.12)$$

in frequency by

$$(\sigma_f)_{min} = 1/\alpha_0 \sqrt{2E/N_0} = \sqrt{3}/\pi\tau \sqrt{2E/N_0}$$
$$= 1/1.81\tau \sqrt{2E/N_0} = B_3/1.61 \sqrt{2E/N_0} \qquad (8.1.13)$$

and in angle by

$$(\sigma_\theta)_{min} = \lambda/\mathcal{L}_0 \sqrt{2E/N_0} = \sqrt{3}\lambda/\pi w \sqrt{2E/N_0}$$
$$= \lambda/1.81w \sqrt{2E/N_0} = \theta_3/1.61 \sqrt{2E/N_0} \qquad (8.1.14)$$

The rms values $\beta_0 = 1.81B$, $\alpha_0 = 1.81\tau$ and $\mathcal{L}_0 = 1.81w$ are those for uniform energy distribution over B, τ, and w, respectively. This distribution gives the minimum 3-dB widths $\tau_3 = 0.886/B$, $B_3 = 0.886/\tau$, and $\theta_3 = 0.886\lambda/w$, but also gives high sidelobes that are not generally acceptable in operational systems. The sidelobes are not of concern in optimum estimation theory for the noise environment.

The expressions for optimum estimation in noise can be compared with (8.1.9). The signal-to-interference ratio $(S/I_\Delta)_{ne}$ can be replaced by

$(S/N)n = E/N_0$, corresponding to reception with a matched filter. The Σ channel noise can be used in the equation because all receiver channels are assumed to have equal noise. The normalized error slope k_z is then equal to $1.81 \times 0.886 = 1.61$ in all coordinates, for optimum estimators.

Sequential and Simultaneous Lobing

Implementation of Figures 8.1.1 and 8.1.2 for angle measurement involves forming beam patterns $f_1(\theta)$ and $f_2(\theta)$, combining them to obtain $\Delta(\theta)$ and $\Sigma(\theta)$, and determining the error signal Δ/Σ. This error signal may be smoothed and used as data, or fed back directly to control the beam position through an antenna servo or beam-steering computer. Although the Δ and Σ symbology may seem to imply monopulse implementation, it applies equally to all measurement and tracking systems. In Figure 8.1.4a, the two beams are formed sequentially, and signals from each are stored over one or more complete switching cycles, after which Δ, Σ, and Δ/Σ are computed from the stored signal voltages. A common type of analog process, Figure 8.1.4b, uses an automatic gain control (AGC) loop with long time constant to perform the normalization, and to extend the dynamic range of signals accommodated by the receiver. In Figure 8.1.4c, the beams exist simultaneously and signals are fed to an RF network, which forms Δ and Σ. Normalization is again performed by an AGC loop, closed on the Σ receiver and giving open-loop control to the Δ channel. If the two beams in (a) and (b) are displaced by $\theta_k = 0.3\theta_3$ (crossing at the -1 dB level), the sequential Δ response is indistinguishable from the response of an optimum monopulse Δ pattern (the derivative of the Σ pattern), as shown in Figure 8.1.5. Even the sidelobe responses are almost identical.

8.2 CONICAL SCAN

The most common form of sequential lobing is *conical scan*, in which a beam is offset from the antenna axis and rotated about the axis at a scan frequency f_s. A block diagram of such a radar is shown in Figure 8.2.1. The provisions for display, synchronization, transmitter, receiver, and duplexing equipment are those which characterize most pulsed radars. The tracking features consist of the mechanically driven scanner, the range-gated automatic gain control (AGC), the error detector, and the two servo channels. The scanner may take the form of a rotating dipole feed, electrically unbalanced to cause an offset in the beam relative to the mechanical axis; a nutating feed device, which maintains constant polarization; or a lens or metal plate, which rotates in front of a fixed feed to provide displacement of the beam. The maximum rate of rotation is limited by the

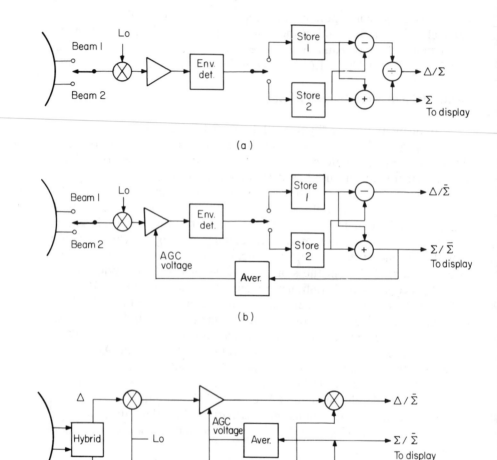

Figure 8.1.4 Sequential and simultaneous lobing systems.

radar PRF, through the requirement for at least four pulses per scan cycle. Scan rates from 1/10 to 1/100 of the PRF are commonly used. The tracking radar display unit provides a full-range presentation of received signals, but the AGC and error-detector channel are controlled by a range gate to operate only during reception of the signal from the selected target. The AGC then maintains constant amplitude of the selected signal, averaged over a period which exceeds the scan cycle, providing consistent performance of the tracking loops, regardless of target size and range.

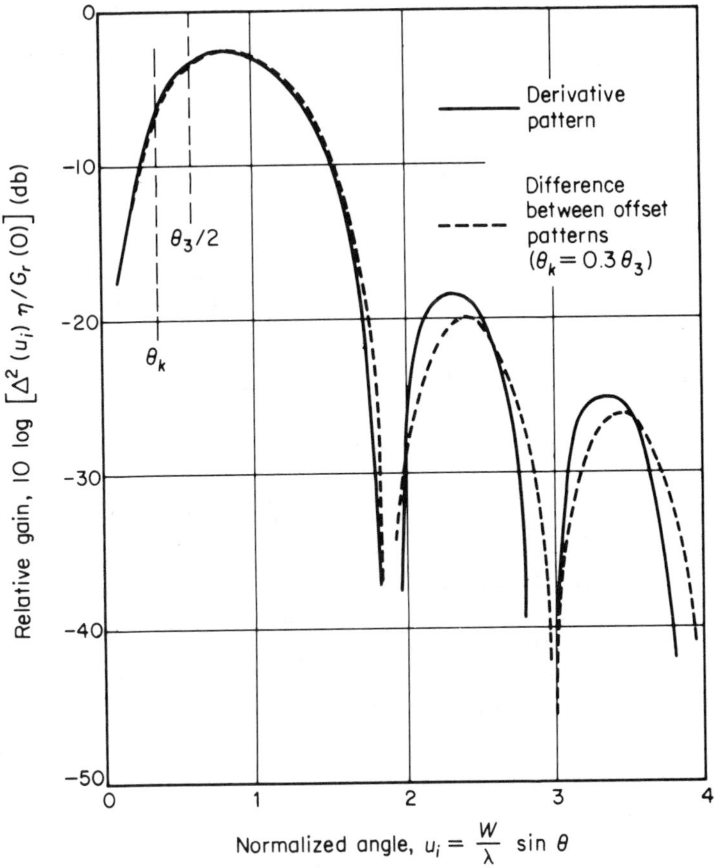

Relative gain, 10 log $[\Delta^2(u_i)\ \eta/G_r(0)]$ (db)
0
-10
-20
-30
-40
-50
$\theta_3/2$
θ_k
Derivative pattern
Difference between offset patterns
$(\theta_k = 0.3\,\theta_3)$
0
1
2
3
4
Normalized angle, $u_i = \frac{W}{\lambda} \sin\theta$

Figure 8.1.5 Similarity of conical-scan and monopulse Δ patterns.

The error detector consists of an envelope detector, which is sensitive to the scan-rate components of signal modulation, and rejects the dc level and components at and above the PRF. The ac output at the scan frequency can be calibrated in terms of off-axis error angle, and its phase relative to the reference-generator voltage indicates the direction of the error. The error demodulator resolves the error voltage into elevation and traverse components, which control the two servos. In an azimuth-elevation pedestal, the traverse error voltage is multiplied by a gain function, approximating the secant of elevation angle, in order to drive the azimuth servo with constant loop gain as the target nears zenith. This is needed because the traverse error signal represents the angle along a great circle passing

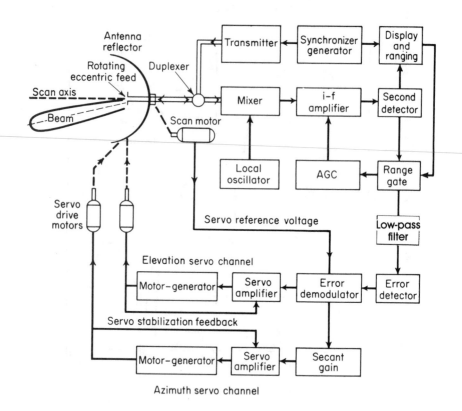

Figure 8.2.1 Block diagram of conical-scan radar.

through the beam axis, while azimuth is the projection of this angle in the horizontal plane.

The relationships between the scanning beam, the target, and the error voltage in a conical-scan system are shown in Figure 8.2.2. The received pulse train is held at an average amplitude E_0 by the AGC loop. The envelope of the instantaneous pulse amplitude then carries the error information in the form of sinusoidal modulation at the scan frequency f_s.

Error Slope in Conical Scan

In conical-scan radar, the individual pulse cannot provide angle data, but (8.1.10) can be applied to find the error caused by thermal noise over the n pulses integrated by the servo loop. Only half of these pulses are

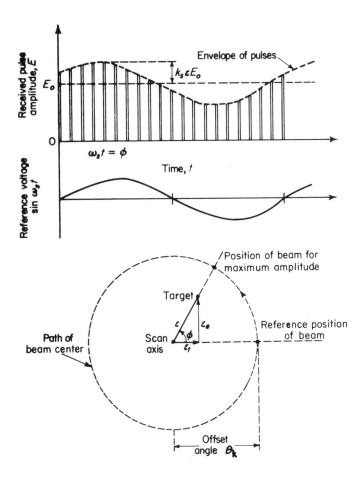

Figure 8.2.2 Error detection in conical-scan radar.

effective in measurement of azimuth, and half in elevation, so the equation for error in each coordinate becomes

$$\sigma_\theta = \frac{\theta_3}{k_s\sqrt{E/N_0}} = \frac{\theta_3}{k_s\sqrt{(S/N)n}} = \frac{\theta_3\sqrt{L_k}}{k_s\sqrt{(S/N)_m n}} \qquad (8.2.1)$$

where k_s is the conical-scan error slope, $(S/N)_m = (S/N)/L_k$ is the single-pulse SNR that would have been received if the beam were directed at the target, and L_k is the *crossover loss* (loss due to beam squint).

The conical-scan error slope k_s and the crossover loss are shown in Figures 8.2.3 and 8.2.4, as a function of squint angle, for both one-way and two-way beam patterns. Data for Gaussian and $(\sin x)/x$ beams are shown, the two curves diverging only for large squint angles. The effective slope $k_s/\sqrt{L_k}$ is maximum at $\theta_k \approx 0.45$ for the two-way case, but the optimum point for system operation gives reduced loss with little reduction in slope at $\theta_k \approx 0.33$, $k_s/\sqrt{L_k} = 1.4$.

For the optimized squint angle with a two-way beam, (8.2.1) becomes

$$\sigma_\theta = \frac{0.7\theta_3}{\sqrt{(S/N)_m n}} \tag{8.2.2}$$

Scintillation Error in Conical Scan

When the target fluctuates during the scan cycle, the error detector of Figure 8.2.1 receives spurious modulation components, which are interpreted as target deviations from the axis. For a power spectral density of target fluctuation $W(f)$, expressed as (fractional modulation)^2/Hz, the ratio of signal to error power within the servo bandwidth will be

$$(S/I_\Delta)n = 1/2\beta_n W(f_s)$$

The corresponding *scintillation error* in the track will be

$$\sigma_s = (\theta_3/k_s)\sqrt{\beta_n W(f_s)} \tag{8.2.3}$$

The rms fractional modulation for a multiple-point target, having a Rayleigh amplitude distribution, is 0.5, giving a total area under the $W(f)$ curve of 0.25. The zero-frequency spectral density can then be expressed in terms of the correlation time t_c' of the signal envelope:

$$W_0 = t_c'/2 \tag{8.2.4}$$

The envelope correlation time is approximately equal to the signal correlation time t_c used in Section 2.4, but may differ by a factor of two depending on the form of the correlation function [8.1, p. 174]. For $f_s t_c' \gg 1$, with exponentially correlated fluctuations, this results in

$$W(f_s) \approx 1/2\pi^2 t_c' f_s^2 \tag{8.2.5}$$

The corresponding scintillation error in a conical-scan radar is

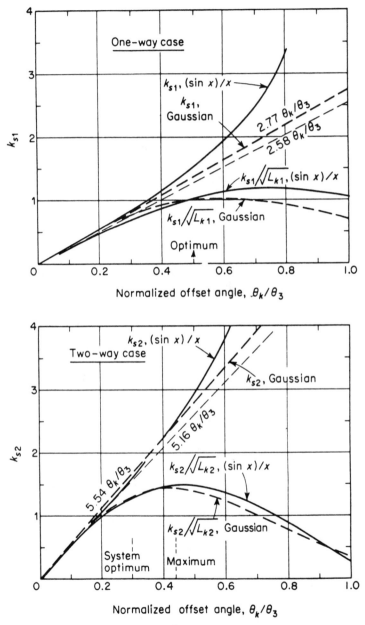

Figure 8.2.3 Conical-scan error slopes.

$$\sigma_s = (\theta_3/k_s)\sqrt{\beta_n/2\pi^2 t_c' f_s^2} = 0.225(\theta_3/f_s k_s)\sqrt{\beta_n/t_c'} \qquad (8.2.6)$$

As an example, assume a target with an envelope correlation time $t_c' = 0.06$ s. This corresponds to an envelope fluctuation spectrum with a half-power bandwidth of 5 Hz. If the scan frequency $f_s = 30$ Hz and the servo bandwidth $\beta_n = 2$ Hz, we have, for optimized squint $\theta_k = 0.33\theta_3$,

$$\sigma_s = 0.025\theta_3$$

This level of error, near $\theta_3/40$, is typical of conical-scan radar in tracking aircraft targets. Note that the scintillation error depends on k_s, and not on L_k, indicating that somewhat greater squint angles are appropriate to minimize scintillation error.

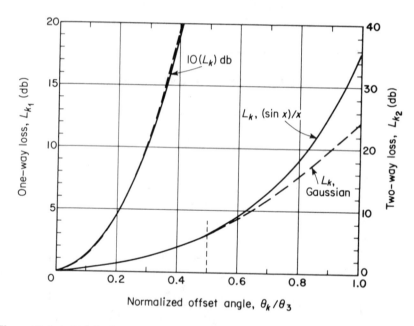

Figure 8.2.4 Conical-scan crossover loss.

8.3 SECTOR SCAN

A sector scanning beam can obtain angle data on targets within its scan sector, and a pair of beams can be used as in Figure 8.3.1 to track multiple targets in both angle coordinates. The scan field may remain fixed, in which case a track-while-scan (TWS) operation will be conducted, or

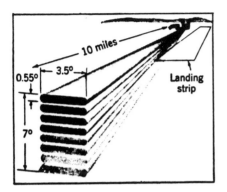 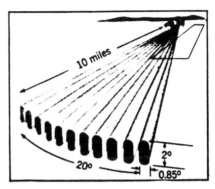

Figure 8.3.1 Antenna scan coverage of typical precision approach radar.

the field may move to follow targets of interest. Angle data on individual targets are provided as offset angles from the center of the scan field.

Measurement of Target Angle

Angle estimates within the scan sector are made by locating the centroid of the received signal envelope (Figure 8.3.2a). The signal from the receiver is the envelope-detected pulse train, modulated by the two-way pattern of the antenna, which scans at a rate ω_a. Were it not for noise disturbances, the centroid could be estimated simply by thresholding the pulses and averaging the times at which the first pulse went above the threshold and the last pulse fell below the threshold. The pulse train is integrated (Figure 8.3.2b) to reduce the effects of noise on both detection and measurement. The times of upward and downward threshold crossings of the integrated output are averaged to locate the centroid (with a delay equal to the integrator time constant). The pulses to be processed must be selected in a range gate before thresholding. When targets over any intervals in range are to be measured, multiple range gates are needed. The centroid estimator gives the time at which the antenna was pointed at the target, and a scan angle encoder reads the beam position continuously to permit that time to be converted to an angle relative to the reference coordinate system.

A closed-loop tracking estimator uses *angle gates* as shown in Figure 8.3.2c. Range-gated pulses from the target of interest are integrated with positive polarity in the first half of the gate, and negative pulses in the second half (Figure 8.3.2d). When the gate is centered on the envelope of the pulse train, the output is zero. The error sensing curve, plotted as

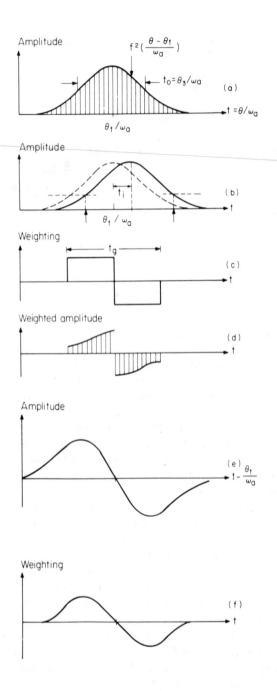

Figure 8.3.2 Centroid estimation on output of a scanning antenna.

a function of target displacement from the center of the gate, appears as in Figure 8.3.2e. The performance of the angle gate estimator can be described by the slope k_p of this error sensing curve, properly normalized for use in the equation for thermal noise:

$$\sigma_\theta = \frac{\theta_3}{k_p\sqrt{2E/N_0}} = \frac{\theta_3\sqrt{L_p}}{k_p\sqrt{2(S/N)_m n}} \tag{8.3.1}$$

where E/N_0 is the ratio of energy in the pulse train to noise spectral density, L_p is the beamshape loss, $(S/N)_m$ is the SNR for the central pulse in the train, and $n = t_o f_r$ is the number of pulses between the half-power points of the one-way beam.

An optimum estimator [8.1, p. 34] would integrate the pulse train with weights proportional to the derivative of the two-way antenna pattern. This would give $k_p/\sqrt{L_p} = 1.4$, for which case,

$$\sigma_\theta = \frac{0.5\theta_3}{\sqrt{(S/N)_m n}} \tag{8.3.2}$$

Equation (8.3.2) is equally valid for Gaussian and $(\sin x)/x$ beams, and for one-way and two-way scanning beam patterns.

When a single transmitter and receiver are time-shared between beams scanning the two coordinates, the energy from the target is divided between the two measurements. Thus, if we take n as the total number of pulses received from the target, we will have

$$\sigma_\theta = \frac{0.7\theta_3}{\sqrt{(S/N)_m n}} \tag{8.3.3}$$

This result is identical to that for a conical-scan system, (8.2.2). In comparing a two-coordinate sector scan system with conical scan, however, we must take into account the broader scan field of the former. The total number of pulses transmitted to cover a sector $\Delta_a \times \Delta_e$ will be

$$n_b = (\Delta_a/\theta_a) + (\Delta_e/\theta_e)$$

Hence, for the same transmitted power and antenna gain, E/N_0 will be greater by $n_b/2$ for the conical-scan system.

The use of rectangular angle gates as an approximation of the optimum estimator causes an increase in error. This can be described as in Figure 8.3.3, in terms of a decrease in k_p relative to its optimum value, or as a corresponding matching loss which reduces $(S/N)n$.

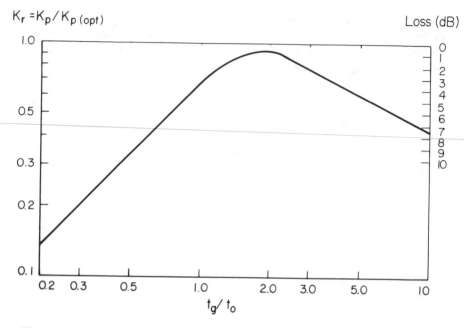

Figure 8.3.3 Measurement slope for rectangular angle gates.

Scintillation Error

Target fluctuation during the time-on-target disturbs the envelope of the pulse train, as shown in Figure 8.3.4. The resulting scintillation error can be calculated, for a Rayleigh fluctuating target and an estimator optimized for thermal noise, with results as shown in Figure 8.3.5. The abscissa is the quantity:

$$n_e - 1 = t_0/t_c' = nt_r/t_c' \qquad (8.3.4)$$

where t_c' is the correlation time of the envelope-detected signal. This time is equal to the reciprocal of the noise bandwidth of the target fluctuation spectrum, measured after envelope detection.

Consider the example of a rigid target with scatterers distributed uniformly across its width L_x, rotating at a rate ω_x about an axis normal to the beam and to L_x. The correlation time is

$$t_c = \lambda/2\omega_x L_x \qquad (8.3.5)$$

The spectrum in this case is uniform over $B_c = 2\omega_x L_x/\lambda = 1/t_c$, and the correlation function has the $(\sin x)/x$ form. After envelope detection, the

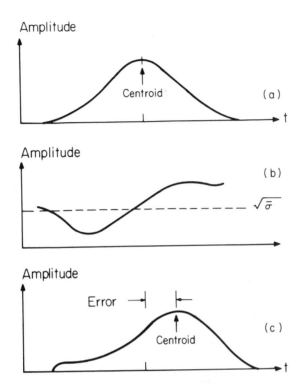

Figure 8.3.4 Effect of target fluctuation on signal envelope.

spectrum is triangular, the correlation function is $(\sin^2 x)/x^2$, and the envelope correlation time $t_c' = t_c$.

Most scanning radars have t_o/t_c' between 0.1 and 10, and hence will have scintillation errors between $0.05\theta_3$ and $0.1\theta_3$. Let the target span be $L_x = 15$ m, for a target flying normal to the radar beam at velocity $v_t = 300$ m/s, range $R = 30$ km. An X-band radar will be assumed, scanning a sector $\Delta_a = 10°$ in $t_s = 0.5$ s, with a beamwidth $\theta_a = 1°$. The following can be calculated:

$$t_o = t_s\theta_a/\Delta_a = 0.5 \times 1/10 = 0.05 \text{ s}$$

$$\omega_x = v_t/R = 0.01 \text{ r/s}$$

$$t_c = 0.03/(2 \times 0.01 \times 15) = 0.1 \text{ s}$$

From Figure 8.3.5, we find

$$\sigma_s = 0.1\theta_a = 0.1° = 1.8 \text{ mr}$$

The scintillation error will normally be the dominant error component in a sector scan system. Scintillation error can be reduced, at the expense of greater thermal noise error, by using a centroid estimator with increased weights beyond the half-power points of the pattern.

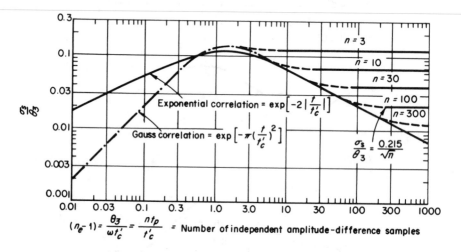

$(n_e - 1) = \dfrac{\theta_3}{\omega t_c'} = \dfrac{n t_p}{t_c'}$ = Number of independent amplitude-difference samples

Figure 8.3.5 Scintillation error in sector scan measurement.

Scintillation Error in Frequency Scan

Frequency scanning systems for height finding and other applications often pass the beam across the target very rapidly, $t_o \ll t_c$, such that the error from Figure 8.3.5 will be very small. In some cases, two beams straddling the target are generated simultaneously by transmitting a pulse with two frequency segments, producing a monopulse estimate. However, as the radar changes frequency by Δf in moving the beam from one side of the target to the other, decorrelation of the target cross section is produced. The frequency decorrelation can be described by

$$n_e - 1 = \Delta f/f_c = 2L_r\Delta f/c \qquad (8.3.6)$$

where L_r is the radial length of the target. The resulting error for a frequency scanning beam is large for most targets. For example, assume a sector $\Delta_e = 30°$, scanned with a frequency shift of 150 MHz. The value for motion across $\theta_e = 1°$ is $\Delta f = 5$ MHz. In locating an aircraft target with $L_r = 15$ m, we find $f_c = 10$ MHz, giving

$$n_e - 1 = 5/10 = 0.5$$

$$\sigma_s = 0.1\theta_e = 0.1° = 1.8 \text{ mr}$$

The only way to reduce this error would be to use a frequency sensitive antenna which would scan the beamwidth with $\Delta f \ll 10$ MHz or $\Delta f \gg 10$ MHz, or to repeat the scan several times at intervals separated at least by t_c, to obtain independent samples of error for subsequent averaging.

8.4 MONOPULSE RADAR

Monopulse is a radar technique in which information concerning the angular location of a target is obtained by comparison of signals received in two or more simultaneous antenna beams (Figure 8.1.4c). The advantages are the greater efficiency of the measurement, the higher data rate, and freedom from the effects of target scintillation error, and reduced vulnerability to jamming.

Amplitude Comparison Monopulse

The conventional amplitude comparison system, using Σ and Δ channels formed at RF, is shown in Figure 8.4.1. Horns are located symmetrically about the axis, and are fed in-phase by the transmitter to produce a single pencil beam along the mechanical axis of the reflector. The Σ channel of the antenna is also connected, through a duplexer, to a receiver channel, permitting reception of echo signals from along the axis with the full gain of the antenna. These signals are used in acquisition and display of targets, in range tracking, and as reference signals for the angle error detectors. Up to this point, the monopulse and conical-scan radars are equivalent, as are the servos required to drive the antenna. The additional receiving channels required in the monopulse radar are indicated by the shaded blocks in Figure 8.4.1. These start with the multihorn feed assembly, which picks up the off-axis components of the received signals and transforms them into elevation and traverse difference signals. These are brought through transmit-receive (T/R) switch tubes and adjustable RF phase shifters to the difference channel mixers. The electrical lengths of the paths between the feed assembly and the mixers must be held approximately equal to preserve the sensitivity of the error detectors and to minimize errors in location of the tracking axis.

After conversion to IF, the sum and difference signals are amplified in separate IF channels, which deliver all signals to the error detectors in the same phase relationships that applied at the feed, and which preserve the amplitude ratios. The AGC loop is based on the output of the sum

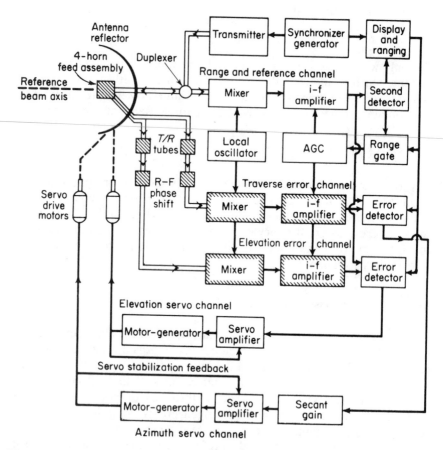

Figure 8.4.1 Block diagram of conventional monopulse radar.

channel, and controls all three amplifiers in the same way to preserve these ratios, while achieving a constant signal level out of the sum channel. The gain of the tracking loop is thus held constant for varying target size and range. The phase-sensitive error detectors now operate on the difference signals, producing bipolar video pulses proportional to the components of tracking error and indicative of its direction. Stretching or sample-and-hold operations, followed by low-pass filtering, now supply dc error inputs to the servo amplifiers. The signals supplied to the error detectors are range-gated to select the desired target.

Monopulse Feed Networks

The operation of a basic monopulse feed network is shown in Figure 8.4.2, which diagrammatically indicates the coupling between the four horns and the three output ports. A target signal received from along the axis will pass entirely into the Σ channel, with essentially perfect cancelation at the two Δ-channel ports and at the unused termination. However, a signal received from a target slightly above the axis will be reflected by the antenna surface more strongly into the lower pair of horns, marked D and C in the figure, and this will cause a Δ signal to appear at the elevation port. Similarly, any traverse error component will appear at the traverse port, with the signal amplitude in each case proportional to the magnitude of the angle offset from the axis.

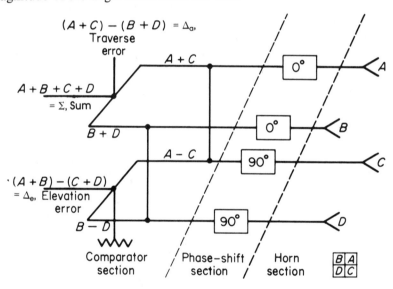

Figure 8.4.2 Configuration of four-horn monopulse feed.

In an amplitude comparison system, the Δ signal will be either in phase or 180° out of phase with the Σ signal, the phase serving to indicate the sense of the error (up or down, right or left). The phase shifts through the RF and IF portions of the system must be carefully equalized to maintain this ability to sense the direction of the error, but need not be held to the tolerance required between the horns and the comparator section of the feed assembly. As can be shown, the energy appearing in the two difference channels represents only that part of the signal which would, in a single-feed system, be lost from the antenna as a result of the offset angle of the target. At the tracking axis, or with small offsets, this energy will represent a negligible portion of the total received signal.

Monopulse Antenna Patterns

The optimum angle estimator for a noise environment would use the Σ and Δ illuminations shown in Figure 8.4.3, with maximum Σ efficiency and Δ slope, but with high sidelobes as shown. When the considering effects of a realistic environment are considered, the designer will taper the illuminations to reduce sidelobes. A typical choice would be $g(x) = \cos(\pi x/w)$ and $g_d(x) = x \cos(\pi x/w)$, as shown in Figure 8.4.4. These illuminations do not give very low sidelobes, the first Σ sidelobe being $G_{sr} = 23$ dB down, and the first Δ sidelobe being $G_{se} = 18$ dB down from the main-lobe peak. Further sidelobes fall off at a rate of 12 dB per doubling of offset angle.

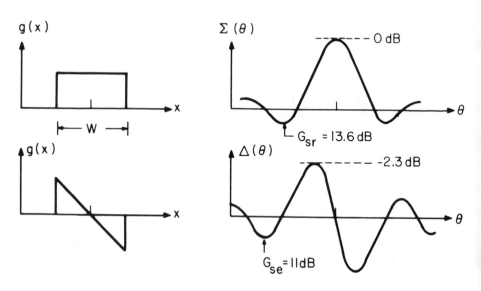

Figure 8.4.3 Optimum monopulse illuminations for noise environment.

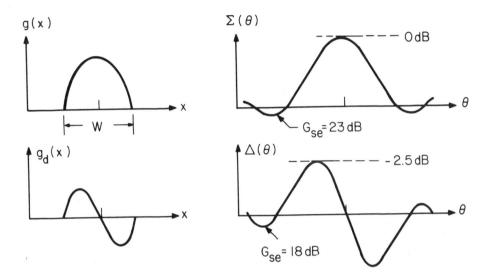

Figure 8.4.4 Typical monopulse illuminations for real environment.

Monopulse Error Slope

The normalized slope (see (8.1.6)) for the patterns of Figure 8.4.4 is $k_m = 1.96$. Data on theoretical values of k_m for different aperture illuminations are shown in Figure 8.4.5a. In comparing the performance of apertures with tapered illuminations to that of the ideal estimator, (8.1.11), several difference slope constants are used. Hannan [8.4] defined a *relative difference slope*:

$$K = \frac{1}{\sqrt{G_0}} \frac{d\Delta}{d\theta}\bigg|_{\theta=0} \tag{8.4.1}$$

where $\Delta(\theta)$ is the difference pattern voltage and $G_0 = G_m/\eta_a$ is the gain of the aperture with uniform illumination. For the optimum difference illumination of Figure 8.4.3,

$$K_0 = \mathcal{L}_0/\lambda = \pi w/\sqrt{3}\lambda \tag{8.4.2}$$

All other illuminations give lower slope, for which a *difference slope ratio* is defined as

$$K_r = K/K_0 \tag{8.4.3}$$

Values of K_r are tabulated in [8.1, p. 25] for a number of illumination functions, and are plotted in Figure 8.4.5b. None of the slopes defined by (8.4.1) to (8.4.3) is related to the actual Σ illumination or pattern. In terms of K, we can write

$$k_m = \frac{d(\Delta/\Sigma)}{d(\theta/\theta_3)} = \theta_3 K/\sqrt{\eta_a} \tag{8.4.4}$$

The slope k_m can be determined easily from measured Σ and Δ patterns, without knowledge of the aperture size or illumination, but it can give an inflated value when θ_3 has been broadened by inefficient Σ illumination.

The precision of a monopulse estimate in a thermal noise environment is

$$\sigma_\theta = \frac{\theta_3}{k_m \sqrt{2E/N_0}} = \frac{\theta_3}{k_m \sqrt{2(S/N)n}} \approx \frac{\theta_3}{2\sqrt{(S/N)n}} \tag{8.4.5}$$

where S/N is measured in the Σ channel with a target on the beam axis. The result, for $k_m = 1.4$, is 3 dB better ($1/\sqrt{2}$ times the error) than for the conical-scan system for the same beam center E/N_0. This reflects the absence of crossover loss or beamshape loss. When using tapered illumination (Figure 8.4.4), with $k_m \to 2$, an additional 3 dB advantage is obtained. The most important advantage, however, is not this greater efficiency, but rather the freedom from scintillation errors, which limit the performance of sequential scan systems.

Monopulse Antenna Options

Discussions of monopulse system implementation customarily divide systems into amplitude, phase, and Σ-Δ types, based on antenna patterns. Sometimes a second set of categories is included, referring to the method of sensing used in the receiver circuits. Rhodes [8.5] showed that these distinctions are arbitrary, and that one type can be transformed into the other in the receiver processing. A more unified view of the subject has resulted from design of phased array radar systems. The entire performance of any antenna, from the viewpoint of its patterns in space, can be

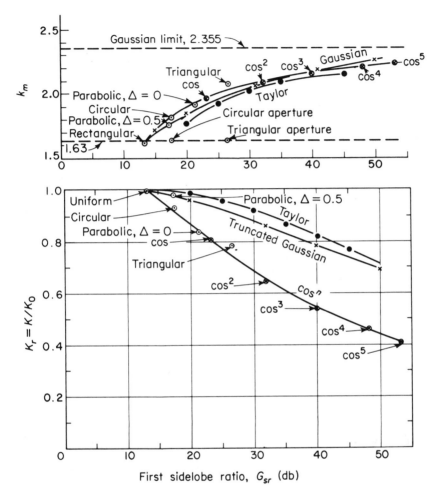

Figure 8.4.5 Difference slope *versus* sidelobe level: (a) normalized monopulse difference slope; (b) difference slope ratio.

derived from the aperture illumination (which need not be measured at the physical surface of the antenna, but on a plane in front of the antenna). We will consider one angular coordinate θ and the corresponding aperture coordinate with illumination function $g(x)$. In the monopulse case, a second illumination function $g_d(x)$ is generated. We might argue that this latter terminology applies uniquely to the Σ, Δ type of system, but the same consideration applies here as to sequential lobing systems: at the tracker output, an error signal appears, which is the result of subtracting the outputs of two channels (and normalization by the sum of these two channels). The response in that error channel varies as a function of target offset from the axis, and that response can be equated to the ratio Δ/Σ of two patterns generated by aperture distributions $g(x)$ and $g_d(x)$. For the Σ channel, $g(x)$ has even symmetry, peaking at the center of the aperture and (for low sidelobes) tapered toward the edges. The phase is uniform across the aperture. For the Δ channel, $g_d(x)$ has odd symmetry, peaking somewhere in each half of the aperture and again (in most cases) tapered toward the edges. These illuminations are illustrated in Figure 8.4.4 for a typical amplitude comparison antenna design.

Phase Comparison Monopulse

The so-called phase comparison monopulse, when implemented with a reflector antenna, places two paraboloids side-by-side with separate feed horns (Figure 8.4.6a). The corresponding patterns are shown in Figure 8.4.6b. The significant difference between this antenna and the one shown in Figure 8.4.4 is the hole in the middle of the Σ illumination, and the resulting high sidelobes in the Σ pattern.

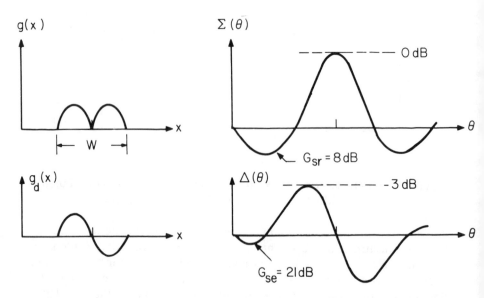

Figure 8.4.6 Illuminations and patterns for phase-comparison monopulse using reflector antenna: (a) aperture illuminations of Σ and Δ channels; (b) beam patterns.

When a planar array is used to implement the phase comparison monopulse system, the Σ pattern is improved, but the Δ pattern suffers, as shown in Figure 8.4.7. More complex feeds, in which $g_d(x)$ may be controlled independently of $g(x)$, must be used to obtain satisfactory sidelobe performance. When this is done, however, the distinction between phase and amplitude monopulse antennas is lost. We will therefore discuss the performance of systems without using this artificial distinction.

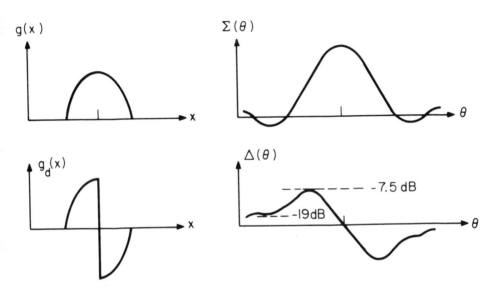

Figure 8.4.7 Illuminations and patterns for phase comparison monopulse using planar array: (a) aperture illumination functions; (b) beam patterns.

Monopulse Feed Horns

The limited control of illumination available from the simple four-horn feed has been discussed by Hannan [8.4] and in Section 4.5. In terms of tracker performance, the unsatisfactory compromises among η_a, K_r, and sidelobes for Σ and Δ channels are summarized in Figure 8.4.8. The most inconvenient connection is between efficiency η_x and Δ-channel sidelobe level G_{se}: to achieve G_{se} better than 20 dB, η_x must be less than 0.5. If this is done in both planes of a two-coordinate tracker, $\eta_a < 0.25$. For this

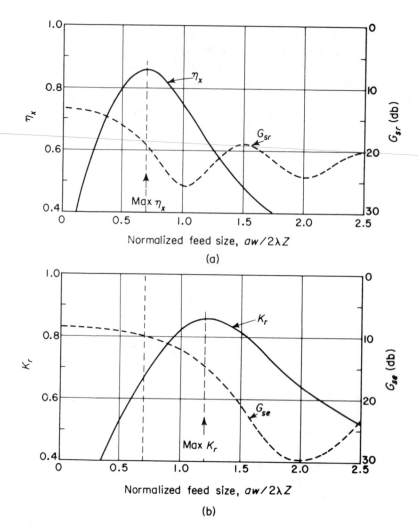

Figure 8.4.8 Aperture efficiency and slope *versus* feed size for four-horn monopulse feed.

reason alone, the simple four-horn feed is seldom used. More complex four-horn feeds, involving mutual coupling between pairs of horns, can ease the efficiency and sidelobe problem, but low-sidelobe performance is not achievable in this class of designs.

In his classic paper, Hannan [8.4] showed how the coupling of additional feed area in different ways to the Σ and Δ channels could produce a satisfactory compromise among the several performance parameters of a horn-fed aperture. Figure 8.4.9 and Table 8.2 summarize these results.

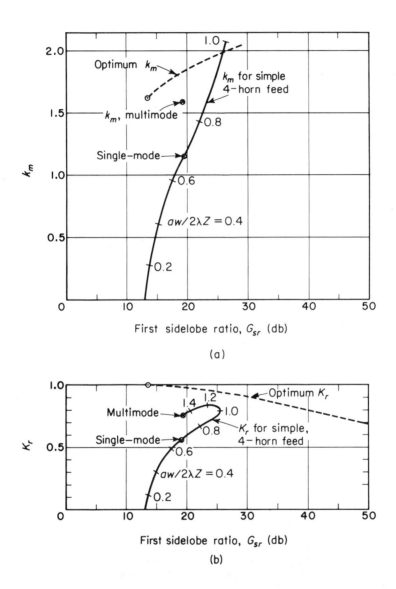

Figure 8.4.9 Comparison of monopulse horn feed performance.

Both Σ and Δ sidelobes can be held to 19 dB below the Σ peak, while maintaining high error slope and efficiency. More recent designs provide lower sidelobes in both channels, at the expense of reduced slope and efficiency, of course. A problem in using these complex feeds is the size of the structure, which must include multiple couplers. For the Nike Hercules tracking radar (Figure 8.4.10), Hannan solved the problem by going to the polarization-twist Cassegrain reflector geometry (Section 4.2). Space-fed arrays of the transmission lens type can use these feeds without any blockage. Very complex feeds, such as the 44-horn focal-plane array of Figure 4.4.4, can be used with these systems, or with offset-fed reflectors to achieve even more precise control of the illumination functions.

Table 8.2 Monopulse Horn Feed Performance

Type of Horn	η_a	H-plane $K_r\sqrt{\eta_y}$	k_m	E-plane $K_r\sqrt{\eta_x}$	k_m	G_{sr} (dB)	G_{se} (dB)	Feed Shape
Simple four-horn	0.58	0.52	1.2	0.48	1.2	19	10	⊞
Two-horn dual-mode	0.75	0.68	1.6	0.55	1.2	19	10	⊟
Two-horn triple-mode	0.75	0.81	1.6	0.55	1.2	19	10	⊟
Twelve-horn	0.56	0.71	1.7	0.67	1.6	19	19	⊞
Four-horn triple-mode	0.75	0.81	1.6	0.75	1.6	19	19	☰

Figure 8.4.10 Nike Hercules tracking radar. (Photo courtesy of Bell Telephone Laboratories.)

Optimum Monopulse Arrays

The optimum array for a real environment is a compromise among G_m, θ_3, and G_{sr} for the Σ channel, and between K_r and G_{se} for the Δ channel. There is no single optimized design, but the general characteristics of such a design include:

(a) independent control of Σ and Δ illuminations;
(b) smooth $g(x)$ with even symmetry, and $g_d(x)$ with odd symmetry;
(c) low edge illumination.

With independent control, optimization of the Σ channel is by conventional means: Taylor or Dolph illuminations. Optimization of the Δ channel has been discussed by Bayliss [8.6], who derived illuminations having maximum K_r for given sidelobe ratio G_{se}.

The array design problem thus becomes a feed network task, and the best performance is obtained when each element is weighted separately for the Σ, Δ_a, and Δ_e channels. This can be done with such designs as the Lopez dual-ladder feed [8.7]. Use of subarrays, and the resulting errors in amplitude of the illumination function, were discussed in Section 4.5.

The complexity of feed networks for separate control of Σ and Δ can be considerably reduced in the space-fed array. Here it is possible to apply directly Hannan's multihorn, multimode designs, and to use larger focal-plane arrays as needed. The complexities of the network for a 44-horn feed are considerable, but not significant when compared to similar problems in the aperture plane with thousands of elements.

8.5 MONOPULSE SIGNAL PROCESSING

Normalization of Error Signal

The conventional monopulse receiver configuration of Figure 8.4.1 uses identical, parallel receivers for the Σ, Δ_a, and Δ_e channels, with a common AGC for normalization. The IF outputs of the Δ channels are applied to a phase sensitive detector for which the Σ channel provides the reference. The error signal outputs are pulses with amplitude:

$$E_\Delta = (|\Delta|/|\Sigma|) \cos\phi \tag{8.5.1}$$

where ϕ is the phase angle between Σ and Δ. Range gating in the IF amplifier confines the AGC and error sensing to the selected target. These pulses of bipolar video are converted to dc by stretching or sample-and-hold circuits followed by low-pass filters.

The characteristics of the AGC loop are critical to performance of the monopulse radar. The principal advantage of monopulse is its freedom from target scintillation error, including the effects of deliberate jammer modulation. The Δ channel from which the error signal is taken is a suppressed carrier channel, having zero amplitude for a target on the axis. Hence, modulation of the carrier signal produces no output, in contrast to sequential systems in which modulation can generate a spurious error signal. The purpose of the AGC is to ensure constant loop gain in the error channel, as target signal amplitude changes with range or target fluctuation. Compensation for change in target range is available from a slow AGC loop, while target fluctuations can be removed only with fast AGC. In the limit, where fluctuations occur from pulse to pulse, a computer operation that divides Δ by Σ can provide instantaneous normalization, at least over the dynamic range of those fluctuations not already removed by the normal fast AGC. Other methods of implementing instantaneous normalization will be discussed.

There is a cost to using fast AGC, as shown in Figure 8.5.1. When the target is on the tracking axis, the error with AGC is greater than without, and with fast AGC it is greater than with slow. The increase results from target glint (Section 10.2), which is greatest during target fades [8.8, pp. 111–122]. The glint error is proportional to the target width L_x, but its peaks can greatly exceed this width.

Conversely, there is a cost to not using AGC, or having a response too slow to follow the target fluctuations. The tracking loop gain will vary, if the Δ signal is not normalized by Σ. For a target which has a motion that can cause tracking lag comparable to or greater than L_x, the random error due to modulated lag in a system with slow AGC will greatly exceed the glint error. Fast AGC holds this random error almost constant, regardless of lag. Instantaneous AGC would produce a larger random error with almost no dependence on lag. The occurrence of lag errors is discussed in Section 10.3.

Detailed analyses of thermal noise error on fluctuating targets have shown that this error component also increases when instantaneous normalization is used [8.10]. In Figure 8.5.2, the thermal noise errors for steady and Rayleigh fluctuating targets are compared. Thermal noise error is shown to increase by a factor of two or more for the fluctuating target, compared to a steady target of the same average signal power. The reason for this is the action of the normalization process in the presence of noise. The instantaneous error signal output will be

$$E_\Delta = \frac{|\Delta + e_{n\Delta}|}{|\Sigma + e_{n\Sigma}|} \cos\phi \qquad\qquad (8.5.2)$$

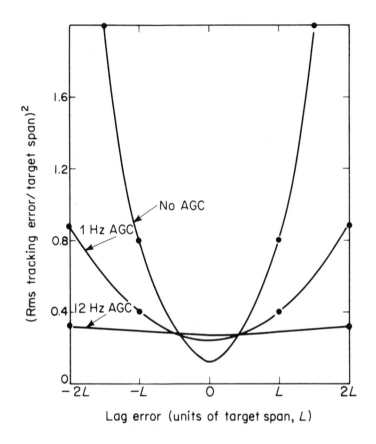

Figure 8.5.1 Radar tracking noise power as a function of tracking lag error for different AGC bandwidths [8.9]. (Copyright 1970 by McGraw-Hill Book Co. Reprinted with permission.)

When Σ is Rayleigh distributed, $\Sigma + e_{n\Sigma}$ is also Rayleigh distributed, and the denominator will occasionally approach zero, producing a noise impulse in E_Δ. The traditional analog processes for implementation of (8.5.1) restrict the amplitudes of these impulses by having limited bandwidth and dynamic range. As a result, experimental data on fluctuating targets has followed the predictions of (8.4.5) rather than the curves of Figure 8.5.2, permitting the thermal noise error to be calculated simply by using the average SNR of any target. Modern digital processing makes possible passing the impulses with much larger amplitude, thereby significantly increasing the error.

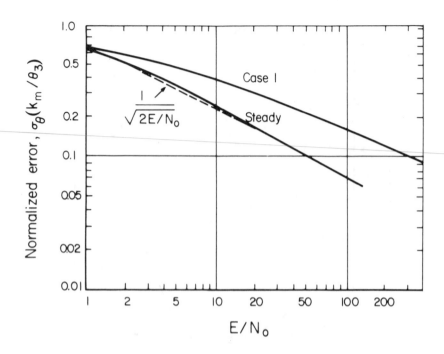

Figure 8.5.2 Thermal noise errors for steady and fluctuating targets with instantaneous normalization (after [8.10]).

The error on fluctuating targets can be reduced essentially to that predicted by (8.4.5), by using some averaging in the normalization process:

$$E_\Delta = \frac{|\Delta + e_{n\Delta}|}{|\Sigma + e_{n\Sigma}|} \cos\phi \qquad (8.5.3)$$

The average must extend over at least two samples of noise to reduce essentially to zero the probability of the denominator tending toward zero. Limiting E_Δ before the error signal in the servo loop will also reduce the error peaks, but the limit level must be set high enough to avoid introduction of bias error in the estimate.

Sensitivity to Errors in Monopulse Implementation

The basic elements of amplitude comparison monopulse are shown in Figure 8.5.3, with two types of error in the electrical elements:

Precomparator error, $d_1 = 1 + a_1 + j\phi_1$,

Postcomparator error, $d_2 = 1 + a_2 + j\phi_2$

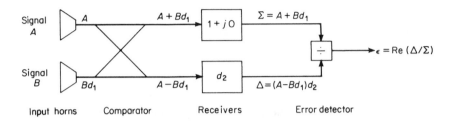

Figure 8.5.3 Monopulse implementation errors.

The imbalance errors for amplitude, a_1 and a_2, are assumed small with respect to unity, and phase imbalance errors ϕ_1 and ϕ_2 are small with respect to one radian. Precomparator error is associated with RF components, and is often stated as a null depth G_n in the Δ pattern, relative to the Σ pattern maximum:

$$G_n = 4/\phi_1^2 \qquad (8.5.4)$$

Restricting our concern to the first- and second-order terms, we may write the shift in null position, relative to the ideal (error-free) system in terms of an error voltage:

$$\begin{aligned}
E_\Delta(0) &= \mathrm{Re}(\Delta/\Sigma)_0 \\
&= (-a_1/2) - (a_1 a_2/2) + (a_1^2/4) + (\phi_1\phi_2/2) - (\phi_1^2/4) \qquad (8.5.5)
\end{aligned}$$

The normalized difference slope, which relates this voltage to angle, is

$$k_m = (\theta_3/\Sigma)\, d\Delta/d\theta \qquad (8.5.6)$$

Thus, an error voltage E_Δ is equivalent to an angle $\Delta\theta = E_\Delta\theta_3/k_m$, and (8.5.5) may be converted to an angle with this scale factor.

The monopulse network error will also affect the difference slope, changing the error-free value k_m to a value k_m', such that

$$k_m'/k_m = 1 + a_2 + (a_1^2/4) - (\phi_1\phi_2) - (3\phi_1^2/4) \qquad (8.5.7)$$

(retaining only first- and second-order terms).

In practice, a major portion of the precomparator error is eliminated by offsetting the feed or data system to center the RF null of the system (as measured, for instance, with receiving equipment carefully adjusted to ideal performance, $a_2 = \phi_2 = 0$). This displaces the zero point by the amount:

$$E_0 = (a_1/2) - (a_1^2/4) - (\phi_1^2/4) \tag{8.5.8}$$

Assuming that the precomparator errors remain fixed, the introduction of postcomparator errors (usually associated with gain and phase variations in active receiver elements) will now cause an error signal:

$$E_2 = E_\Delta(0) - E_0 = \phi_1\phi_2/2 \tag{8.5.9}$$

The corresponding angle error for this case is

$$(\Delta\theta)_2 = \theta_3\phi_1\phi_2/2k_m = \theta_3\phi_2/k_m \sqrt{G_n} \tag{8.5.10}$$

The rms error σ_θ caused by postcomparator error after setting the RF null to zero is found from (8.5.10) by inserting the rms value of ϕ_2. This postcomparator phase variation can result from any of the following factors:

(a) tuning of the receiver such that the signal is not centered in the IF passband;

(b) change in the signal power level at the receiver input;

(c) temperature effects on receiver components.

In addition, there may be changes in precomparator errors a_1 and ϕ_1, occurring after calibration and setting to zero. These will cause the null point to shift in accordance with (8.5.5), and can result from any of the following:

(d) change of operating frequency within the band;

(e) ambient temperature or solar heating effects on the RF structure;

(f) polarization changes in the received signal (see below).

We can see from the above that active elements in the signal paths are more likely to increase the gain and phase errors a and ϕ, as compared to passive elements. Therefore, most monopulse systems place the comparator at RF, as close as possible to the antenna ports. Active amplifiers in the postcomparator portion of the receiver are essential, and relations (8.5.8) and (8.5.9) can be used to establish the required phase matching ϕ_2 between Σ and Δ amplifiers. Another source of error is dc offset in the phase-sensitive error detector. This produces a fixed bias error in boresight axis position. To reduce this error, a periodic phase reversal (commutation) at a submultiple of the PRF, f_r/m, is often introduced in the Δ channel, in the RF or early in the IF stages. The error signal output is then a square wave at f_r/m, and, after sufficient amplification of this audio signal, it is decommutated to provide a dc error signal for the servo amplifier.

Polarization Effects

Antennas are designed to respond to input waves having a particular polarization (e.g., vertical), and there is always one orthogonal polarization (or cross-polarization) for the antenna (e.g., horizontal). The antenna patterns for the cross-polarization will differ from those for the intended polarization, and targets which scatter both polarizations will generate spurious responses at the antenna output. Figure 8.5.4 shows typical Δ and Σ patterns for the intended polarization, and Δ_c and Σ_c for the cross-polarization.

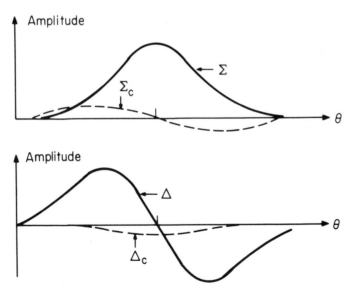

Figure 8.5.4 Antenna patterns for intended polarization and cross-polarization.

In most cases of symmetrical antennas, Σ_c becomes zero on the axis, where Δ_c has nonzero value. The monopulse error voltage will then be

$$E_\Delta = \frac{e\Delta + e_c\Delta_c}{e\Sigma + e_c\Sigma_c} \approx \frac{e\Delta + e_c\Delta_c}{e\Sigma} \qquad (8.5.11)$$

Here, $e = \sqrt{\sigma}$ is the target field scattered at the intended polarization, and $e_c = \sqrt{\sigma_c}$ is the cross-polarized field. The polarization interference component, in the Δ channel, will be $I_\Delta = \sigma_c\Delta_c^2$, and the error, for σ_c having random phase relative to σ, will be

$$\sigma_\theta = \frac{\theta_3(\Delta_c/\Sigma)}{k_m\sqrt{2\sigma/\sigma_c}} \qquad (8.5.12)$$

In many cases, the cross-polarized target cross section averages about -6 dB relative to the intended polarization. Then if the cross-polarized antenna response is, for example, -30 dB relative to the intended polarization ($\Delta_c/\Sigma = 0.032$), we have

$$\sigma_\theta = \theta_3 \times 0.032/2\sqrt{4} = 0.008\theta_3$$

When higher tracking precision is required, the cross-polarized response of the antenna must be reduced below -30 dB.

Phase Processors for Monopulse Error Channels

An alternative processor for monopulse error sensing is shown in Figure 8.5.5. Phase processing refers not to the antenna, but to the way in which Δ and Σ signals are converted into an error signal.

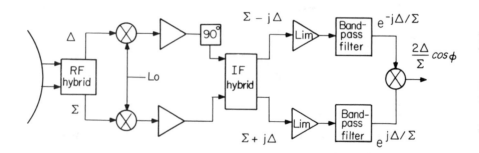

Figure 8.5.5 Phase processing monopulse system.

The antenna is assumed to provide Δ and Σ outputs, which are down-converted to IF and applied as inputs to a hybrid network to produce Σ $\pm$ jΔ. These combined signals are hard-limited, bandpass filtered, and converted to an error signal by a phase detector with an output of bipolar video pulses. The hard-limiting ensures instantaneous normalization, with the time constant equal to the reciprocal of the bandwidth of the preceding amplifier stages. Symmetrical limiting of the combined signals eliminates bias error, and the characteristics of the phase detector reduce impulses in output error on a fluctuating target. Commutation may be used to eliminate dc error in the phase detector. When both angle coordinates are to be processed, the Σ channel must be amplified before being divided into signals for the two IF hybrids, one for combination with Δ_a and one for Δ_e.

Advantages of the phase processor include the ability to provide data on many targets at different ranges within the beam, without the complexity of multiple AGC loops. However, great care is needed to ensure sampling of the error outputs at times near the signal peaks. The limiting action will produce random noise outputs at high level when no signal is present. The limiting processor is also disadvantageous when any type of doppler processing is required, because the clutter spectrum will be spread, as discussed in Section 5.3.

Processing in Logarithmic Amplifiers

Noncoherent radars can perform monopulse processing in logarithmic IF amplifiers, as shown in Figure 8.5.6. The Σ and Δ channels are combined after amplification to produce $\Sigma + \Delta$ and $\Sigma - \Delta$ channels for log amplification. The log outputs are subtracted and added to form

$$\log(\Sigma + \Delta) - \log(\Sigma - \Delta) = \log\frac{\Sigma + \Delta}{\Sigma - \Delta}$$

$$\approx \log(1 + 2\Delta/\Sigma) \approx 2\Delta/\Sigma \qquad (8.5.13)$$

$$\log(\Sigma + \Delta) + \log(\Sigma - \Delta) = \log(\Sigma^2 + \Delta^2) \approx 2 \log\Sigma \qquad (8.5.14)$$

These are the desired signals, and the error output has been normalized with the time constant of the IF amplifiers.

Monopulse with Doppler Processing

When doppler processing (including MTI) is to be used in tracking radar, the receiver channels must remain linear to a point beyond the

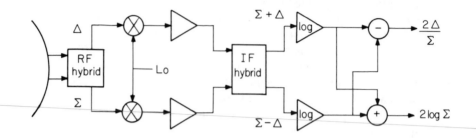

Figure 8.5.6 Logarithmic monopulse processor.

clutter rejection filter. In the conventional system, this means that parallel doppler processors must be used in IF or baseband channels, preserving matched amplitude and phase response for the Δ and Σ channels. In an analog processor, these matching requirements are difficult to meet unless a doppler tracking loop (Section 9.3) is used to center the target signal in the filter passband. Modern digital techniques make possible an exact match through the doppler processing after A/D conversion. To reduce complexity, the monopulse processing need be carried out only in a single range gate or sample taken at the target range.

Multiplexed Monopulse Channels

Many options are available for multiplexing the Σ with the Δ channels, after initial RF or IF amplification has established to SNR and channel bandwidth.

(a) *Time Multiplexing*. Range gates are used to sample the three channels after preamplification. The channels are then added with small time delays $\tau_{da} > \tau$ and $\tau_{de} > \tau + \tau_{da}$, after which a single IF amplifier with AGC can provide most of the receiver gain. Demultiplexing and postamplification occurs prior to error detection. Only the preamplifiers, postamplifiers, and short delay lines can introduce errors in amplitude and phase matching between channels. Since the multiplexing process is linear, doppler filtering may be included in the common amplifier channel.

(b) *Frequency Multiplexing*. After preamplification and bandpass filtering to the signal bandwidth B, the Δ channels may be converted to IFs which differ from that of the Σ channel, f_1, by an offset greater than B:

$$f_{da} - f_1 > B, \quad f_1 - f_{de} > B$$

The three frequency channels are then summed and passed through a common amplifier having bandwidth at least $3B$, and providing most of the receiver gain, with AGC or limiting for normalization. Demultiplexing occurs before error detection, using filters at f_{da} and f_{de} to select the error signal data. Since the frequency conversion to f_1, f_{da}, and f_{de} is a linear process, doppler filtering may be accomplished on each of the three frequency channels before the common amplifier, or after it, if AGC is used.

(c) *Multiplexing of RF Difference Signals.* Greater economy of receiver components is achieved with the configuration of Figure 8.5.7. The two Δ channels are time multiplexed by sampling in a microwave resolver or switch. A combined Δ channel is thus formed, containing half of the information from Δ_a, and half from Δ_e. This is combined with the Σ channel in a hybrid to produce $\Sigma \pm \Delta$ signals, which are processed in any convenient manner (logarithmic processing is shown). The resulting combined error voltage E_d is resolved into ϵ_a and ϵ_e by phase demodulation with the resolver reference signal. Commutation is accomplished in the course of the sampling, and decommutation is during the phase demodulation.

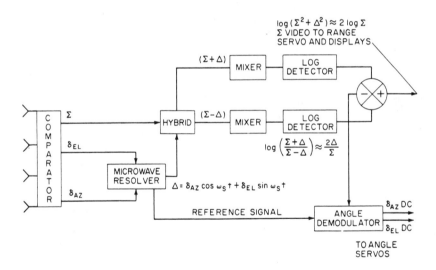

Figure 8.5.7 Two-channel monopulse system [8.11].

When sinusoidal sampling is used (e.g., with a microwave resolver), the monopulse receiving patterns will consist of two beams, offset on opposite sides of the antenna axis and rotating about the axis at the sampling frequency f_s (Figure 8.5.8). The monopulse estimate is then formed

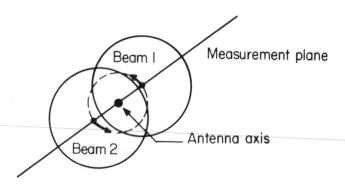

Figure 8.5.8 Two-beam interpretation of time-multiplexed monopulse.

in a plane which rotates in space for sequential sampling of azimuth and elevation. The estimate is still made on the basis of simultaneous lobe comparison, eliminating the effects of target fluctuation (in an ideally implemented system). This concept has also been discussed in Soviet literature [8.12] under the name *scan with compensation,* and in more recent US literature [8.13] as *conopulse.* Unlike the three-channel multiplexing processors, the use of RF multiplexing reduces the efficiency of the estimation by 3 dB, since half of the input Δ signal energy is terminated in a load resistor. It becomes no more efficient than conical scan, although it preserves (at least to first order) the ability of the monopulse system to cancel target fluctuation components. When used in doppler radars, the sampling must either be sinusoidal, at a rate low enough to avoid shifting the target modulation into the clutter spectrum, or use square waves with batch-processed bursts of pulses. An appropriate linear processor must then be used to recover the error signals.

The configuration shown in Figure 8.5.7, with a monopulse feed providing signals to the resolver, can also be used in a single-channel, conical-scan system known as *conical scan on receive-only,* or COSRO. Only one of the two receiver channels is used following the hybrid, and one of the two beams of Figure 8.5.8 is produced. However, if the transmitter is coupled through a duplexer to the Σ channel of the feed, the target will be illuminated with a steady beam on the axis, and one-way scanning will be produced. In many cases this is the preferred configuration for conical scan, since the lobing rate is not broadcast for use in controlling deception jammers. Even when two-way conical scan is to be used, the monopulse feed, resolver, and hybrid may be preferred to mechanical rotation or nutation of the feed, for reasons of reliability.

Processing of Elementary Beams

In amplitude or phase monopulse systems that do not form Δ and Σ channels at RF, the four elementary beams are received and amplified separately, before combination in error detectors. The combination can be performed at IF, at baseband (bipolar video), or after envelope detection. The resulting amplitude and phase tolerances on the receiver channels are much more stringent, for a given angle accuracy, than in Σ and Δ processing because all receiver errors (and detector errors as well, in some cases) are precomparator errors (d_1 in Figure 8.5.3), capable of causing shifts in the tracking axis. For this reason, the Σ and Δ processing methods are almost universally preferred, except in stacked-beam 3D radar, where the emphasis is on detection, rather than measurement accuracy.

Off-Axis Angle Estimation

The estimate of target angle may be made when the angle is offset from the measurement axis, but at the expense of greater error than for tracking on the null. This can occur in tracking with dynamic lag error, or in a stacked-beam system, where the target can appear at any angle relative to the beams. A second component of thermal noise error will appear, proportional to the offset θ from the null axis:

$$\sigma_2 = \theta / \sqrt{2E/N_0} \tag{8.5.15}$$

The increased error, as a function of θ/θ_3, is plotted in Figure 8.5.9. When estimates over $\pm \theta_3/2$ are to be made, some form of fast normalization is essential to avoid large errors. If the normalization time constant t_n greatly exceeds the signal correlation time t_c' of a Rayleigh distributed target, the scintillation error component will be

$$\sigma_s = \theta \sqrt{2\beta_n W_0} = \theta \sqrt{\beta_n t_c'} \tag{8.5.16}$$

where the fluctuation correlation time and spectral density W_0 are defined as in Section 8.2. In the limit, when all of the fluctuation power falls within the servo bandwidth,

$$\sigma_s \rightarrow \theta/2$$

This error will be reduced directly by the open-loop voltage gain of the AGC loop within the servo bandwidth β_n.

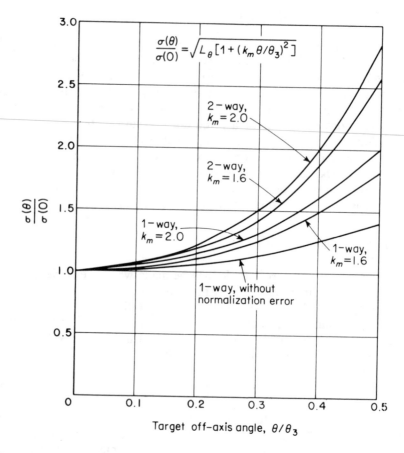

$$\frac{\sigma(\theta)}{\sigma(0)} = \sqrt{L_\theta \left[1 + (k_m \theta/\theta_3)^2\right]}$$

Target off-axis angle, θ/θ_3

Figure 8.5.9 Ratio of off-axis error to on-axis error for a target of given amplitude.

REFERENCES

[8.1]   D.K. Barton and H.R. Ward, *Handbook of Radar Measurement*, Artech House, 1984.

[8.2]   P.M. Woodward, *Probability and Information Theory, with Applications to Radar*, Artech House, 1980.

[8.3]   M.I. Skolnik, Theoretical accuracy of radar measurements, *IRE Trans.* **ANE-7**, No. 4, December 1960, pp. 123–129.

[8.4]   P.W. Hannan, Optimum feeds for all three modes of a monopulse antenna, *IRE Trans.* **AP-9**, No. 5, September 1961, pp. 444–461.

[8.5] D.R. Rhodes, *Introduction to Monopulse,* Artech House, 1980.

[8.6] E.T. Bayliss, Design of monopulse antenna difference patterns with low sidelobes, *Bell System Technical J.* **47,** No. 5, May-June 1968, pp. 623–650.

[8.7] A.R. Lopez, Monopulse networks for series feeding an array antenna, *IEEE Trans.* **AP-16,** No. 4, July 1968, pp. 436–440.

[8.8] R.V. Ostrovityanov and F.A. Basalov, *Statistical Theory of Extended Radar Targets* (W.F. Barton, tr.), Artech House, 1985.

[8.9] J.H. Dunn and D.D. Howard, Target Noise, Chapter 28 in *Radar Handbook* (M.I. Skolnik, ed.), McGraw-Hill, 1970.

[8.10] T.E. Connolly, Statistical prediction of monopulse errors for fluctuating targets, *IEEE 1980 Int. Radar Conf. Record,* pp. 458–463.

[8.11] P.A. Bakut, *et al., Problems in the Statistical Theory of Radar* (Vol. 2), Soviet Radio, 1964; translation: DDC Doc. AD645775.

[8.12] H. Sakamoto and P.Z. Peebles, Jr., Conopulse radar, *IEEE Trans.* **AES-14,** No. 1, January 1978, pp. 199–208.

Chapter 9

Range and Doppler Measurement

9.1 RANGE MEASUREMENT

The time delay t_d between transmission and reception of a radar signal is related to the target range R by

$$R = t_d c/2 \qquad (9.1.1)$$

Measurement of the time delay is accomplished in two steps: the centroid, or other definable point on the received waveform, is located; the delay between that point in the transmission and its reception is then measured and converted to output data. Estimation of the signal centroid involves the implementation of Figures 8.1.1 and 8.1.2 in the time coordinate: formation of time weighting functions $f_1(t)$ and $f_2(t)$, combining them into $\Delta(t)$ and $\Sigma(t)$, applying them to the input signal $a(t)$, and determining the resulting error signal Δ/Σ. These are the same operations as those with beam patterns for angle measurement, but the process is much more easily performed in the time domain, using range gates, correlators, or digital operations on samples in time.

Optimum Estimator

The optimum estimator for the centroid of a signal consists of a matched filter, a differentiator, and a zero-crossing detector (Figure 9.1.1). A narrow impulse, formed when the differentiator output passes through zero, marks the peak of the matched-filter output, which is delayed by a fixed amount from the centroid of the input signal. The time delay of this impulse, relative to an equivalent event in the generation of the transmitted signal, can be determined with a high-speed counter or equivalent device, and converted to range data using (9.1.1).

425

The output of the matched filter can be identified with the Σ channel of the generalized measurement response, Figure 8.1.2. This output is applied to a threshold for target detection, and the threshold output is used to select the zero-crossing output corresponding to the peak of the signal, rejecting sidelobe responses and noise crossings. The rms error in the time estimate, caused by thermal noise, is

$$\sigma_t = \frac{1}{K_0 \sqrt{2E/N_0}} = \frac{1}{\beta \sqrt{2E/N_0}} \approx \frac{\tau_o}{1.61 \sqrt{2E/N_0}} \qquad (9.1.2)$$

where τ_o is the half-power width of the output pulse from the matched filter, and β is the rms bandwidth of the signal spectrum, defined by (8.1.15). The width τ_o is about 6% less than the effective width τ_n used in Section 1.5 for the width of the clutter cell.

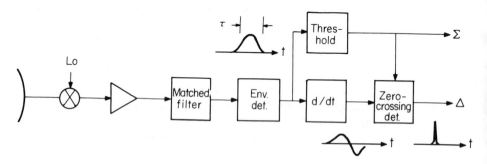

Figure 9.1.1 Block diagram of optimum centroid estimator.

The rms bandwidths for different signal spectra having total width B are given in Table 9.1. For matched-filter processing, these values of β can be used directly in (9.1.2) to find the thermal noise error.

Mismatched Estimator for a Single Pulse

The receiver need not be matched perfectly to the input signal to achieve reasonable precision in range measurement. In processing a single pulse, the process illustrated in Figure 9.1.2 can be used to approximate the matched filter.

Table 9.1 Bandwidth Parameters of Different Signal Spectra

Spectrum	$A(f)$	β/B
Rectangular	1	1.81
Triangular	$1 - \lvert 2f/B \rvert$	0.99
Parabolic	$1 - 2(f/B)^2$	1.53
	$1 - (2f/B)^2$	1.19
Cosine	$\cos(\pi f/B)$	1.14
Cosine²	$\cos^2(\pi f/B)$	0.89
Cosine⁴	$\cos^4(\pi f/B)$	0.67
Gaussian	$\exp(-f^2/2\sigma_f^2)$	
(assume $B = 6\sigma_f$)		0.74

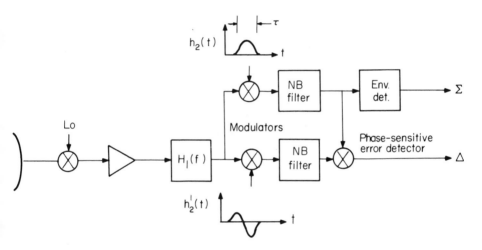

Figure 9.1.2 Block diagram of practical range estimator.

The cascaded response $H_1(f)\,H_2(f)$, or $h_1(t) \otimes h_2(t)$ of the filter and correlator is approximately matched to the echo pulse, and $h_2'(t)$ forms the derivative response. The phase-sensitive error detector converts the IF Δ signal to baseband. Normalization can be accomplished by AGC or by mathematical division of output Δ by Σ. The functions $h_2(t)$ and $h_2'(t)$, which multiply the IF waveform, implement the functions $\Sigma(t)$ and $\Delta(t)$ required by the generalized estimator of Figure 8.1.2. The curve of dc output *versus* displacement of the target from the center of the correlator function reproduces the S-curve shown in Figure 8.3.2e for the angle gate in a sector scanning radar. This dc output serves as the input to a loop which controls the timing of the correlator functions $h_2(t)$ and $h'_2(t)$. When $H_1(f)$ is closely matched to the input signal, $h_2(t)$ should be reduced to a narrow gate sampling a single IF cycle, and $h'_2(t)$ to a double gate sampling two successive cycles with opposite polarity. At the other extreme, when $H_1(f)$ is wideband, $h_2(t)$ must more closely match the signal waveform.

Pulse compression waveforms are usually processed in filters, $H_1(f)$, the phase response of which is the conjugate of the signal spectrum $A(f)$. The amplitude response $|H_1(f)|$ may be weighted to reduce sidelobes. This approximation to the matched filter permits the Σ and Δ correlator functions to be done with minimal loss at video, as in Figure 9.1.3.

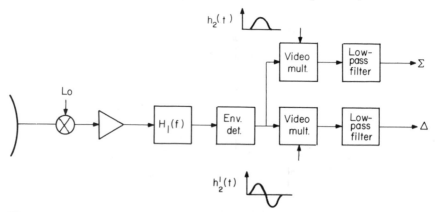

Figure 9.1.3 Block diagram of range estimator with video correlation.

The mismatched processor is characterized by reduced Σ channel SNR:

$$S/N = E_1/N_0 L_m$$

and by reduced error slope:

$$K < \beta$$

The slope K is used in place of K_0 in (9.1.2), and is defined as

$$K = \frac{1}{\Sigma_0}\frac{d\Delta}{dt} \qquad (9.1.3)$$

where Σ_0 is the sum channel signal that would have been produced by a matched filter. This definition is analogous to (8.4.1) for angle measurement. As in the case of monopulse angle estimation, a normalized slope can be defined as

$$k_t = \frac{d\,(\Delta/\Sigma)}{d\,(t/\tau_o)} \qquad (9.1.4)$$

The time delay error for a single pulse is then given by

$$\sigma_{t_1} = \frac{\tau_o}{k_t\,\sqrt{2\,(S/N)}} \qquad (9.1.5)$$

The slope k_t may be larger than the matched-filter value 1.61, if the increase in $\tau_o/\sqrt{\eta_f}$ resulting from filter mismatch exceeds the decrease in slope K.

For pulse compression signals having uniform spectrum over B, the slope k_t for delay measurement is exactly analogous to the monopulse slope k_m of Figure 8.4.5a. This is because the uniform received spectrum is analogous to the uniform field incident on an antenna aperture, with weighting applied both to the signal and the noise samples. The analogy cannot be applied, however, to signals received with nonuniform spectra. Procedures for evaluation of the error slope and resulting accuracy for signals with nonuniform spectra are discussed in [9.1]. Separate rms bandwidths must be found for the signal voltage spectrum, the filter voltage response, and the filter power response. Hence, the estimation error can be expressed as a function of input energy ratio, using these bandwidths and the filter matching loss. In general, the single-pulse error in delay estimation, for nonuniform signal spectra, can be expressed by substituting, in (9.1.2), the slope $K \le K_0$ and the single-pulse energy ratio E_1/N_0:

$$\sigma_{t_1} = \frac{1}{K\,\sqrt{2E_1/N_0}} \qquad (9.1.6)$$

The results for rectangular pulses are discussed in Section 9.2. For Gaussian pulses and filters (i.e., filters formed by several cascaded amplifier stages tuned to the same frequency), the results are shown in Figure 9.1.4, with K normalized to the input pulsewidth τ_3. Also shown in the figure are the filter efficiency $\eta_f = 1/L_m$, and the ratio of input to output pulsewidth. Note that the maximum value of $K\tau_3 \approx 1$ for the matched filter, for which the output pulse is broadened by $\sqrt{2}$.

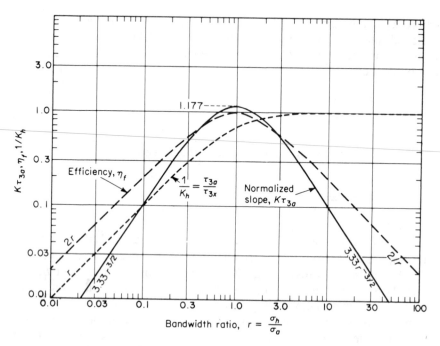

Figure 9.1.4 Error slope, efficiency, and pulse broadening for Gaussian pulse and filter.

Ranging on Pulse Trains

The optimum estimator of Figure 9.1.1 is applicable to coherent pulse trains, if the matched filter is implemented as a comb filter passing all lines of the signal spectrum. However, for noncoherent pulse trains, the envelope detector must precede integration. The optimum system then consists of an IF filter matched to the spectral envelope $A(f)$ of Figure 5.3.1, followed by the envelope detector, a video integrator, differentiator, and zero-crossing detector as in Figure 9.1.5. The matched filter integrates over each pulse of width τ, the video integrator operates from pulse to pulse over t_o, and the differentiator operates within the pulsewidth τ of the integrated output.

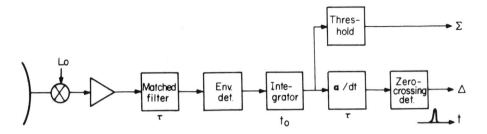

Figure 9.1.5 Optimum estimator for noncoherent pulse train.

The pulse-to-pulse matched filter for the coherent pulse train, and the wideband video integrator for the noncoherent pulse train, are not easily implemented, and are unnecessary for a tracker that is intended to obtain range on a single target. Instead, for coherent pulse trains, the correlator form of the matched filter is often used, which is identical to Figure 9.1.2, but with the narrowband filters integrating over the entire pulse train of duration t_o, rather than over τ. Here, the cascaded response $H_1(f)\,H_2(f)$, or $h_1(t) \otimes h_2(t)$ of the filter and correlator is matched to the echo pulse, and h_2' forms the derivative response. The phase-sensitive error detector converts the IF Δ signal to baseband, with proper polarity to serve as the error signal for the tracking loop. In the case of noncoherent pulse trains, the narrowband filters are restricted to integration over τ, and low-pass filters are placed after the phase detector for Δ and envelope detector for Σ to integrate over the pulse train.

The thermal noise errors for these processes on pulse trains are given by (9.1.2) for the matched filter. For the mismatched filter and the non-coherent pulse train, assuming adequately high single-pulse SNR, the error is

$$\sigma_t = \frac{1}{K \sqrt{2E/N_0}} = \frac{\tau_o}{k_t \sqrt{2\,(S/N)(n/L_m)}} \qquad (9.1.7)$$

When the ratio S/N at the error detector drops below about four, additional noise components are generated, as discussed in Section 10.4.

Range Measurement in a Digital Signal Processor

Modern digital signal processors normally perform approximate matched filtering prior to the final down-conversion to baseband or envelope detection. If the matched filtering and down-conversion are performed by analog circuits, the A/D converter will sample the detected output at a rate near B (perhaps somewhat higher to reduce straddling loss). The matched filter, in pulse compression, may also be performed digitally on IF signals sampled at a rate equal to twice the IF, or at rate B on signals converted to baseband after bandpass filtering to B. Whichever processes are used, digital samples are available at a rate of at least $1/\tau_n$, where τ_n is the width of the output pulse. These samples can be used to obtain range estimates. The process is similar to off-axis angle estimation, since the samples are taken at fixed intervals and the target delay may be at any point within the sampling interval.

The time response of a given digital sample at the receiver output is given by the impulse response of the preceding circuits (and signal processing steps), centered on that sample:

$$f_1(t) = f(t_1) = a_0(t_1)$$

where $a_0(t)$ is the waveform resulting from the receiver and the approximately matched filter, which precedes the digital sampling. The adjacent sample is

$$f_2(t) = f(t_2) = a_0(t_2)$$

The estimator, which forms the sum and difference of f_1 and f_2, is the time analog of the monopulse processor in angle. The Σ and Δ responses are formed according to (8.1.3) and (8.1.4), and normalized to Δ/Σ. The maximum accuracy is obtained for signals midway between two sampling points, corresponding to a target on the axis of the monopulse angle tracker. When the signal is near one of the sample times, two estimates can be made, one using the previous sample and another using the following sample. This helps to reduce the error for off-axis measurement, and if the samples are spaced somewhat more closely than τ_n (as will be the case for a low-sidelobe weighting) the error remains almost constant for any signal delay, and is given by (9.1.6).

9.2 RANGE MEASUREMENT ON A RECTANGULAR PULSE

The rectangular pulse requires special discussion because, for an ideal pulse, the rms bandwidth is infinite. This implies that the matched filter, followed by a differentiator, will have infinite noise output. As a result, some compromise must be made, either by assuming a band limit to the input signal or a mismatched filter in the receiver.

Optimum Estimation on a Rectangular Pulse

Information theory [9.2] gives a lower limit to the time-delay measurement error on the ideal pulse as

$$(\sigma_t)_{min} = \frac{\tau}{E/N_0} \qquad (9.2.1)$$

The inverse relationship with energy ratio, rather than with the square root of this ratio, is unusual, but has been explained [9.1, p. 67] by the requirement that an optimum processor include a band-limiting filter, which is adapted to the actual input energy ratio.

The practical transmitter limits the signal bandwidth to B_a, giving a trapezoidal pulse with rise time $\tau_e = 1/B_a$. In a classic study, Skolnik [9.3] showed that the rms bandwidth for the band-limited rectangular pulse could be approximated as

$$\beta_a = \sqrt{2B_a/\tau} \qquad (9.2.2)$$

For this pulse, using a matched receiver, the error in time delay measurement becomes

$$\sigma_t = \sqrt{\tau/4B_a\,(E/N_0)} \qquad (9.2.3)$$

This corresponds to the substitution in (9.1.2) of $\beta = \beta_a$ from (9.2.2).

This thermal noise error is shown in Figure 9.2.1 as a function of the input energy ratio and bandwidth limit. The minimum error predicted by Manasse [9.2] can only be achieved by using the adaptive band-limiting process of Figure 9.2.2. Regardless of the transmitted bandwidth B_a, the

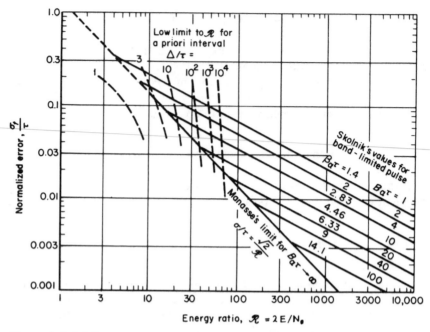

Figure 9.2.1 Measurement accuracy *versus* energy ratio for a rectangular pulse.

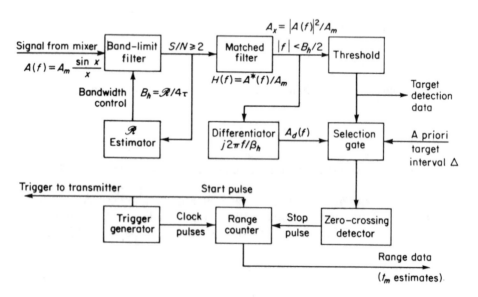

Figure 9.2.2 Adaptive estimator for rectangular pulse.

receiver bandwidth B_h must be limited to ensure an input $S/N > 2$ to the matched-filter and differentiator combination. This requires

$$B_h < (E/N_0)/2\tau \tag{9.2.4}$$

If a fixed value of $B_h < B_\alpha$ is used, it must be set according to (9.2.4) with the minimum expected energy ratio, E_{min}/N_0. The error for greater SNR will then follow the lines on Figure 9.2.1 identified as constant $B_\alpha\tau$, varying inversely with $\sqrt{E/N_0}$. Since the restriction on B_h leads, in general, to a mismatched filter, not all the transmitted signal energy will be useful in measurement. In cases where $B_h\tau < 1.5 < B_\alpha\tau$, a significant fraction of the signal energy will be excluded by the filter, and the value of E/N_0 used in (9.2.3) must be reduced by the matching loss.

Split-Gate Range Tracker

The correlator functions used in Figure 9.1.2 to produce the Σ and Δ outputs can be approximated, for rectangular pulses and other real signals, by the rectangular gates of Figure 9.2.3. Such gates can also be used on the output pulse after application of pulse compression to complex signals. The gates can be applied as modulator functions at IF, or at video as in Figure 9.1.3. The Σ gate controls the input to the receiver AGC, if used, and also the angle error channels of the radar.

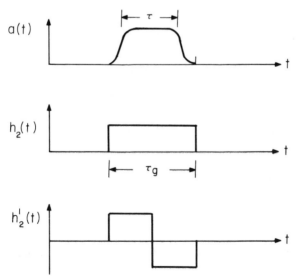

Figure 9.2.3 Waveforms for split-gate range tracker.

The gates constitute mismatched video correlator functions for the Σ and Δ channels of the range estimator, and the processor responses, which should approximate the matched filter and its derivative, are those of the gates in cascade with the IF filter $H_1(f)$. As with the IF or video correlator, a matched $H_1(f)$ should be followed by narrow gates (e.g., impulse sampling for an A/D converter). In this case, the split gate with $\tau_g \rightarrow 0$ represents a differentiator for the matched-filter output. If $H_1(f)$ has wideband response, τ_g should be matched closely to the half-power width τ_3 of the pulse at its input. The passage of the signal with wideband noise through the envelope detector will introduce a detector loss, increasing the error of the estimator (see Section 10.4).

The noise error in a split-gate estimator is described by

$$\sigma_{t_1} = \frac{1}{K\sqrt{2E_1/N_0}} \tag{9.2.5}$$

where the slope K, normalized to the 3-dB width of the input pulse, is shown in Figure 9.2.4. For a rectangular pulse with wideband IF filter, the optimum gate width is near $\tau_g = \tau$, with $K\tau_3 \approx 2$. The error sensitivity drops steeply if $\tau > \tau_g$. Because the echo from an extended target may be stretched beyond τ_3, it is customary to use $\tau_g \approx 1.5\tau_3$ to ensure good centroid tracking. For the rectangular pulse with a matched filter, the signal reaching the gate will be triangular. The optimum gate is then very narrow, responding to the slope reversal at the peak of the output pulse. The high values of $K\tau_3$ shown in Figure 9.2.4 represent approximations to the matched-filter estimator of Figure 9.2.2.

When a rectangular split gate is used on pulses with Gaussian shape, the results, as shown in Figure 9.2.4, can approach those for the rectangular pulse processed with a wide gate. The Gaussian shape characterizes the output of a low-sidelobe pulse compression system. The optimum gate width τ_g is about twice the half-power pulsewidth τ_3, but when the IF filter is matched to the Gaussian pulse a narrow gate should be used. In either case, the maximum error slope is near unity.

When using the split-gate estimator on a pulse train, the error is reduced by $\sqrt{n}$ relative to the single-pulse value.

Leading-Edge Tracker

The leading-edge tracker is a device that measures range to the leading edge of a near-rectangular pulse, ignoring the flat top and the trailing edge. The tracker is used primarily to select the closest target of a formation or a mass of distributed echoes, as in the case of an aircraft dropping confusion reflectors. Analysis of this tracker is simple when the signal bandwidth exceeds the bandwidth of the receiving filter, as will normally be the case. The simplified block diagram is shown in Figure 9.2.5. The pulse is passed through the filter $H_1(f)$, which introduces a sloped leading edge of rise time $\tau_e \approx 1/B_h$. This filtered signal $a_1(t)$ is differentiated once to produce a short pulse waveform $a_x(t)$ of width τ_e. Then, a second differentiator is used to locate the center of the short pulse. (A split-gate discriminator with $\tau_g \approx \tau_e$ may replace the second differentiator). The resulting error slope [9.1, p. 79) is

$$K\tau = \sqrt{5B_h\tau/9} \qquad (9.2.6)$$

This slope is worse by a factor of two than that of the optimum delay estimator for the rectangular pulse, reflecting the nonoptimum weighting and the loss of information in the trailing edge.

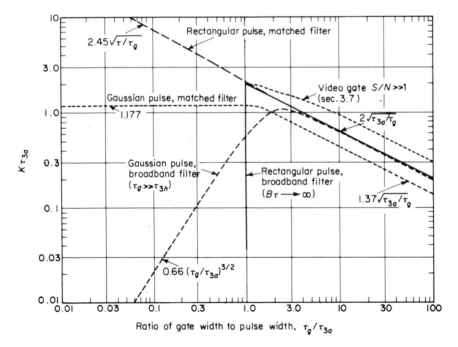

Figure 9.2.4 Normalized slope for split-gate tracker.

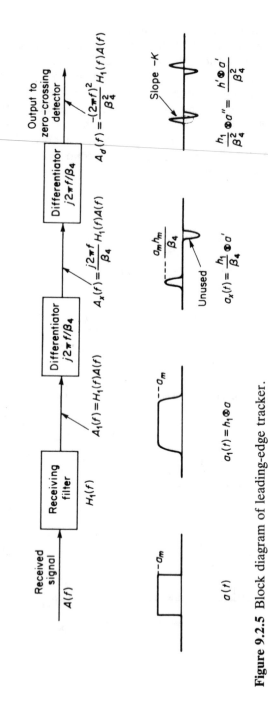

Figure 9.2.5 Block diagram of leading-edge tracker.

9.3 DOPPLER MEASUREMENT

Measurement of target doppler is directly analogous to range measurement, with time and frequency functions interchanged. In doppler measurement, however, two time and frequency scales must be considered:

(a) the pulsewidth τ and corresponding spectral envelope width B;
(b) the time-on-target t_o and corresponding width B_f of fine lines in the spectrum of the coherent pulse train.

Radars using CW or long-pulse transmissions may achieve meaningful resolution and measurement on the spectral envelope, but in most cases it is the fine-line spectrum of the coherent pulse train transmission that provides the data.

When doppler shift of the signal has been measured, it is related to the radial velocity of the target by

$$v_r = f_d \lambda / 2 \tag{9.3.1}$$

The purpose of doppler tracking is not always to perform this velocity measurement, but often to achieve resolution of the desired target from clutter and other targets. However, when the doppler filter is used for resolution there will be data available on radial velocity, and it is often advantageous to exploit these data in the radar system.

Doppler Measurement on a Single Pulse

The optimum estimator for the doppler shift of a single pulse or sample of a signal is shown in Figure 9.3.1. The two filters, offset by an amount $< B/2$ from the expected signal frequency, generate the two responses called for by the generalized measurement diagram of Figure 8.1.1. These responses are subtracted to obtain Δ and added to obtain the Σ channel output. The Δ channel is the familiar frequency discriminator.

Normalization of the output error can be by AGC using analog or digital division of the Δ by the Σ outputs, or by limiting in the preceding IF amplifier. The use of limiting depends on the other purposes of the Σ channel data, and whether there are interfering signals present in the stages prior to the doppler filters.

For the ideal estimator, two matched filters will be used, with small offset from the center frequency of the signal, and the estimation error in

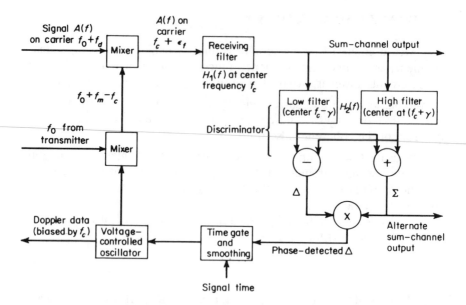

Figure 9.3.1 Basic doppler estimator.

thermal noise will be

$$(\sigma_f)_{\min} = \frac{1}{\alpha_0\sqrt{2E_1/N_0}} \tag{9.3.2}$$

Here, E_1 is the energy of the signal pulse or sample, and α_0 is its rms duration defined by

$$\alpha_0^2 = \frac{\displaystyle\int_{-\infty}^{\infty} (2\pi t)^2 \, |a(t)|^2 \, dt}{\displaystyle\int_{-\infty}^{\infty} |a(t)|^2 \, dt} \tag{9.3.3}$$

In most cases, a pulsed signal waveform will be rectangular (from the output of a saturated transmitter). The corresponding rms duration is

$$\alpha_0 = (\pi/\sqrt{3})\tau = 1.81\tau = 1.61/B_3 \tag{9.3.4}$$

where the final portion of the expression applies only to the rectangular pulse without phase modulation or coding. Substitution in (9.3.2) gives

$$(\sigma_f)_{min} = \frac{1}{1.81\tau \sqrt{2E_1/N_0}} \qquad (9.3.5)$$

In CW radar, the signal amplitude envelope and spectrum are often Gaussian, resulting from the scanning of a beam across the target position. If the time between one-way, half-power points of the beam is t_o, the Gaussian function describing the signal envelope will be

$$a(t) = \exp(-t^2/2\sigma_t^2) = \exp(-2.77t^2/t_0^2) \qquad (9.3.6)$$

The rms signal duration is

$$\alpha_0 = \sqrt{2}\,\pi\sigma_t = 1.89t_o = 1.18/B_3 \qquad (9.3.7)$$

where B_3 is the half-power width of the spectrum. This may be substituted into (9.3.2) to find the optimum doppler measurement accuracy.

The filters used to develop the discriminator function in Figure 9.3.1 need not be perfectly matched to the signal spectrum. For mismatched filters a reduced slope will replace α_o in the equation for thermal noise:

$$\sigma_{f_1} = \frac{1}{K_f \sqrt{2E_1/N_0}} \qquad (9.3.8)$$

where K_f, normalized to the half-power bandwidth B_3 of a Gaussian signal spectrum, is shown in Figure 9.3.2. The abscissa is the ratio of the widths of the signal spectrum and filter bandwidth. The optimum slope is $1.18/B_3$, as given in (9.3.7).

The discriminator is sometimes implemented as a nearly triangular function, band-limited to B_h as in Figure 9.3.3. When this function is applied to the spectrum of a rectangular pulse, the results are as shown in Figure 9.3.4. Again, the optimum slope is very near $1/B_3$, for discriminator bandwidths near $B_h = 2/\tau$.

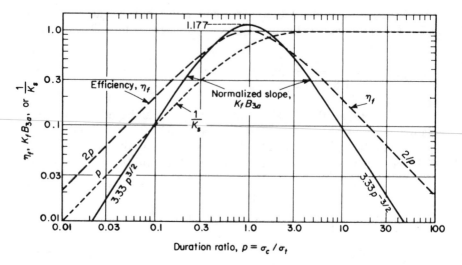

Figure 9.3.2 Error slope for Gaussian signal and filter.

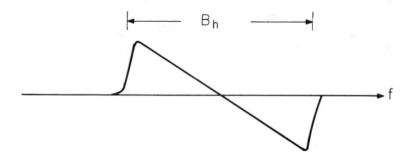

Figure 9.3.3 Triangular discriminator response.

Doppler Measurements on Pulse Trains

Measurements of doppler on noncoherent pulse trains are made by simply averaging the results of n single-pulse measurements. Equation (9.3.8) becomes

$$\sigma_f = \frac{1}{K_f \sqrt{2\,(E_1/N_0)n}} \tag{9.3.9}$$

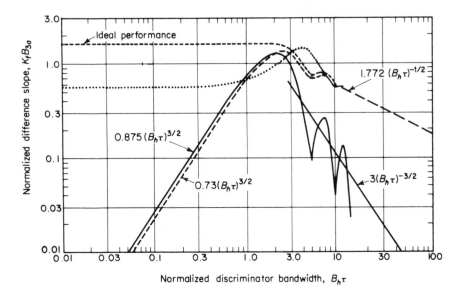

Figure 9.3.4 Error slope for triangular discriminator on rectangular pulse.

The slope K_f is calculated for the IF filter or correlator, using as B_3 the spectral width of the individual pulse or signal sample.

Doppler tracking on coherent pulse trains requires a system such as that shown in Figure 9.3.5. A voltage-controlled oscillator (VCO) offsets the receiver local oscillator signal to place the received signal at the intermediate frequency f_c. After filtering matched to the individual pulse (the cascaded combination of a spectral envelope filter and a range gate), the signal is passed through the fine-line filter for the Σ channel, and a discriminator for the Δ channel. The discriminator output controls the VCO to keep the signal centered in the passband of the processor. Doppler data, offset by f_c, may be extracted by a counter on the VCO output. The correlator equivalent of this circuit is discussed in [9.1, p. 117].

The processor is now matched to the entire pulse train, extending over the time t_o. Equation (9.3.2) now uses

$$\alpha_0 = 1.81 t_o = 1.18/B_f$$

where B_f is the 3-dB width of the fine spectral line. For mismatched processors, the curves of Figures 9.3.2 and 9.3.4 can be applied, using B_f in place of B_3 and (for a uniform burst of pulses) t_o in place of τ. In general,

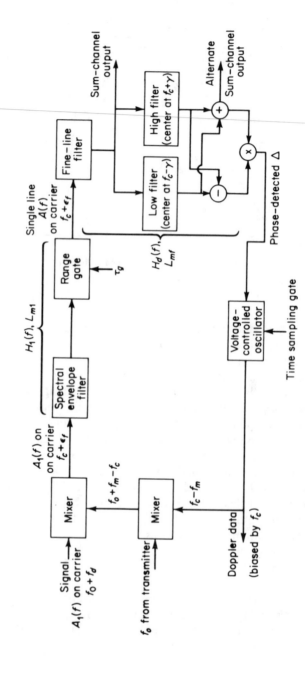

Figure 9.3.5 Block diagram of pulsed doppler tracker.

the slope K_f is now approximately the reciprocal of the fine-line width B_f, rather than of the width of the spectral envelope. The energy ratio is that of the entire pulse train, rather than of a single pulse. Further advantages of the fine-line processor are the following.

(a) The SNR presented to the envelope and error detectors at the receiver output is now that of the entire pulse train, coherently integrated, and improved by a factor approaching $n = f_r t_o$ relative to that of the single pulse. The effects of operation at low SNR (Section 10.4) are avoided.
(b) Improvement in the signal-to-clutter ratio is also obtained when the target doppler is resolvable from that of the clutter.

Resolution of Doppler Ambiguities

In processing of coherent pulse trains, there will be ambiguity as a result of the presence of nearly identical lines in the spectrum, spaced by f_r. Within the main lobe of the spectrum of a rectangular pulse, there will be $2/\tau f_r = 2/D_u$ lines. To resolve the ambiguity, the VCO in Figure 9.3.5 must be designated to within $\pm f_r/2$ of the central line in the spectrum. This can be done in any of the following ways:

(a) The signal frequency may be known *a priori,* on the basis of external measurements or physical constraints, to lie within a particular ambiguous interval f_r;
(b) The signal frequency may be measured noncoherently over the pulse train, in addition to the coherent measurement, with sufficient accuracy to identify the proper interval;
(c) The frequency may be computed from measured range rate, as measured over the period of the pulse train;
(d) The ambiguity may be resolved by measurements at two or more PRFs.

To apply the second option, with high probability of success (e.g., 99.7%), we require that the rms error in the noncoherent measurement be

$$\sigma_{fn} = \frac{\sqrt{3}}{\pi\tau\sqrt{2E/N_0}} < f_r/6$$
$$\sqrt{E/N_0} > 2.34/D_u$$

(9.3.10)

where D_u is the duty factor of the waveform. This is an acceptable constraint for values of D_u that are not too small, but for short-pulse radars ($D_u < 0.01$), it is usually preferable to use the third option.

The requirement for option (c) is that the error in range rate, estimated for a series of n pulses over time t_o, be less than $v_b/6$. This gives

$$\sigma_{\dot{r}} = (\sigma_{r_1}/t_o)\sqrt{12/n} = \sqrt{12}\,\sigma_r/t_o < f_r\,\lambda/12$$

$$\sqrt{E/N_0} > 15\,f_0/nB_3 \tag{9.3.11}$$

where $f_0 = c/\lambda$ is the radar frequency and $K_f = 1/B_3$ has been assumed. For example, in an S-band radar with $B_3 = 1$ MHz, $f_r =$ kHz, and $t_o = 1$ s, there will be $n = 1000$ pulses. The requirement is for $\sqrt{E/N_0} = 45$, or $E/N_0 = 2000$, which is only 3 dB per pulse. In most short-pulse and pulsed doppler tracking radars operating on a single target, the use of differentiated range data provides a rapid means of resolving the PRF ambiguity in doppler data.

The use of PRF diversity to resolve doppler ambiguity is often a prerequisite for satisfactory acquisition or identification of the target, before tracking begins. In such cases, it can ensure unambiguous velocity data without waiting for further measurements. If the target is detected on two different PRFs, and measurements are obtained on the apparent radial velocity for each detection, the two values will be

$$v_1 = v_r - iv_{b_1} + \epsilon_1$$

and

$$v_2 = v_r - jv_{b_2} + \epsilon_2$$

where i and j are integers, v_{b1} and v_{b_2} are the two blind speeds, and ϵ_1 and ϵ_2 are the errors in measurement. The problem is to find unique values of i and j that correspond to a v_r within the physical limitations of an actual target. Since values of ϵ are unknown, they may be replaced by $\pm 3\sigma_v$ and solution for i and j attempted. Figure 9.3.6 shows a typical case in which the measurement error is low enough to permit a unique solution. The true $v_r = v_1 + 4v_{b_1} = v_2 + 3v_{b_2}$. If errors are neglected, the problem can be solved by using the Chinese remainder theorem. Other solutions are often more convenient when digital processing is available. In some cases, three or more PRFs must be used to obtained unambiguous results. This procedure also can resolve range ambiguities in medium- and high-PRF waveforms, and is likely to be used in those systems when PRF switching to avoid blind regions generates streams of data with ambiguities in both coordinates.

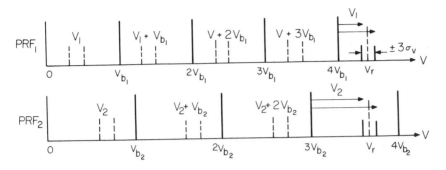

Figure 9.3.6 Resolution of ambiguities with two PRFs.

Four-Coordinate Tracking Radar

Range and doppler tracking can be combined with monopulse angle tracking to implement the four-coordinate tracking radar, diagrammed in Figure 9.3.7. The receiver uses three input channels, the Σ, Δ_a, and Δ_e channels, the signals of which are formed by the monopulse antenna. These signals are down-converted to IF by a local oscillator derived from the same synthesizer that generates the transmitter frequency. The range tracker operates on the Σ channel, generating Δ_r by applying a split gate to the Σ channel at IF. The Σ range gate is applied to the two angle error channels as well as to the Σ channel. The four resulting channels, now including Δ_r, are then down-converted to a second IF, where narrowband filters are applied to integrate over the pulse train. After this filtering, the Σ channel is further divided to serve as a reference phase input to the angle error detectors and the range error detector, and as an input to the discriminator (the Δ_v channel). The discriminator output controls the VCO, which is the input to the down-conversion to the second IF. The range error detector controls the ranges gates applied to the first IF. The angle error detectors control the antenna servomechanisms.

In four-coordinate tracking, operation depends on having all four loops locked to follow the target. In target acquisition, it is necessary to have designation data from an external source to avoid prolonged search in all four coordinates. This subject will be discussed further in Section 10.1.

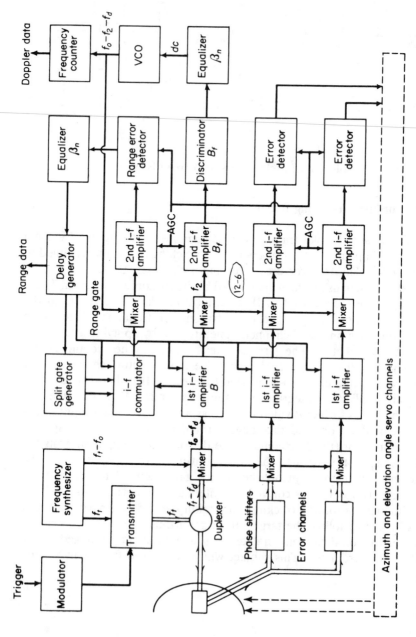

Figure 9.3.7 Four-coordinate monopulse tracker.

REFERENCES

[9.1]   D.K. Barton and H.R. Ward, *Handbook of Radar Measurement,*
        Artech House, 1984.

[9.2]   R. Manasse, Range and Velocity Accuracy from Radar Measure-
        ment, MIT Lincoln Laboratory Rep. 312–26, February 3, 1955,
        DDC Doc. AD 236236.

[9.3]   M.I. Skolnik, Theoretical Accuracy of Radar Measurements, *IRE
        Trans.* **ANE-7,** No. 4, December 1960, pp. 123–129.

Chapter 10
Tracking Radar Acquisition and Target Considerations

10.1 TARGET ACQUISITION

A special application of search-radar theory is concerned with the problem of target acquisitions by the narrow-beam tracking radar. This section will review the application of search-radar theory to this special case, considering the type of designation data made available to the tracking radar, the errors in this data, and the requirements for reliable target acquisition.

Application of the Search Radar Equation

During acquisition the tracking radar must operate in a search mode over a limited volume of space. The search radar equation discussed in Section 1.3 is the basic relationship which determines the range at which a tracker can acquire its target:

$$R_m^4 = \frac{P_{av}A_r t_s \sigma}{4\pi\psi_s k T_s D_0(1) L_s} \qquad (10.1.3)$$

In the case of the tracker, the search solid angle ψ_s must be large enough to ensure a high probability that the desired target lies within the scan. The scan frame time t_s and the detection probability per scan are specified so that the required cumulative probability of acquisition P_{ca} is obtained within a time t_a. As the search angle is reduced, by high-quality designation data, the acquisition range R_m increases. In the limit, the tracker beam is pointed at a known target position, and signal integration over the entire time t_a can be used to detect targets at low SNR. The acquisition capability of the radar is increased if the signal processor has the ability to integrate signals in many range and doppler cells, covering the uncertainty volume in those coordinates.

Sources of Designation

Designation of the target can come from different sources of differing accuracies in each radar coordinate.

(a) *Optical Designation*. When the target is visible to a collocated optical instrument, designation in both angle coordinates is available with high accuracy. Scanning in angle will be unnecessary, but range and velocity of the target will be unknown. The solid angle ψ_s will be equal to the beam angle $\psi_b = \theta_a\theta_e$, and the detection range can be found directly from (1.2.26), with $t_o = t_a$:

$$R_m^4 = \frac{P_{av}t_a G_t G_r \lambda^2 \sigma F^4}{(4\pi)^3 k T_s D_x(1) L_t L_\alpha} \tag{10.1.2}$$

This is the equation on which the Blake chart and RGCALC were based.

(b) *2D Search-Radar Designation*. A collocated 2D search radar can supply azimuth and range designation data to the tracker. Elevation (and usually radial velocity) are unknown. The tracker must scan in elevation over the sector which is covered by the search radar, and in azimuth by an amount sufficient to cover the $\pm 3\sigma_x$ error of the designation data (often about half the azimuth beamwidth of the search radar). Extensive range search may not be needed, relieving the tracker of the requirement for a processor with many range gates. Mechanically steered trackers may be limited in the speed with which the elevation sector can be searched, and the scan may be inefficient because of the time requirement to reverse the scan at the upper and lower limits. Hence, the uniform distribution of search energy, used in deriving (10.1.1), may not be realized and an additional loss factor may be needed in the denominator. In particular, when only elevation scanning is needed, the distribution of energy may actually peak at the ends of the scan where there is the lowest probability of finding the target.

The search radar is often located at some distance from the tracker, requiring that the designation data be converted to Cartesian coordinates, translated to those of the tracker site, and reconverted to tracker spherical coordinates. When accompanied by a time lag in transferring the designation data, this will increase the region of uncertainty over which the radar must search. When the separation between the search radar and the tracker is large, and the target is at high altitude, a further error may be introduced by the inability of the 2D radar to identify the ground range $(G = R \cos E)$, causing errors in the Cartesian coordinates in the ground plane. Search over significant intervals in all three spatial coordinates may then be required.

(c) *3D Search Radar Designation.* The availability of elevation data from the 3D search radar eliminates most of the uncertainties in designation data to the tracker, even when the radars are not collocated. The tracker scan volume should normally be small for this type of designation.

(d) *Jammer Strobe Designation.* In the jamming environment, targets carrying jammers may be designated to the tracker for engagement, based solely on angle data. Designation from a collocated 2D radar will consist of an azimuth jammer strobe, usually quite accurate. The tracker will have no difficulty in locating the same target if it is also emitting on the tracker frequency. If the jammer is quiet at the tracker frequency, an elevation sector scan will be needed. When triangulation from 2D radars is available, range data are also available, subject to the limitations of ambiguity when multiple jammers are present. Jammer designation from collocated 3D radar will eliminate the need for angle scanning, as will triangulated data from two 3D radars.

Establishing the Scan Volume

The probability of acquisition will be product of two probabilities: P_v, the probability that the target lies within the scan volume; P_d, the probability of detection of a target if it lies within this volume. When those probabilities can be stated for each of n_v resolution elements, the overall probability of acquisition can be written as

$$P_a = \sum_{i=1}^{i=n_v} P_{vi} P_{di} \tag{10.1.3}$$

The first step in establishing the acquisition scan for a tracker is to identify the sector width in azimuth and elevation which must be scanned in order to obtain adequate P_v for the designation data available. It will be assumed that range and doppler acquisition intervals will play a small role in the probability of acquisition, because modern signal processors can be designed to cover at small cost the entire unambiguous area in range and doppler.

The designation data in azimuth and elevation will be characterized by standard deviations σ_a and σ_e, and the scan will normally extend over sectors A_m and ΔE, which are proportional to the corresponding values of σ. If we let X/σ_x be the ratio of scan sector to rms designation error in one coordinate, Table 10.1 gives the probabilities that the target will lie within the scan volume for linear, rectangular, and elliptical scans (Figure 10.1.1).

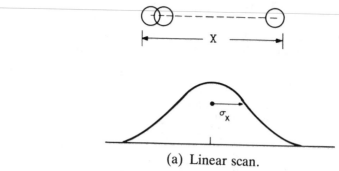

(a) Linear scan.

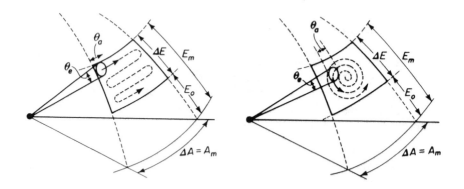

(b) Rectangular raster scan.   (c) Elliptical scan.

Figure 10.1.1 Acquisition scan parameters.

Table 10.1 Probability of Target's Lying in Scan Volume

X/σ_x	P_x	P_2	P_r
0.5	0.20	0.04	0.03
1.0	0.38	0.14	0.10
2.0	0.68	0.46	0.35
3.0	0.87	0.76	0.66
4.0	0.955	0.91	0.86
5.0	0.988	0.976	0.95
6.0	0.997	0.994	0.99
7.0	0.9995	0.999	0.998
8.0	0.9999	0.9998	0.9997

The probability of failing to acquire the target because it lies outside the scan volume will normally be assigned about one-half the allowable probability of failure to acquire. Thus, if $P_a = 0.9$ is required, $P_v = 0.95$ will be assigned. For a linear scan (uncertainty exceeding the tracker beamwidth in one coordinate), the scan sector should then be $4\sigma_x$ in that coordinate. For a two-coordinate rectangular scan (e.g., raster scan centered on the designated point), the limits would be almost $5\sigma_x$, while for an elliptical (spiral) scan about the designated point a diameter of $5\sigma_x$ would be used.

Number of Scans for Acquisition

The required cumulative probability of acquisition P_{ca} is normally achieved by making m scans centered on the designated point, with the probability of acquisition on each scan given by

$$P_a = P_v P_d = 1 - (1 - P_{ca})^{1/m} \tag{10.1.4}$$

This equation is based on the assumption that the probability P_a is independent from scan to scan, as will be the case if the designation error and the target signal are both independent over t_a/m. If the designation error remains constant over the m scans, and only P_d is independent from scan to scan, the equation becomes

$$P_d = 1 - (1 - P_{ca}/P_v)^{1/m} \tag{10.1.5}$$

Unless frequency diversity is used to reduce target fluctuation loss, the normal procedure is to scan the volume about $m = 4$ times in achieving high P_a. Consider as an example the requirements for scan volume and single-scan P_d to achieve $P_{ca} = 0.99$ in four scans. If the designation error is independent from scan to scan, we find, from (10.1.4),

$$P_a = P_v P_d = 0.68 \text{ and } P_v \approx P_d \approx 0.83$$

The scan would extend over a sector about $3.5\sigma_x$ wide in each of two angle coordinates, and for $P_d = 0.83$ the fluctuation loss would be 6 dB. If the designation error remains constant over the four scans, (10.1.5) requires $P_v > P_{ca}$. Assuming $P_v = 0.995$, we find

$$P_d = 0.73 \text{ and } P_a = P_v P_d = 0.726$$

The scan must now extend just beyond $6\sigma_x$, and the fluctuation loss for $P_d = 0.73$ would be 4 dB. In the first case, the probability of acquisition would increase rapidly toward unity as the scanning continued, whereas in the second the probability would increase rapidly to 0.995 and no further because of the nonzero probability that the target lies outside the scan sector.

The division of the allowable acquisition time t_a among m scans can be optimized by balancing the reduction in $t_s = t_a/m$ against the reduction in detectability factor and fluctuation loss for the reduced P_d in (10.1.4) or (10.1.5).

Example of Acquisition Range Calculation

In Chapter 1, a radar example was given for a horizon-scanning search radar using a $1.3° \times 2°$ beam. For a comparison of tracker acquisition range, these same radar parameters can be considered in a mode where the antenna performs a small raster scan about the designated target position. Assume that the designation errors from a remote 2D radar are

$$\sigma_a = 2°, \quad \sigma_r = 3 \text{ km}$$

and that the target is restricted to altitudes below $h_m = 15$ km. The elevation sector scan must cover

$$E_m \approx \sin E_m = h_m/R_m$$

Since the acquisition range R_m is unknown, this equation for E_m will be substituted in the search radar equation.

Assume that $P_{ca} = 0.99$ is required within $t_a = 10$ s from the time of target designation. Calculations will first be made with

$$m = 4 \text{ scans}, \quad t_s = 10/4 = 2.5 \text{ s}$$

$$P_v = 0.995, \quad P_d = 0.73$$

$$A_m = 6\sigma_a = 12°, \quad \Delta R = 6\sigma_r = 18 \text{ km}$$

To determine the detectability factor $D_0(1)$, a false-alarm probability must be assigned. We will assume that an MTD processor is used, in which the number of doppler filters is

$$n_f = n = f_r t_o$$

From (2.3.19) we can write

$$P_{fa} = t_o/n_t n_f t_{fa} = 1/n_t f_r t_{fa}$$

where $n_t = 2\Delta R/\tau c = 120$. The false-alarm time t_{fa} will be set at 40 s, allowing one false alarm per four acquisition attempts. The false-alarm probability can then be calculated:

$$P_{fa} = 1/(120 \times 1108 \times 40) = 2 \times 10^{-7}$$

The reasoning for choice of t_{fa} is that a false alarm will interrupt the acquisition scan for about 2 s, reducing the average time available for scanning to 9.5 s. This will have negligible effect on the probability of acquisition, and if necessary the effect can be reduced further by a slight reduction in P_{fa}. The radar loss budget will be as shown in Table 10.2.

Table 10.2 Loss Budget for Acquisition Calculation

Loss Factor	Value (dB)	Source of Data
Transmitter loss, L_t	1.0	Fig. 1.2.5
Antenna loss, L_a	0.2	Fig. 1.2.5
Antenna beam factor, L_n	2.0	Sec. 1.3
Atmospheric loss, L_α	1.2	Fig. 1.2.5
Matching factor, M	0.8	Fig. 1.2.5
Beamshape loss, L_p	2.5	Sec. 1.3
Miscellaneous loss, L_x	7.1	Table 5.4.2
Integration loss, L_i	0	Assume coherent MTD
Fluctuation loss, L_f	4.0	Fig. 2.4.2
Total L_s	18.8 = 75.9	

Equation (10.1.2) now gives

$$R_m = 144 \text{ km}$$

This is somewhat beyond the unambiguous range of the waveform, and in the absence of clutter a reduction of PRF would be appropriate.

The acquisition scan for this example can now be specified, based on a maximum range equal to the unambiguous range $R_u = 134$ km. Thus, we have

$$\sin E_m = 15/134 = 0.11 \approx E_m$$
$$E_m = 6° = 3.2\theta_e, \quad A_m = 12° = 9.2\,\theta_a$$

The number of search beam positions is

$$n_s = A_m E_m / \theta_a \theta_e = 30$$

An appropriate scan pattern would consist of four bars at elevations separated by $1.6° = 0.8\theta_e$, overlapping at the -2 dB (one-way) level. Each bar would be traversed in 0.5 s at $\omega = 24°/s$, with 0.125 s allowed for reversal of the scan (and moving to the next elevation) at the end. The time-on target would be

$$t_o = \theta_a / \omega = 0.054 \text{ s}$$

This allows for transmission of two bursts of 32 pulses per beamwidth, adequate for diversity coverage of blind speeds. The loss in useful energy at the ends of the scans would be $0.625/0.5 = 1.25$, or 1 dB. This would reduce the acquisition range to 136 km.

In the presence of rain, use of circular polarization would introduce a target loss of about 3 dB, reducing the range to $R_m = 115$ km. In chaff, the low-PRF S-band radar example would not be able to operate at so long a range, for reasons discussed in Section 7.2.

To test for optimization of the number m of scans during the 10 s acquisition time, calculations can be run for $m = 5$. This reduces the time $t_s = t_a/m$ by a factor $1.25 = 1$ dB. The required detection probability from (10.1.5) becomes $P_d = 0.65$, reducing the required $D_0(1)L_f$ by 1.2 dB. However, a somewhat greater fraction of the time will be lost in turning at the end of each bar, so the 0.2 dB advantage cannot be realized. The conclusion is that $m = 4$ is optimum for a mechanical scanning antenna, although $m = 5$ may be 0.2 dB better for a system with electronic scan.

10.2 DYNAMICS OF TRACKING

Tracking radar data are used to reconstruct the trajectories of moving targets, and to predict the future trajectories of targets. The target dynamics are crucial factors in establishing tracker requirements and available accuracies in trajectory measurement. Data on straight-line trajectories, or others such as satellite orbits which are tightly constrained by laws of motion, can be smoothed over long periods to reduce random errors. Targets that have random acceleration capabilities must be tracked with short smoothing times (wide data bandwidths), and predictions are limited to time not too far in the future.

Target Dynamics

The target trajectory is characterized, at any given time, by its present position and its derivatives in three spatial coordinates. The limits on the extent of the position coordinates to be tracked are set by radar coverage and line of sight, and different classes of targets have limits on their altitudes, speeds, accelerations, and higher derivatives. For example, Table 10.3 shows typical limits for several classes of radar target.

Table 10.3 Trajectory Limits for Typical Targets

Target Type	Altitude h_m (km)	Speed v_t (m/s)	Acceleration a_t (m/s^2)
Aircraft (subsonic)	15	300	60
(supersonic)	30	1000	80
Missiles (short range)	100	2000	150
(ICBM)	1000	7000	600
Satellites	40000	10000	10
Ships	0	30	2
Land vehicles	0	50	4
Troops	0	2	2

When viewed in radar (spherical) coordinates, the target will have velocities limited by v_t:

$$\dot{R}_{max} = v_t, \quad \dot{A}_{max} = v_t/R, \quad \dot{E}_{max} = v_t/R$$

Accelerations in the spherical coordinates are not limited by a_t, being composed of two components. The real accelerations are

$$\ddot{R}_{max} = a_t, \quad \ddot{A}_{max} = a_t/R, \quad \ddot{E}_{max} = a_t/R$$

The *geometrical accelerations* and higher derivatives are caused by the nonlinear relationships between Cartesian and spherical coordinates. The geometry of the classical pass-course problem is illustrated in Figure 10.2.1. The corresponding geometrical accelerations and other derivatives, for straight-line trajectories at constant speed, are shown in Tables 10.4 to 10.6 and Figures 10.2.2 to 10.2.4. In the angular coordinates, these derivatives are all normalized to $\dot{A}_{max} = \omega_m = v_t/R_c$, where $R_c = R_a \cos E_{max}$ is the ground range at crossover (the point of closest approach to the radar). In range, they are normalized to $R_a = R_{min}$ at the point of closest approach.

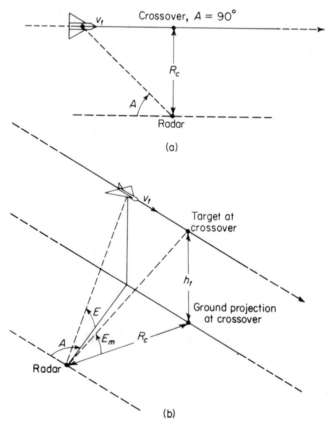

Figure 10.2.1 Geometry of the pass-course problem: (a) ground projection of flight path; (b) pass course with target in level flight.

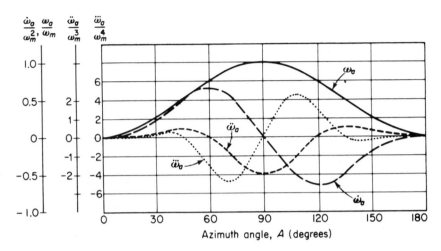

Figure 10.2.2 Derivatives of azimuth for pass course.

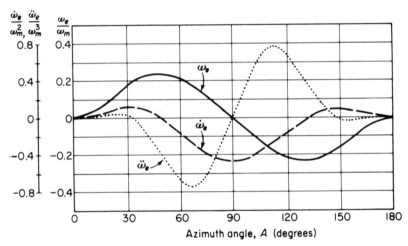

Figure 10.2.3 Derivatives of elevation angle for pass course ($X = 1$).

For example, a subsonic aircraft or missile at 3 km altitude, crossing at a range of 3 km, has the following peak magnitudes of velocity and acceleration in radar coordinates:

$$\omega_m = 0.1 \text{ r/s}, \quad \dot{\omega}_a = 0.0065 \text{ r/s}^2$$

$$\omega_e = 0.024 \text{ r/s}, \quad \dot{\omega}_e = 0.005 \text{ r/s}^2$$

$$\dot{R}_{\max} = 300 \text{ m/s}, \quad \ddot{R}_{\max} = 21 \text{ m/s}^2$$

Table 10.4 Derivatives of Azimuth Angle for Pass Course

Derivative	Equation	Peak Value	Peak Azimuth
$\dot{A} = \omega_a$ $\omega_m \sin^2 A$		$\omega_m = \dfrac{v_t}{R_c}$	90°
$\dot{\omega}_a$ $2\omega_m^2 \sin^3 A \cos A$		$\pm \dfrac{3\sqrt{3}\,\omega_m^2}{8} = \pm 0.65\omega_m^2$	60°, 120°
$\ddot{\omega}_a$ $2\omega_m^3 \sin^4 A(4\cos^2 A - 1)$	$-2\omega_m^3$		90°
$\dddot{\omega}_a$ $24\omega_m^4 \sin^5 A \cos A \cos 2A$	$\mp 4.5\omega_m^4$		72°, 108°

Table 10.5 Derivatives of Elevation Angle for Pass Course

First Derivative	Peak Values	Peak A
$\dot{E} = \omega_e = X\omega_m \dfrac{\sin^2 A \cos A}{1 + X^2 \sin^2 A}$	$\pm 0.38X\omega_m$ $\pm 0.24\omega_m$ $\pm \omega_m/X$	55°, 125° $(X \ll 1)$ 48°, 132° $(X = 1)$ 0°, 180° $(X \gg 1)$

Second Derivative	Peak Values	Peak A
$\dot{\omega}_e =$ $X\omega_m^2 \sin^3 A \dfrac{2 - 3\sin^2 A - X^2 \sin^4 A}{(1 + X^2 \sin^2 A)^2}$	$-X\omega_m^2$ $-0.5\omega_m^2$ $-\omega_m^2/X$	90° $(X \ll 1)$ 90° $(X = 1)$ 90° $(X \gg 1)$

NOTE: $X \equiv h_t/R_c$.

Table 10.6 Derivatives of Range for Pass Course

$$\ddot{R} = -\frac{v_t^2}{R_a}\sin^3\alpha \qquad \left(\text{max: } \frac{-v_t^2}{R_a}\right)$$

$$\dddot{R} = -\frac{3v_t^3}{R_a^2}\sin^4\alpha \cos\alpha \qquad \left(\text{max: } \pm 0.85\frac{v_t^3}{R_a^2}\right)$$

$$\ddddot{R} = -\frac{3v_t^4}{R_a^3}(5\cos^2\alpha - 1)\sin^5\alpha$$

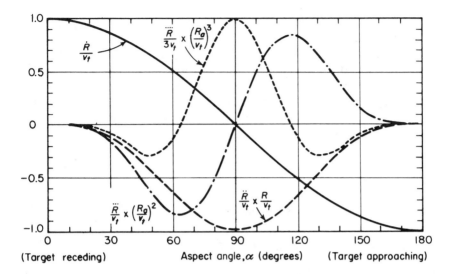

Figure 10.2.4 Derivatives of range for pass course.

The peak geometrical accelerations are all equivalent to a target with about $20 \text{ m/s}^2 = 2g$ maneuver, even though the target moves in a straight line with constant speed.

Tracking Error Coefficients

Conventional servomechanism theory can be used to predict the lag errors in the radar tracking loops, as a function of servo bandwidth and other design parameters. In one formulation, assuming that a track has been established long enough to remove initial transients, the lag error is written (for example, in azimuth) as

$$\epsilon_a = \frac{\omega_a}{K_v} + \frac{\dot{\omega}_a}{K_a} + \frac{\ddot{\omega}_a}{K_3} + \cdots \tag{10.2.1}$$

The coefficients K_v, K_a, K_3 . . . are the servo error coefficients, the values of which increase with increasing loop gain and bandwidth.

Servos are classified according to the first coefficient that is finite in the loop design: the type 1 servo has a finite K_v (but infinite position error constant K_0); type 2 has finite K_a, but infinite K_v. Infinite K_a and higher coefficients are not available. Even the type 1 servo can be designed with such a high value of K_v that the error is negligible for the highest target velocity, so it is primarily K_a and the target acceleration (real and geometrical) which determine the dynamic lag error of the radar.

The acceleration error coefficient is intimately connected to the closed-loop bandwidth of the servo. If this bandwidth is expressed in terms of the equivalent noise bandwidth β_n (measured positive from zero frequency), we may write

$$K_a \approx 2.5\beta_n^2 = 0.63/t_o^2 \qquad (10.2.2)$$

where t_o is the equivalent averaging time of the tracking loop. The acceleration lag error can thus be written as

$$\epsilon_a = \dot{\omega}_a/K_a = \dot{\omega}_a/2.5\beta_n^2 = 1.6\dot{\omega}_a t_o^2 \qquad (10.2.3)$$

It is this relationship, which applies also in elevation and range coordinates with the corresponding acceleration components, which sets the maximum averaging or smoothing of data in the radar spherical coordinates.

Optimum Servo Bandwidth

The optimum servo bandwidth can be found for minimum total tracking error. If the important error components consist of thermal noise and dynamic lag, (8.4.5) and (10.2.3) can be combined to give the variance of the total error:

$$\sigma_\theta^2 = \frac{\theta_3^2\beta_n}{k_m^2 f_r B\tau(S/N)} + \frac{a_t^2}{6.3R^2\beta_n^2} \qquad (10.2.4)$$

The optimum bandwidth is found as

$$\beta_n = \left[\frac{a_t^2 k_m^2 f_r B\tau(S/N)}{1.6\theta_3^2 R^2}\right]^{1/5} \qquad (10.2.5)$$

For example, assume the following parameters:

$a_t = 20$ m/s^2 (real or geometrical 2g), $k_m = 1.6$

$f_r = 1000, \ldots B_\tau = 1, \ldots S/N = 10,$ $\theta_3 = 0.02\ r$

The optimum bandwidth is 2.8 Hz, giving

$\sigma_\theta = 0.2$ mr for thermal noise;

$\epsilon_a = 0.1$ mr for acceleration lag;

$\sigma_\theta = 0.22$ mr total error.

Equation (10.2.5) applies equally to azimuth and elevation servo loops. For the range tracker, the term $R\theta_3$ in the denominator is replaced by the range resolution cell $\tau_n c/2$, and k_m in the numerator is replaced by k_t or $K\tau_3$. The noise is greater than the lag, for optimum bandwidth, because a slight reduction in bandwidth would cause lag to increase much faster than the decrease in noise. If the remaining error components are independent of servo bandwidth, the total error may increase but the optimum bandwidth will not change. If other errors vary directly with bandwidth or with its square root, the optimum bandwidth will be less than given by (10.2.5).

Bandwidth in Sampled-Data Systems

In sampled-data systems, such as those used in track-while-scan (TWS) operations on data from scanning radars, the loop bandwidth is often unstated. The α-β tracker [10.1] produces, on the observation $x_o(k)$, smoothed output estimates $x_s(k)$ for position and $v_s(k)$ for velocity, and a predicted position $x_p(k + 1)$ for the next observation t_t seconds later, given by

$$x_s(k) = x_p(k) + \alpha[x_o(k) - x_p(k)]$$
$$v_s(k) = v_s(k - 1) + (\beta/t_t)[x_o(k) - x_p(k)] \qquad (10.2.6)$$
$$x_p(k + 1) = x_s(k) + t_t v_s(k)$$

The coefficients α and β determine the dynamic performance of the loop, and also the noise outputs in x_s and v_s. The α-β tracker is a type 2 servo, having zero lag for constant target velocity. The servo's lag for constant target acceleration is

$$\epsilon_a = (1 - \alpha)t_t^2 a_t/\beta \qquad (10.2.7)$$

The error σ_s of the random noise in the output can be expressed in terms of the input noise σ_1 by

$$\sigma_s = \sigma_1 \left[\frac{2\alpha^2 + \beta(2 - 3\alpha)}{\alpha[4 - \beta - 2\alpha]} \right]^{1/2} \tag{10.2.8}$$

In [10.1, App. 2A], Washburn has related these TWS loop parameters to those of the conventional tracking servo:

$$\beta_n = \sigma_s^2/\sigma_1^2 t_t \tag{10.2.9}$$

For typical parameters $\alpha = 0.6$, $\beta = 0.43$, the ratio of output noise to input is $\sigma_s/\sigma_1 = 0.75$, $\beta_n = 0.56/t_t$, and the acceleration error coefficient is $K_a = 0.78/t_t^2 = 2.5\beta_n^2$. These relationships are based on data obtained on samples spaced t_t apart (unity detection probability per scan, in TWS).

Phased array trackers are often designed to minimize the required tracking data rate $f_t = 1/t_t$. Equations (10.2.7) to (10.2.9) may be used to determine the rate required for a given accuracy. For example, assume that a monopulse phased array achieves an angle precision $\sigma_1 = 0.01\theta_3$ on each track dwell. For a beamwidth $\theta_3 = 1.7° = 30$ mr, this corresponds to $\sigma_1 = 0.3$ mr. When processed in the α-β filter with the typical parameters given above, the output data will have $\sigma_s = 0.23$ mr, corresponding to 2.3 m error at a target range of 10 km. If the track data rate $f_t = 4$ Hz, $t_t = 0.25$ s, the acceleration error constant will be

$$K_a = 0.78/t_t^2 = 12.5$$

The error on a target acceleration $a_t = 2g = 20$ m/s^2 will be 1.6 m. The allowable lag error and the target acceleration will determine the required tracking data rate for phased array radar systems. The duration of each dwell will be determined by the required single-dwell precision σ_1, the single-pulse energy ratio, and the requirements for clutter rejection.

Characteristics of type 3 trackers and those based on the Kalman filter are also discussed in [10.1], along with adaptive filters and comparison of the various techniques. Proper application of the Kalman filter, and conversion of radar coordinates into Cartesian coordinates before application of smoothing and tracking filters, can eliminate the effects of geometrical accelerations and other predictable effects, reducing the required tracking data rate to that needed to accommodate unpredictable target accelerations. This is an important consideration in multiple-target tracking radars, where time may not be available for high-rate tracking of all targets.

10.3   TRACKING AT LOW S/N

In deriving the equations for thermal noise error in tracking (e.g., (8.2.1) for conical scan, (8.4.5) for monopulse), it was assumed that the ratio S/N (in the Σ channel, for monopulse) was high enough to ensure linear operation of the error detector. The Δ channel noise then caused a simple additive output, which was balanced by motion of the tracking servo within the linear portion of the error sensing curve. The instantaneous output of the error detector in the monopulse case is then given by

$$E_\Delta = \frac{|\Delta + e_{n\Delta}|}{|\Sigma + e_{n\Sigma}|} \cos\phi \approx |(\Delta/\Sigma) + (e_{n\Delta}/\Sigma)| \cos\phi \qquad (10.3.1)$$

The Δ/Σ portion is the desired target error output, and the $e_{n\Delta}/\Sigma$ term is the noise error. The random output error voltage is thus $(e_{n\Delta}/\Sigma) \cos\phi$, and its rms value is $1/\sqrt{2S/N}$. The appearance of N in the expressions is simply the result of the assumption that the Δ channel noise is equal to the Σ channel noise, the latter having no effect on the output.

When S/N is near or below unity, a small-signal suppression loss or detector loss appears, similar to the loss C_x, which was encountered in target detection (Section 2.2). The loss is not identical to C_x, but is given, for monopulse error detection, by

$$C_a = (S + N)/S = 1 + N/S \qquad (10.3.2)$$

for conical-scan by

$$C_d = (2S + N)/2S = 1 + N/2S = 1 + L_k/2(S/N)_m \qquad (10.3.3)$$

and for linear scan by

$$C_a = (S + N)/S = 1 + N/S = 1 + L_p/(S/N)_m \qquad (10.3.4)$$

These detector losses could normally be ignored for $S/N > 4$, but would cause the thermal noise error to increase steeply for $S/N < 1$, if the tracking loop gain and bandwidth could be held constant. However, the circuits which perform normalization (division of Δ by Σ) can no longer maintain this function when the Σ channel becomes dominated by noise, causing the voltage gain of the error detector to vary inversely with C_a^2 (or with $C_a C_d$ for conical scan). The decrease in gain causes the thermal noise output error, in terms of random motion of the beam, to increase more slowly at low S/N, reaching a maximum value when only noise is present. The total tracking error continues to increase until the target is lost, because the servo no longer has sufficient gain to follow the moving target.

Equations for Thermal Noise at Low S/N

The equations for output thermal noise, in terms of n', the number of pulses integrated in the reduced servo position loop bandwidth at low S/N, and β_{no}, the servo bandwidth at high S/N, are as follows:

(a) for monopulse,

$$\sigma_\theta = \frac{\theta_3\sqrt{1 + S/N}}{k_m(S/N)\sqrt{2n'}} = \frac{\theta_3}{k_m\sqrt{(1 + S/N)(f_r/\beta_{no})}} \tag{10.3.5}$$

(b) for conical scan,

$$\sigma_\theta = \frac{\theta_3\sqrt{L_k[L_k + 2(S/N)_m]}}{k_s(S/N)_m\sqrt{2n'}} = \frac{\theta_3\sqrt{L_k}}{k_s\sqrt{[L_k + (S/N)_m](f_r/2\beta_{no})}} \tag{10.3.6}$$

For the usual values k_m or $k_s/\sqrt{L_k} \approx 1.4$, the maximum rms noise errors are $\theta_3/\sqrt{2f_r/\beta_{no}}$ for monopulse and $\theta_3/\sqrt{f_r/\beta_{no}}$ for conical scan. Plots of error *versus* S/N are shown in Figures 10.3.1 and 10.3.2 for monopulse and conical-scan radars. Also shown on these plots is the ratio of servo bandwidth β_n to the value β_{no} for high S/N.

The errors for low S/N no longer represent the antenna motion required to produce a balancing target response, but rather are the random motions of the tracking axis about its short-term mean position, which, in turn, will tend to follow the target with lag errors determined by the reduced servo gain. The random motions can be small, for f_r/β_{no} large, and useful data can be provided on targets with low accelerations, as long as the total energy ratio E/N_o is sufficient to ensure recognition of target presence and maintenance of a servo error signal which can overcome dc drifts in the low-level circuits.

Range and Doppler Tracking at Low S/N

The action of range trackers at low S/N is similar to that of angle tracking loops, with detector losses given by C_a in (10.3.2). The equations are the same as those for monopulse tracking, with error slopes K or

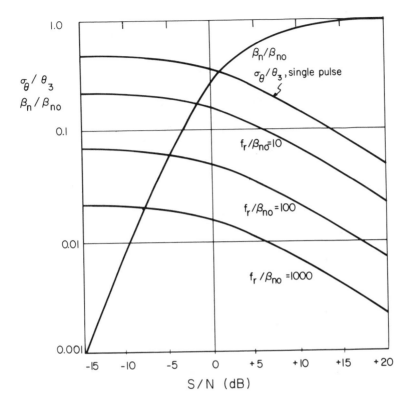

Figure 10.3.1 Thermal noise error and servo bandwidth *versus S/N* ratio for monopulse radar.

k_t/τ_o replacing k_m/θ_3. Thus, for typical $K\tau_{3a} = 1$,

$$\sigma_t = \frac{\tau_3\sqrt{1 + S/N}}{(S/N)\sqrt{2n'}} = \frac{\tau_3}{\sqrt{(1 + S/N)(f_r/\beta_n)}} \qquad (10.3.7)$$

As with angle tracking, the noise output at low *S/N* no longer represents target position errors which produce a balancing error signal, but a deviation of the tracker position from its short-term mean value. The tracker can be following a constant-velocity target, if the rate was established before the signal faded to low *S/N*.

In doppler tracking, the presence of the narrow doppler filter will normally prevent the occurrence of low $(S/N)_f$ at the error detector or discriminator, until the signal has faded to the point where its existence is uncertain.

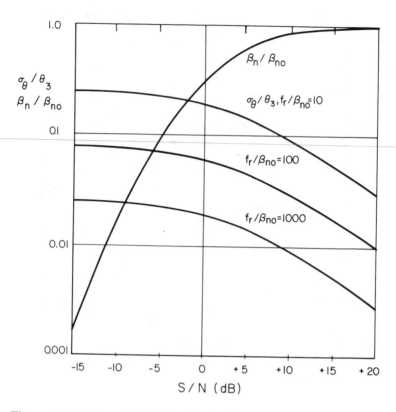

Figure 10.3.2 Thermal noise error and servo bandwidth *versus* *S/N* ratio for conical-scan radar.

Loss of Track

Loss of track will occur when the displacement of the target from the beam, gate, or discriminator response has gone beyond the peak of the error curve (normally, about one resolution cell width from the tracking null). The error signal will decrease rapidly beyond this point, and unless a noise peak happens to restore the tracker to a point nearer the target position, the tracker will coast on noise in all three coordinates until a reacquisition attempt is made. Loss of track can result from target dynamics in excess of those for which the tracker was designed, from fading of the target, from capture by an interfering signal, or some combination of these factors. The usual cause, in the absence of severe interference, is a target fade during acceleration, causing reduction in loop bandwidth and acceleration error constant to the point that the tracker lags one resolution cell behind the target. Imbalance in the servo loop can also cause the tracker to drift away from a faded target.

The range coordinate is the most likely to be affected first, in radars using short pulses or compressed pulses. For a pulsewidth of 0.5 μs, an error of 75 m will be sufficient to initiate loss of track. Even at short ranges (e.g., 10 km), the usual angle resolution cell (e.g., 20 mr) will be 200 m wide, and the target is less likely to reach this error level. At the long ranges where loss of track is more likely to occur, the angle resolution cell is much wider than that is range. The acceleration error coefficient is given by (10.2.2) as

$$K_a = 2.5\beta_n^2$$

where the loop bandwidth β_n varies according to

$$\beta_n = \beta_{no}/(1 + N/S)^2$$

where β_{no} as before is the loop bandwidth for high S/N. The acceleration lag error at any given S/N is

$$\epsilon_a = a_t(1 + N/S)^4/2.5\beta_{no}^2 \tag{10.3.8}$$

Consider, for example, a target with $a_t = 20 \text{ m/s}^2$, under track by a radar with a range servo bandwidth $\beta_{no} = 2$ Hz. The lag error will be 2 m for high S/N. For $\tau = 0.5$ μs, $f_r = 1000$, $k_t = 1$, the thermal noise error at $S/N = 4$ will be 1.7 m, and the bandwidth will be reduced to 1.28 Hz, increasing the lag error to 5 m. A target fade to $S/N = 0.7 = -1.5$ dB, lasting for 2 to 3 s, will cause an error:

$$\epsilon_a = 20(1 + 1.42)^4/10 = 70 \text{ m}$$

which is sufficient to initiate loss of range track. Once the range gate has left the target, angle tracking will also be interrupted, and reacquisition of the target in all three coordinates will be necessary.

Note that the energy ratio, under conditions that cause loss of track, is

$$E/N_0 = nS/N = (f_r/2\beta_{no})S/N = 100, \text{ or } +20 \text{ dB}$$

This is more than sufficient for target detection, but the inability of the tracking loops to maintain adequate bandwidth has caused loss of track. The signal should be visible to an operator as it leaves the range gate. If pulsed doppler processing were used, the single-pulse S/N would no longer be the criterion in establishing detector loss and bandwidth reduction. Instead, the ratio $(S/N)_f$ at the output of the doppler filter will be used in (10.3.2) to (10.3.8). In most cases, $(S/N)_f$ will not drop to low enough values to introduce the nonlinear effects which are discussed in this section.

10.4   SEARCH, ACQUISITION AND TRACKING WITH PHASED ARRAY RADARS

The phased array radar is ideally suited to tracking of multiple targets within its electronic scan field, as no time need be lost in reorienting the antenna from one target to the next. The target track dwells may be interlaced in such a way as to obtain a high data rate on each of many targets, the number depending only on the dwell times required to obtain the target echoes and reject clutter. Additional dwells for acquisition of new targets may also be interlaced with existing target tracks. Search of at least a limited volume can be conducted, when there are not too many targets to be tracked. The successful application of phased array technology requires that the functions be properly assigned, with due regard for the environment in which the radar is to operate, so that a single radar frequency can be used with a set of beam patterns and waveforms to execute the radar mission.

Choice of Frequency

Choice of RF is governed primarily by the accuracy and resolution requirements of the tracking mode, given a limit to the available antenna size. The achievable tracking accuracy is set by the antenna beamwidth and the environment in which the radar must operate. In the ideal case, when only thermal noise sets the accuracy limit, the energy received from critical targets can be increased to achieve the desired accuracy. For example, if an error on $0.01\theta_3$ is allowed, (8.4.5) gives a requirement for $E/N_0 > +31$ dB. In practice, however, several other factors will produce error components proportional to the beamwidth, some of which can be reduced by averaging over multiple dwells if time permits. To achieve a total error of less than $0.01\theta_3$, the designer might assign to each of four major error sources an allowance of $0.005\theta_3$. In a radar having no major mechanical limitations, these critical error components might be as follows.

(a) *Target fluctuation increase in thermal noise error* (Figure 8.5.2). The reliance on a single coherent burst to extract tracking data on a fluctuating target increases by about 10 dB the required energy ratio, as compared to a steady target or a fluctuating target tracked with several bursts. To meet a requirement of $\sigma_\theta < 0.005\theta_3$, $E/N_0 > +49$ dB will be needed.

(b) *Monopulse implementation error* (Equation (8.5.10)). The required quotient of monopulse RF null depth divided by the square of IF phase error must exceed $2 \times 10^4 = +43$ dB to hold this error component below $0.005\theta_3$. For example, with $G_n = 33$ dB, the rms phase error allowed is $\phi_2 = 0.31 = 18°$.

(c) *Polarization error* (Equation (8.5.2)). The cross-polarized response of the array Δ channels must not exceed -35 dB relative to the Σ main-lobe gain, to maintain this error component below $0.005\theta_3$ on a typical aircraft or missile target.

(d) *Multipath error* (Section 11.2). This error will be in the order of $0.05\theta_e$ whenever the tracked target is at an elevation angle below θ_e. At higher elevations, the Δ-channel sidelobes on the surface must not exceed -35 dB relative to the Σ main-lobe gain, to maintain this error component below $0.005\theta_e$.

As a result of these and other errors (see Chapter 11), the tracking function normally requires $\theta_3 < 2° = 0.035$ radian, to provide data of sufficient accuracy for missile guidance or gunfire control, and in critical applications $\theta_3 < 1° = 0.018$ radian will be required. The resulting maximum wavelength will depend on the available aperture diameter D, according to

$$\lambda_{max} = \theta_3 D / K_\theta = (0.015 \text{ to } 0.03)D \tag{10.4.1}$$

The required wavelengths and frequencies for different applications will be approximately as shown in Table 10.7.

Table 10.7 Wavelengths and Frequencies for Tracking (based on $\theta_3 = 1°$ to $2°$)

Application	D_{max} (m)	λ_{max} (m)	f_{min} (GHz)
Airborne radar	1	0.015 to 0.03	10 to 20
Ground mobile	3	0.05 to 0.1	3 to 6
Shipboard	6	0.1 to 0.2	1.5 to 3
Ground fixed	30	0.5 to 1	0.3 to 0.6

The frequency dictated by the beamwidth and available aperture then combines with the clutter environment to control the unambiguous range of the waveform. Target tracking can be carried out successfully when the blind speed exceeds the total spread of the clutter spectrum by a factor of about two, if a pulse doppler waveform and processor are used. For MTI processing with 35 dB improvement, the blind speed must be about 10 times the clutter spread, using as a measure of spread the $4\sigma_v$ width of a Gaussian spectrum. When the clutter consists of rain, chaff, birds, or unwanted land vehicle echoes, its spectral spread will be from 20 to 40 m/s (see, for example, Figure 7.3.3). The required blind speed of the waveform is then from 40 to 80 m/s if PD processing is used. The available unambiguous range is expressed, for any given wavelength, by (5.3.6):

$$R_u = c\lambda/4v_b$$

Figure 10.4.1 shows the unambiguous range for radars in the different bands, with blind speeds of 50 to 200 m/s.

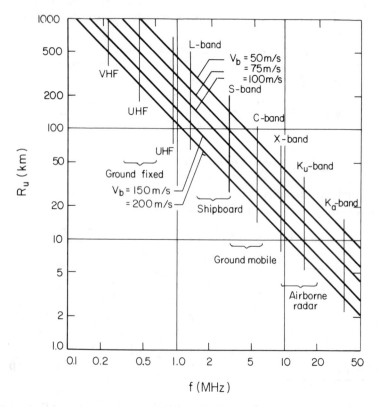

Figure 10.4.1 Available unambiguous range as a function of radar frequency.

From the figure, it can be seen that an S-band radar can track targets at unambiguous ranges of 90 to 180 km, while rejecting moving clutter. At C-band, the unambiguous range will extend to between 50 and 100 km, while at X-band 30 to 60 km can be expected. The dwell times for targets at lower elevations, where the maximum clutter spread will be encountered, will permit bursts of 16 pulse or more to be transmitted, after an allowance for reception of clutter from maximum range (5.4.6). In the case of chaff or ground clutter received from a ducted path, the maximum range will be about 450 km, requiring a delay $t_{dc} = 3$ ms between the beginning of a coherent burst and the processing of target data. Table 10.8 shows the resulting dwell times, the number of targets which can be tracked at 4 Hz data rate, and the loss resulting from the 3 ms transient time.

Table 10.8 Tracking Capacities *versus* Unambiguous Range (based on 16-pulse bursts after 3 ms initial transient)

Unambiguous Range (km)	Repetition Interval (ms)	Dwell Time (ms)	Number of Targets	Loss (dB)
180	1.2	22.2	11	0.6
90	0.6	12.6	20	1.2
45	0.3	7.8	32	2.1
30	0.2	6.2	40	2.9

To avoid the large loss in efficiency for the shorter unambiguous ranges, most designers would increase the coherent dwells to 32 pulses or more. The dwell times will be decreased and the target capacity increased as the targets appear at higher elevation angles or the clutter environment is less severe. Dwells not needed for tracking can be applied to acquisition of new targets or to search functions.

Unambiguous Range for Target Acquisition

Target acquisition may be delayed if the waveform supports a blind speed only twice the width of the clutter spectrum. The region Δ_v of inadequate target detection is related to the clutter width Δ_c, blind speed v_b, and the number of pulses processed coherently by (5.4.3):

$$\sigma_v = \Delta_c + v_b/n_c$$

If $\Delta_c = v_b/2$, use of $n_c = 16$ leaves only 37% of the target velocities within the passband of the sensitive doppler filters. Multiple bursts with PRF diversity will be required to cover all target velocities, and some of the required PRFs will not support the full unambiguous range. This means that the probability P_v of the target lying within the acquisition volume, in (10.1.4), will be low, requiring many scans or many bursts per scan in order to achieve high probability of acquisition.

Target acquisition in the clutter environment is carried out more expeditiously if the designation data include fairly accurate range and radial velocity. These data permit the tracker to select PRFs for the acquisition scan that will not be blind in doppler, and in many cases a range-ambiguous PRF can be selected at which the target need not compete with clutter from too short a range.

Unambiguous Range for Search

The ability to search with a narrow pencil beam at the higher radar frequencies used in tracking is very dependent on the clutter environment. Radar time becomes the limiting problem, and it is important to use a waveform which makes efficient use of the available time. No *a priori* data on target range or velocity are available in the search mode, so all ranges and velocities out to the maximum for the radar and target must be covered. Dual PRF diversity bursts can cover all target velocities to v_m if the blind speed v_b meets the requirement of (7.2.19):

$$v_b \approx \Delta_v \left(\sqrt{v_m/\Delta_v} + 1/4 \right)$$

where Δ_v is the width of the notch at which target detection is inadequate (-6 dB). This requirement may be combined with that of (5.4.3):

$$\Delta_v = \Delta_c + 2v_b/n_c$$

where Δ_c is the width of the clutter spectrum and n_c is the number of coherently processed pulses, to express the requirement for blind speed as a function of clutter spectral width and number of coherently processed pulses in the burst. Figure 10.4.2 shows the resulting blind speed requirements for $v_m = 600$ m/s, as a function of clutter spectral width, for several values of n_c. The requirement for efficient search is more stringent than for tracking, because only a small fraction of blind velocities can be permitted. In the case of airborne radars, the motion of the radar platform is the dominant factor in clutter spectral width, often forcing the adoption of high-PRF waveforms. Ground and shipboard radars see a clutter spectrum determined primarily by the wind speed and shear.

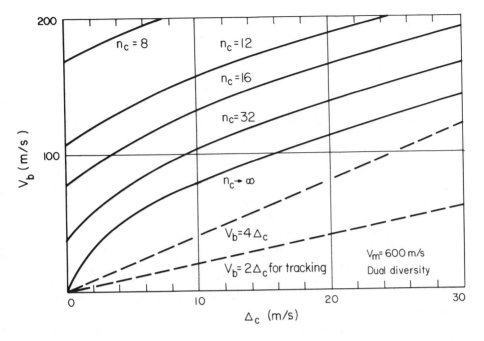

Figure 10.4.2 Required blind speed in search mode as a function of clutter velocity spread.

The available unambiguous range for a given blind speed, at different radar frequencies, has been plotted in Figure 10.4.1. Because of the much higher blind speeds required for effective search, typically 150 m/s or greater, the unambiguous ranges are smaller for each radar band: about 150 km for L-band, 50 km for S-band, 25 to 30 km for C-band, and near 15 km for X-band. Only the large antennas typically used in fixed ground installations can provide the beamwidths needed for accurate tracking at the wavelengths needed for search with unambiguous ranges beyond 100 km in clutter.

Multifunction Array Radar

The multifunction radar is intended to combine the search and tracking functions, as well as other possible modes such as missile guidance and, in airborne radars, terrain mapping. Two critical design factors in multifunction radar are:

(a) Choice of radar frequency, high enough to provide the narrow tracking beam and low enough to permit rejection of clutter during volume search modes;

(b) Budgeting of radar time, to provide the dwell times necessary for clutter rejection in both search and tracking modes.

The budgeting of average power among the several functions and coverage regions has been considered one of the major problems, but increasing the total average power to solve this problem is easier than finding time for transmission and reception of the required waveforms.

In the case of long-range missile or satellite surveillance, clutter problems are minimized by blanking the ranges of atmospheric and ground clutter. The dwell time required is that needed to propagate a pulse to maximum range and back, in each beam position, and sometimes to propagate a burst of pulses for target discrimination. Dwells lasting for multiples of the maximum target delay time are seldom required. This freedom from clutter limitations, along with the very high average powers available from large phased arrays, makes space surveillance a primary application for multifunction radars. In terrestrial applications, the problems are more difficult, as shown below.

The multifunction phased array radar is basically a 3D surveillance radar with provision for random access to the surveillance volume. This permits the radar to interlace tracking and other special modes with the routine search function. Once a target is detected, it can be tracked either with specially assigned beam dwells, or by track-while-scan (TWS) operation on the output data from the search raster.

Budgeting of Radar Time

The typical multifunction radar time budget will allocate about 50% for search, with the remaining 50% shared among tracking and other functions. The minimum time requirements for search, for waveforms matched to the csc^2 coverage envelope (e.g., Figure 7.3.1), may be determined from (1.3.7). Consider, for example, a system for which the unambiguous range is to be a twice the detection boundary of Figure 7.3.1: $(R_u)_{max} = 150$ km, $h_m = 25$ km. The equivalent elevation sector, for which $t_r = 1$ ms is required, is

$$\theta'_m \approx \sin\theta'_m = \sin\theta_1(1 + \ln\sin\theta_2 - \ln\sin\theta_1)$$
$$= 0.17(1 - 0.69 + 1.75) = 0.35 \text{ radian} = 20°$$

For a phased array covering a 90° sector, there will be 1800 square degrees in the equivalent coverage, or 800 beam positions for a 1.5° pencil beam. The beam will actually broaden to 2° at the extremes in azimuth scan, but this effect can be used to reduce beamshape loss in those regions, and will be ignored in establishing the time budget. To obtain one pulse per beam position requires 0.8 s, or 1.6 s at 50% search usage, with no lost time between pulses. To allow for beam switching and use of pulse compression, these times will increase to about 1 and 2 s, respectively. If $n_c = 16$ pulses were to be used, the search frame time would be 32 s, an excessive value for tactical surveillance.

Solutions to the time budget problem requires that MTD, MTI, and frequency diversity waveforms be used only to the extent that clutter appears in each elevation beam position.

Multifunction Radar Example for Air Defense

To see what effect chaff and rain have on multifunction radar design, consider the example used in Section 7.3. The rain clutter problem at S-band could be solved with a combination of circular polarization and pulse compression ($B = 15$ MHz). The chaff problem could not be solved at S-band while providing the required unambiguous range $R_u = 135$ km, but the combination of pulse compression and frequency diversity offered a solution. This approach also eliminates the requirement for use of a circularly polarized array, with its target loss and complication of phase shifter design problems. The use of three-pulse MTI for ground clutter then required the transmission of eight three-pulse bursts in the lowest beam position (and possibly in the second beam as well). Referring to Figure 7.3.2, it can be seen that clutter will be reduced by 24 dB if it lies more than one full beamwidth off the center of a beam position. Note that solution of the airborne clutter problem has required the use of a wideband waveform, with the processing of 15,000 range cells in each of the seven full-range beams.

Assume the use of $\theta_e = 1.5°$, to support adequate accuracy in the tracking mode, with seven full-range beams (instrumented to 150 km) in the search raster at 0.5° elevation and 1.5° intervals up to 9.5°. All these beams will have rain and chaff over most of the search range interval, requiring the eight-pulse diversity waveform, with the lowest two beams requiring eight three-pulse MTI bursts. The dwell time requirements for elevation coverage in a single azimuth beamwidth will be as follows:

Beams 1 and 2:	(8 × 3 ms/beam) × 2 = 48 ms
Beams 3–7:	(8 × 1 ms/beam) × 5 = 40 ms
Upper beams:	(8 × 1 ms/beam) × 6 = 48 ms
Total per azimuth:	136 ms
Total for 60 azimuths:	8.16 s
Elapsed time at 50% search usage:	16.3 s

The beams covering 10° to 30° are equivalent to six full-range beams, according to (1.3.7). The search loss for the beam arrangement used here would be high, becuase no overlap has been allowed between beam positions (except in azimuth as the beam is scanned off broadside). Solution of the clutter problems by using doppler waveforms, at L-band for unambiguous range coverage or at higher frequencies with ambiguous waveforms, requires even greater time allocations.

Search frame rates in excess of about 6 s are unsuitable for support of TWS operations on high-performance targets or targets in a dense environment. Hence, if the example search process were used, tracking files supported by dedicated tracking beams would be required on all objects detected during search. As shown below, this requirement further complicates the time budgeting process by increasing the number of tracking beams far beyond the number of interesting targets present in the surveillance volume.

The solution to the time budget problem, for multifunction air defense radar, requires either the use of multiple search beams, in parallel, for each 90° sector, or the elimination of rain and chaff from the environmental model.

Multifunction Radar Example for Landing Control

A successful application of multifunction phased array radar is the AN/TPN-19 Precision Approach Radar discussed in [10.2]. The principal radar parameters are listed in Table 10.9.

The radar operates only to 35 km in range at low altitudes, and the major environmental problems are land and rain clutter, plus high attenuation in the maximum rainfall rate of 50 mm/h. Circular polarization, frequency diversity and MTI (in the search mode) and high-resolution pulse compression (in the track mode) are all used in combination with the narrow beam to overcome the strong rain clutter. The basic blind speed of 58 m/s is adequate for the relatively low clutter spread encountered at low altitudes and short ranges.

The success of the search-track combination in this case results from the very limited coverage required: 20° in azimuth, 15° in elevation, and 35 km in range. There are only 280 beam positions in the search sector, and with overlap the system scans about 500 beam positions at a rate of 2/s, with a three-pulse MTI burst in each beam (a total of 3000 pulses per second). Dual-frequency diversity is provided by dividing each pulse in half and transmitting the two RFs in the same pulse. Overlapping beams

Table 10.9 AN/TPN-19 PAR Parameters

Functions	Raster scan, and monopulse track
Frequency	9.0 to 9.2 GHz
Detection range	37 km (clear weather)
(T-33 Aircraft)	28 km (50 mm/h, rain)
Scanned sector	$\pm$ 10° azimuth
	$-1°$ to $+ 14°$ elevation
Data rate	2 scans/s—scan mode
	22/s per target—track mode
Transmitter:	
Type	TWT/CFA
Peak power	320 kW (10kW with CFA off)
Average power	1 kW (25W with CFA off)
Pulsewidth	1µs
Pulse rate (average)	3500 Hz
Modulation: Scan	4-frequency diversity, 0.5 µs subpulses
Track	Chirp, bandwidth = 120 MHz
Receiver:	
Type	3 monopulse channels plus 2 scan channels
Noise figure	3.3 dB
Video, scan mode	Linear ⎫ to digital video integrator Dicke-Fix ⎭
	Coherent MTI ⎫ to DMTI Noncoherent MTI ⎭
Antenna:	
Type	Limited-scan phased array
Gain	43.5 dB (at boresight)
Beamwidth-Azimuth	1.4°
Elevation	0.75°
Polarization	Circular (20 dB integrated cancelation ratio)
Number of array elements	824
Bits of phase control	3
Reflector size	2.8 m $\times$ 3.6 m
Tracking accuracy: Azimuth	0.143°
Elevation	0.072°

use different pairs of frequencies for fourfold diversity at each target location. The remaining 500 pulses per second are more than enough to track the targets at a data rate of 22 Hz per target. Because of the short range and high attenuation, rain clutter from the second and subsequent range ambiguities does not pose a problem. In reduced rainfall and attenuation rates, the narrow beam and circular polarization are adequate to avoid problems from ambiguous clutter.

In all respects other than the limited scan field, the AN/TPN-19 operates as a dual-mode phased array radar for search and tracking, with monopulse channels for optimum angle accuracy and a very narrow elevation beamwidth for minimum multipath error. The antenna, shown in Figure 10.4.3, uses a small array of 824 elements to illuminate a 2.8 × 3.6 m reflector, producing the narrow beam with electronic scan capability over the scan field. Circular polarization is provided by a grid on the reflector. Monopulse channels are provided by space-feeding of the lens array.

Figure Figure 10.4.3 AN/TPN-19 precision approach radar antenna.

Sequential Detection

A potential advantage of the multifunction radar is that a false alarm in the search mode does not carry great cost in terms of system reaction. A standard procedure is to validate each alarm, using a second dwell, possibly of higher energy, or a series of beam dwells, which can determine whether a target is actually present. This process is called *sequential detection*. The cost of an alarm is then the assignment of additional energy (and time) to the search beam position that originated the alarm. A minimum assignment requires only a second dwell on the beam position of the original alarm. A threshold on the validation dwell is set lower than during search (e.g., 10^{-2}), and validation alarms are accepted only in the range cells immediately surrounding the cell that gave the first alarm. The probability of successful validation on an actual target is thus increased to near 99%, with small probability of false validation on noise. If this validation requirement occurs on 10% of the search beam positions, the total time and energy requirements are increased by only 10% and 0.4 dB. In return, the probability of a false alarm in each search beam can be increased to 10%. The gain in search energy depends on the number of range and doppler cells in the search beam.

Consider a system with unambiguous range $R_u = 150$ km, with range resolution $\tau_n c/2 = 150$ m, and using a 16-pulse MTD. If target detection is permitted (at full sensitivity) from 150 to 30 km, there will be

$$n_d = n_t n_f = 800 \times 16 = 12,800$$

opportunities for an alarm in each beam. The same requirement applies when pulse compression is used instead of MTD to reduce clutter. A false-alarm probability of 10^{-5} will lead to false alarms in 13% of the beams, at a cost of 0.5 dB in validation dwells, while a probability of 10^{-6} will cause only 1.3% of the beams to have an alarm, with a cost of 0.05 dB. The change in SNR requirement between $P_{fa} = 10^{-5}$ and 10^{-6} is about 0.7 dB (at $P_d = 0.7$, see Figure 2.2.2). The higher false-alarm rate, requiring 0.36 dB more validation power and 12% more time, saves 0.34 dB on the original search allocation. Were it not for the time problem, it would be possible to save about 0.4 dB through this process of sequential detection. The potential gain from sequential detection is greatest when the number of detection decisions per beam, n_d, is small and the detection probability is low. However, in systems using high resolution in range and doppler to reduce clutter, n_d is large and there is no significant gain.

Tracking Capacity

The 50% of the multifunction radar's time budget not allocated to search will be used for tracking targets of interest, and other operations required by the particular system. Waveforms used in tracking can be selected in accordance with the target range and required accuracy, minimizing the tracking dwell times. Few targets would require the full accuracy of the radar, as most would require only sufficient accuracy to maintain tracks and to recognize changes that might render them more threatening than when they were originally classified. All newly detected targets require track initiation, followed by tracking with sufficient accuracy to be classified as to their level of interest. The track initiation procedure, for systems in which dedicated beams are used for tracking, typically requires about 10 dwells on the position of a validated alarm.

The conventional wisdom in multifunction radar engineering would limit the number of target tracking channels to the number of actual aircraft or missiles in the surveillance region, plus a small margin for false alarms. Other detectable objects (e.g., land vehicles or birds), which fail to meet the criteria for aircraft or missiles, would be dropped from the track files. More recent experience, however, indicates that every object which is detectable must be carried in a track file to avoid repeated validation and track initiation attempts. The severity of this problem depends on the environment in which the radar must operate. When the false detections result entirely from random noise, the validation procedure serves to reduce the probability of false track initiation to a small fraction of $P_{fa}n_d$ per beam. However, when the alarm results from an unwanted target (e.g., ground clutter, land vehicle, or a bird that passes the doppler filter), the probability of false validation and subsequent track initiation action is greatly increased.

Consider, for example, a radar attempting to acquire targets as small as 0.1 m^2 in a volume also occupied by birds. In Section 3.6 the cross-sectional distribution of birds was given as a log-normal function with a standard deviation of 6 dB about a median of -30 dBm^2. With 10^5 birds in a region within 50 km of the typical ground radar, there will be 130 birds having cross sections above -12 dBm^2, and these will be detectable with essentially the same probability as the desired target, in the absence of doppler rejection. If the altitude of the target to be detected is below that of the average bird, the received signal from the bird may even be larger than that of the target. Each bird detection will be validated with high probability, leading to a track initiation action over the assigned sequence of 10 dwells. The true velocity of the bird will be found lower than that of a target. If the track is dropped, the next search scan will again start the process. Hence, large numbers of low-velocity objects must be maintained in track files to avoid the expenditure of validation and track initiation energy. Furthermore, larger numbers of random detections will call forth validation dwells, and many of these will start a track initiation sequence that fails because of inadequate SNR. The system loading from such undesired targets can be large, both with respect to radar time and energy, and loading of the data processing equipment. The only solutions are:

(a) The reliance on doppler filtering to reject almost all birds, land vehicles, helicopters (if they do not constitute legitimate targets), and other sources of correlated false alarms; or

(b) The carrying of TWS files on large numbers of detectable objects which do not qualify as targets.

This latter solution imposes a loading on the data processing, but not on the radar, which is executing the search routine without regard to the number of detections that result.

Track Initiation

Initiation of track after detection by a multifunction radar can be made relatively certain, because no angle scanning is required and only a small range interval need be searched. The acquisition range is thus essentially equal to the detection range, and little time delay is experienced between detection and availability of tracking data. This can be important in cases where a high-velocity target is detected at short range (e.g., after unmasking at low altitude).

Summary of Multifunction Radar

The key issues in the operation of multifunction radar are as follows:

(a) There is an inherent conflict between the frequencies that are suitable for search and target detection, at ranges beyond about 30 km, and those suitable for tracking with specified accuracy;

(b) There is an inherent limit in the rate at which search can be conducted with the narrow beams needed for adequate tracking accuracy, unless multiple beams are used for search;

(c) Unless the search mode of the multifunction radar can support TWS operation on most targets, there is a severe problem of track overload caused by correlated alarms from unwanted targets;

(d) If these difficulties can be overcome, the track function of the multifunction radar can produce data almost from the moment of initial target detection, minimizing the reaction time of the system;

(e) The most significant advantage of the multifunction radar is its ability to carry out the several functions from one antenna. This is of great concern in airborne applications, and has also played a role in shipboard and land-based systems.

10.5   TRACKING RADAR ECM AND ECCM

Discussions of ECM against tracking radars [10.3, 10.4] and ECCM used in tracking radars [10.5, 10.6] have been concerned largely with listing and describing a number of jamming techniques, rather than with analysis of their operating principles and effects. In recent years, however, two Soviet textbooks have become available [10.7, 10.8] in which considerable engineering analysis is presented. The material in [10.8] is particularly valuable, as it lays the foundation for consideration of the basic objectives and effectiveness of ECM against tracking radars. Much of this material has been covered in US reports, but these are not generally available and cannot serve as the basis for discussion in the open literature.

Objective of ECM against Tracking Radar

Before discussing particular ECM techniques, and means of countering them, it is important to identify the objective of the ECM, and the approaches used in accomplishing these objectives. The following are the objectives to which ECM might be addressed:

(a) Preventing the tracker from acquiring its target;
(b) Delaying acquisition of the target;
(c) Preventing the tracker from obtaining range or doppler data on the target;
(d) Introducing large errors in range or doppler tracking;
(e) Breaking lock in range or doppler;
(f) Introducing large errors in angle tracking;
(g) Breaking lock in angle;
(h) Introducing additional false targets in the tracker;
(i) Destroying the tracking radar.

The achievement of each of these objectives will have some effect on the system which uses the tracking radar data, but the effects of some are more severe than others. For example, (a) and (i) are likely to leave no residual capability in the system, while (c) and (d) may have very little effect, depending on the system requirements for range and doppler data.

Evaluation of an ECM, and of ECCM to counter it, then depends on the data requirements of the using system. In the following discussions, the denial of adequate data to the using system will be the criterion of jamming effectiveness.

Prevention of Acquisition

The tracking radar is most vulnerable to jamming in its acquisition mode. During this period, it operates as a search radar, scanning some region in four-dimensional space to find its target. The discussion of search radar jamming, Section 7.5, is directly applicable to this mode of tracker operation.

The robust techniques of barrage noise jamming and chaff deployment are primary means of preventing target acquisition. The noise jamming, however, must not come directly from the target, as that would provide a beacon for easy lock-on in angle. Active jammers on stand-off or escort platforms, or on decoys launched from the target, may be effective. The calculation of jammer noise temperature, (1.4.2) or (7.5.1), provides insight into the problem caused in the tracker. The difference between this problem and that of the search radar lies in two areas:

(a) The tracker is scanning a much smaller volume in its acquisition mode, and may have a greater excess in SNR to be overcome by the jammer;

(b) The tracker beamwidth is usually smaller than that of the search radar, and if the sidelobe levels are low, it will be more difficult for the jammer to mask a large sector. The tracker will also have a narrow elevation beam, requiring the jammer to be more nearly in line with the target or to penetrate the elevation sidelobes.

Example of Stand-off Noise Jamming

To illustrate the use of stand-off noise jamming as a means of preventing target acquisition, let us assume that the S-band tracking radar used as an example in Section 10.1 (and based on the narrow-beam search radar example of Chapters 1 and 5) is upgraded to 2 kW average power, and is required to acquire 1-m^2 targets at $R = 80$ km with high probability in $t_a = 10$ s. The actual tracker beamwidths would be 1.6° × 1.6°, instead of 1.3° × 2°, but the gain and other calculations would remain the same. The margin of performance in a benign environment is +21 dB: single-pulse S/N (at beam center) is +30 dB, where only +9 dB is required. Acquisition will be prevented if the stand-off jammer (SOJ) noise temperature is 31 dB above the receiver noise, reducing the single-pulse S/N to −1 dB:

$$T_j = 1.26 \times 10^3 \times 10^3 = 1.26 \times 10^6 \text{ K}$$

The detection probability on a Case 1 target for a 32-pulse burst with $S/N = -1$ dB per pulse, integrated coherently with losses as shown in Table 5.4, is only 0.02. From (1.4.2), the required jamming effective radiated power density (ERPD) for a jammer at $R_j = 200$ km is

$$P_j G_j / B_j = 1.5 \times 10^{-7} / F_j^2 \text{ W/Hz} = 0.15 / F_j^2 \text{ W/MHz}$$

If the jammer could operate directly behind the penetrating targets, it could cover the entire 300 MHz tuning band of the radar with an ERP = 45 W.

The tracker will have low-sidelobe and sidelobe cancelation requirements similar to those of the search radar, and these issues are discussed in Sections 4.4. and 7.5. If the radar receiving sidelobe levels over a sector $\pm 3°$ from the beam axis are $F_j^2 = -30$ dB, the SOJ would have to have an ERP = 45 kW to prevent acquisition of targets at $R = 80$ km within that sector. Targets at low elevation would tend to be masked not only by the direct jammer noise entering the sidelobes, but by ground-reflected jammer power entering the main lobe (see Section 11.2).

Example of Chaff to Prevent Acquisition

The use of chaff to prevent target acquisition can be very effective against tracking radars, because they must operate in the higher microwave bands, where MTI and doppler techniques are more difficult to apply (at least at long ranges). The discussions of Section 10.4 have covered the issues of unambiguous range, blind speed, and chaff rejection, and it has been shown that chaff rejection may not be possible with waveforms used for unambiguous tracking at long ranges. The MTI and PD processing material covered in Chapter 5 is applicable here.

(a) *Chaff at Target Range.* In order to be effective at preventing acquisition of the target, the chaff must cover a considerable volume around the target position, or in a region such that the radar ambiguity function will superimpose the chaff on the target. If the chaff were located at the target range, it would produce $S/C = 0$ dB on a 1.0-m^2 target when its reflectivity equaled the reciprocal of the radar resolution volume. Assume that the radar uses pulse compression to produce $\tau_n = 0.2$ μs. The resolution volume at $R = 80$ km is

$$V_c = (80 \times 28/1.33)^2 (30) = 8.5 \times 10^7 \text{ m}^3$$

A chaff reflectivity of 1.2×10^{-8} m²/m³ would produce $S/C = 0$ dB. At this range, however, the velocity spread of the chaff, from (3.6.15), is only

$$\sigma_v = 0.3 \times 2 \times 80 \times 0.028 = 1.3 \text{ m/s}$$

using a wind shear coefficient $k_{sh} = 2$ (m/s)/km, and assuming a wind direction along the beam azimuth. Since the radar blind speed would be 55 m/s, even an MTI canceler with wind compensation could achieve about 35 dB improvement, and very large chaff density would then be needed to prevent acquisition.

(b) *Stand-off Chaff.* For targets at low elevation angle, the ambiguous range regions may extend for hundreds of kilometers beyond the target itself, making it possible to use *stand-off chaff* as an effective screen. This will be effective when the radar is forced to use a coherent waveform for rejection of land or rain clutter. If the target itself is in a clear environment, the use of a low-PRF waveform with pulse-to-pulse frequency agility will remove interference from range ambiguities. However, if the coherent waveform is required to reject land clutter, rain, or chaff at the target range, then the stand-off chaff can be extremely effective. In the example, the range for such chaff would be

$$R_c = R + R_u = 80 + 134 = 214 \text{ km}$$

$$V_c = 6 \times 10^8 \text{ m}^3$$

The actual chaff cross section at this range, in order to reduce S/C to 0 dB, must be

$$\sigma_c = (214/80)^4 = 51 \text{ m}^2$$

$$\eta_v = \sigma_c/V_c = 8.5 \times 10^{-8} \text{ m}^2/\text{m}^3$$

The chaff in this second range interval would be at altitudes from 6 to 10 km, and would be moving at speeds near the optimum response speed of the MTI or doppler processor, and hence could not be canceled. Thus, with a reflectivity of 10^{-7} m²/m³ the S/C for targets at 780 km would be reduced to -1 dB and acquisition would be prevented. The chaff cloud required to mask targets from 40 to 100 km, over a 10° sector, would extend approximately from 175 to 235 km in range, 40 km in width, and to 10 to 12 km in altitude. This volume is 2.4×10^{13} m^3, requiring a total chaff cross section of 2.4×10^6 m^2. According to (3.6.14), this would require about 1000 kg of chaff, far less than to overcome the MTI or doppler improvement on chaff at the target range.

Delay of Acquisition

The tracker must stop its scan to evaluate each alarm as a potential target. Delay of acquisition can be produced by false targets which arrive from angles other than that of the target itself. Less average jammer power is required to generate false targets in the sidelobes of the antenna than to mask the real target with sidelobe noise jamming.

(a) *Repeater Jamming.* For a repeater jammer which transmits the radar waveform, the required ERP is given by

$$P_{ja}G_j = P_{av}\, G_t\, (\sigma/4\pi)\, (R_j^2 F^4/R^4 F_j^2)\, (n_j/L_t L_{aj})\, (J/S) \qquad (10.5.1)$$

where P_{ja} is the average jammer power, n_j is the number of false targets to be generated within t_o, and J/S is the required ratio of jammer to signal power at the processor output. For example, with the 2-kW S-band tracker discussed above, the jammer average ERP need be only 1 mW per false target, for $J/S = 1$ with the jammer in the main lobe of the radar. To generate 10 false targets through a single sidelobe $F_j^2 = -30$ dB, $P_{ja}G_j = 10$ W is required. Even in -50 dB sidelobes, $P_{ja} = 10$ W, $G_j = 20$ dB is sufficient. If the radar duty factor is $D_u = 0.01$, the jammer duty factor is $D_j = n_j D_u = 0.1$ to produce 10 false targets per radar beamwidth.

(b) *ECCM against Repeater Jammers.* To overcome false target confusion, the tracker must have available the sidelobe blanking techniques discussed in Section 7.5. The technique of pulse-to-pulse frequency agility or other waveform changes can be used to reject false targets from within the range of repeater jammers, if clutter rejection is not required. If rejection of range-ambiguous clutter is not required, PRF stagger can also prevent false targets from being generated within the jammer range. This provides another reason for the use of chaff, in combination with active jamming, since many ECCM techniques against one are inconsistent with protection from the other. Only the antenna techniques of low sidelobes and blanking can reject false repeater targets, while maintaining the ability to reject clutter as well.

(c) *Cover-Pulse Jamming.* The self-screening jammer (SSJ) has a more difficult task in delaying acquisition, since most emissions will permit the radar to lock immediately in angle on the source. False target generation is of little value in this case. Slowly modulated noise waveforms can be used effectively against improperly designed CFAR detection circuits. These cover pulse jammers cause a local increase in the noise level, raising the CFAR threshold in a way that prevents the target echo from being detected without giving evidence of the jamming. Modern CFAR systems monitor the threshold level for such increases, initiating angle lock on the noise source if it is within the acquisition scan region.

(d) *Chaff as a Deceptive Measure.* Chaff has been used as a deceptive target generator against noncoherent tracking radars, causing false alarms and trial locks on multiple echo sources remote from the target. Self-protection chaff projectiles can be carried on the penetrating target for this purpose. This technique is of little value against tracking radars equipped with MTI or doppler processors.

(e) *Rapid Acquisition as ECCM.* A basic tracker ECCM against false target generators that impede acquisition is a rapid acquisition sequence, which can evaluate many potential targets and select the real target. Once the false targets are evaluated as such, their locations must be kept in a file to prevent further attempts to lock on them during the remainder of the acquisition scan.

Denial of Range and Doppler Data

One of the easiest ECM tasks is to deny range and angle data to the tracker. The self-screening jammer needs only minimal noise power to mask the echo signal of its own platform, since the full gain of the radar antenna will be brought to bear on the jammer. The burn-through range calculation of Section 1.4 (1.4.3) showed that a noise jammer with ERP of 10 kW, barrage jamming a 300 MHz band, would mask the 1 m^2 target at a range of 192 m, against the example radar (which had only 110 W average power, but 40 dB antenna gain). Against the upgraded radar with 2 kW average power, this range would be increased to 820 m.

(a) *Burn-through Mode.* In a phased array radar, the routine search and track operations can be interrupted to apply a special *burn-through mode* to a beam containing a noise jammer. Let us assume, for example, that a long-range multifunction phased array radar has a 10 kW transmitter, and can allocate this entire power for $t_o = 0.1$ s to the burn-through mode on a single target, with coherent integration over 10 ms and noncoherent integration of 10 coherent outputs. The loss budget of Table 5.4 will be assumed, but with beamshape loss eliminated and an equivalent amount added for the additional noncoherent integration. The burn-through range equation, in terms of average power, is

$$R_{bt}^2 = \frac{P_{av}t_o G_t \sigma F^2 B_j}{4\pi P_j G_j L_t L_{\alpha j} D_x} \quad (1)$$

$$= 1.36 \times 10^7 \qquad\qquad (10.5.2)$$

$$R_{bt} = 3700 \text{ m}$$

If burst-to-burst frequency diversity were used to reduce target fluctuation loss, a further gain of 7.5 dB could be realized, increasing R_{bt} to 8800 m.

(b) *Alternatives to Burn-through.* It does not appear productive to allocate the entire resources of a multifunction radar for several 0.1 s intervals, as the target approaches, only to determine the target range when it has arrived within 4 km. Systems that use tracking radar data must be designed to counter SSJ targets without direct measurement of range or doppler. Triangulation from two or more sites offers one solution, and use of homing missiles in a home-on-jam mode offers another.

Introduction of Range and Doppler Errors or Break-Lock

Just as it is easy to deny range and doppler data, the introduction of false targets in these coordinates can readily be accomplished by the SSJ. For example, having raised the noise level to mask the echo signal and reduce the tracking receiver gain, a spurious signal can be introduced (by a repeater) to pull the range (or velocity) tracker away from the echo position. In itself, this has little value beyond that of the masking noise, as the angle track will continue on the jammer.

(a) *Objective of Range and Velocity Pull-off.* The objective of the range or velocity pull-off jammer is not simply to generate errors in those coordinates, but to lead the gate to a position where it cannot readily reacquire the echo target when the jammer is abruptly turned off. When this is done, all tracking loops are opened, and they tend to coast at their previous rates until some control action intervenes. A properly designed tracking radar will coast in angle for several seconds before the target leaves the beam. During this time, the radar will attempt to reacquire the echo target, using the same process that originally acquired the target. If this can be done while the beam remains on the target, little data are lost.

(b) *Combined Angle and Range or Velocity Deception.* The potential importance of range and doppler deception jamming is in combination with angle deception techniques, to be discussed below. Basically, range or doppler deception can be successful if, at the moment the jammer is turned off, an angle rate error has been established, which is sufficient to carry the beam off the target before reacquisition takes place. The time available for reacquisition will be inversely proportional to the angle rate error.

Introducing Large Errors or Breaking Lock in Angle

Since the jammer can readily deny range and doppler data, only angle data are reliable in the radar output. Modern systems are designed to intercept targets successfully when the angle data are reasonably accurate. The techniques of angle jamming may be divided into categories in various ways:

Victim: Scanning radars, or monopulse radars;
Sources: Single-point, or multiple-point jamming;
Approach: Exploitation of hardware deficiencies, or generic.

We will discuss first the techniques used against scanning radars, and for exploitation of design defects in any radar, for which a single-point jammer can often be made effective.

(a) *Scan-Frequency AM Jamming.* Against conical-scan and other sequential scanning radars, large errors can be generated by amplitude modulation (AM) of a noise or repeater jammer. Equation (8.2.3) gives the error for a conical-scan tracker as a function of the AM components falling within one servo bandwidth on each side of the scan frequency f_s. The error is

$$\sigma_s = (\theta_3/k_s) \sqrt{\beta_n \, W(f_s)}$$

Normally, on a fluctuating target, these components contribute a small fractional modulation to the target signal: $2\beta_n \, W(f_s) \ll 1$, and the components are equally divided between the two angle channels. When a strong jammer is 100% amplitude modulated by a sinusoid at or near the scan frequency, the fractional modulation power within the servo bandwidth can approach $\beta_n \, W(f_s) = 0.5$. The error slope, for one-way scanning over the strong jammer, becomes $k_{s_1} \approx 1$. The resulting rms error approaches $0.7\theta_3$, which is excessive for most systems. Indeed, the radial error (the vector sum of azimuth and elevation errors) will approach θ_3, leading to almost certain breaking of the angle track.

The radial error for *spot scan-frequency jamming,* in which the jammer AM is within β_n of the scan frequency f_s, is given [10.7, p. 50] by

$$\epsilon / \theta_3 = 0.36 \, (\theta_3/\theta_k) \, q_j m_j / (2 + q_j) \approx q_j m_j / k_{s1} \, (2 + q_j) \qquad (10.5.3)$$

where q_j is the jammer-to-signal voltage ratio and m_j is the fractional modulation of the jammer. The use of $2 + q_j$ in the denominator results from the fact that the two-way error slope on the target is twice that on the jammer [10.9, p. 35]. For $q_j \gg 1$, this becomes simply

$$\epsilon = m_j \theta_3$$

as derived from (8.2.3) for $m_j = 1$. The corresponding rms error in each angle coordinate, ignoring nonlinearity and probable breaking of lock for $m_j \to 1$, becomes

$$\sigma_\theta = (m_j \, \theta_3 / \sqrt{2} \, k_{s_1}) \, q_j / (2 + q_j) \qquad (10.5.4)$$

When the conical-scan frequency is unknown (e.g., COSRO, Section 8.5), the jammer AM components must be spread over the band $B_s = (f_s)_{max} - (f_s)_{min}$, reducing the power of components which lie within the servo bandwidth by the factor $B_s/2\beta_n$. This can be termed *barrage scan-frequency jamming*. The rms error for each angle coordinate becomes

$$\sigma_\theta = (m_j \, \theta_3 / k_{s_1}) \, \sqrt{\beta_n / B_s}) \, q_j / (2 + q_j) \qquad (10.5.5)$$

The large errors and nonlinearities which lead to breaking track in spot scan-frequency jamming will not occur in this case. Methods of producing barrage AM noise over the scan frequency band include direct amplification of bandpass audio noise, and sweeping of a sinusoidal frequency over B_s. A rapid sweep will produce the effect of uniform noise over the band.

A slow sweep will cause brief periods of large error as the jammer modulation frequency f_j coincides with f_s (Figure 10.5.1). This mode of jamming can be termed *swept scan-frequency jamming*.

The duration of the large error will be

$$t_j = t_s (2\beta_n / B_s) \qquad (10.5.6)$$

where t_s is the sweep period of the jammer. In order for the error to be large, $t_j \geq 1/\beta_n$ is required, or

$$t_s \geq B_s / 2\beta_n^2 \qquad (10.5.7)$$

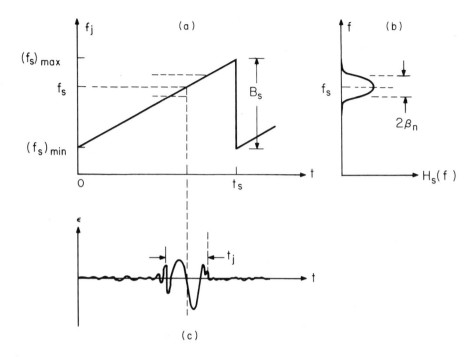

Figure 10.5.1 Swept scan-frequency jamming effect: (a) jammer modulation frequency *versus* time; (b) servo response *versus* modulation frequency; (c) servo error *versus* time.

The fraction of the time when the radar is subject to large error is $2\beta_n/B_s$, leading to the same rms error as for the barrage case (10.5.5).

By designing the radar for a large ratio $B_s/2\beta_n$, the effect of barrage scan-frequency jamming can be kept small, and the sweep period of swept scan-frequency jamming can be increased to give long periods of undisturbed data. If the receiver output can be blanked when the jammer sweeps through the actual scan frequency, good data can be obtained with gaps of length t_j at intervals t_s.

Against monopulse radar, the use of AM jamming is ineffective, unless the radar uses receiver commutation or multiplexing (Figure 8.5.7) to overcome receiver gain imbalance. The errors that can be generated in this way are not large, however, unless the receiver channels are seriously unbalanced. The analysis of [10.8, p. 237] established that jamming at the commutation frequency counteracts the advantages of employing commutation.

(b) *Intermittent Jamming.* The intermittent jammer, or single-source blinking jammer, makes use of rectangular-wave modulation to force the radar receiver into nonlinear operation. Many radars, and especially those using sequential scanning for angle measurement, use a slow AGC to hold the receiver output within the linear dynamic range of the amplifiers and the detection and error sensing circuits. If the AGC time constants (for following increasing and decreasing signals) are known, the jammer can generate on and off cycles which force the receiver to operate with the signal alternating between saturation and circuit noise level (Figure 10.5.2). In the case shown, the jammer cycles have been successfully set to deny almost all useful output data.

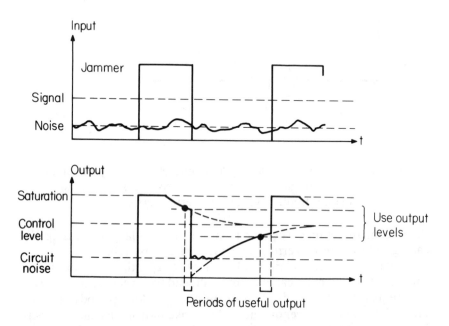

Figure 10.5.2 Intermittent jamming against receiver with AGC.

The radar ECCM solution to this type of jamming is to extend the dynamic range over which useful output can be obtained, and to use fast AGC or logarithmic amplifiers to minimize the time spent in saturation or at the circuit noise level. This is the normal practice in monopulse radars, which can be made immune to intermittent jamming. In conical-scan radar, fast AGC can be used if the scan modulation is extracted from the AGC voltage itself. In MTI and doppler radars, the use of logarithmic amplifiers and fast AGC prior to clutter filtering would cause spreading of the clutter spectrum, so care must be taken in arrangement of the control loops to ensure that the combination of jamming and clutter does not suppress the signal.

(c) *Skirt-Frequency Jamming.* The monopulse Σ and Δ channels must be well matched in phase so that the error detector can preserve the polarity of the target error voltage (8.5.1):

$$E_\Delta = (|\Delta|/|\Sigma|) \cos\phi$$

An error ϕ_2 in receiver phase matching causes a shift in the tracking point (8.5.10), and if the error exceeds 90° the tracking loop may settle on the previously unstable null, more than θ_3 away from the desired tracking axis. A jamming technique that exploits this problem is the *skirt-frequency jammer*, in which a strong signal is offset to the edge of the receiver bandpass curve (Figure (10.5.3).

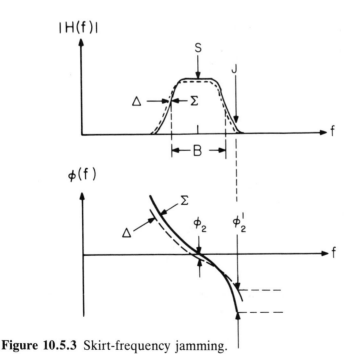

Figure 10.5.3 Skirt-frequency jamming.

A signal in the center of the IF filter produces a small phase-matching error, ϕ_2, because it lies in the region of response for which the Σ and Δ responses can be maintained almost equal. If the jammer at an offset beyond $B/2$ has sufficient power to dominate the output, skirt-frequency jamming can produce Σ and Δ voltages which differ by a much larger ϕ_2'. This is because the slope of phase *versus* frequency at the skirt of the response in much greater than near the center, and it is more difficult to match the two channels in this region.

The radar ECCM for this technique is either (1) to use automatic frequency control, bringing the dominant signal into the center of the filter, or (2) to use a single-tuned circuit (Figure 10.5.4) within the broad response band of the multiple-stage IF, restricting the phase change to $\pm\pi/2$ over the band of significant amplitude response, and making it easier to match the Σ and Δ channel phase response in this band.

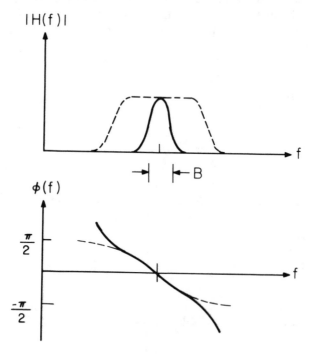

Figure 10.5.4 Single-tuned filter to reduce skirt-frequency effects.

(d) *Image-Frequency Jamming.* Many superheterodyne receivers, operating without RF preselection filters, have equal responses above and below the local oscillator frequency. The undesired response is considered to be the *image* response. Monopulse radars can track on signals at the image frequency, provided that the RF phase paths for the signals and the local oscillator are properly matched (and not compensated by adjustment of IF phase paths). If differences in the RF phase paths are compensated with IF phase shift, the image response may arrive at the error detector with incorrect phase, canceling or reversing the contribution of the desired signal. This permits a jammer, offset by twice the IF, to introduce large errors or break the track.

A radar which is designed and adjusted to track on broadband barrage jamming will obtain valid error signals on both the intended and the image response [10.10]. Use of an RF preselector is an alternative ECCM.

(e) *Cross-Polarized Jamming.* The use of a jamming signal with polarization orthogonal to the radar receiving antenna is regarded as a generic type of angle deception ECM against all radars, including monopulse [10.8, pp. 238–253]. The response of the monopulse antenna to cross-polarized target components was discussed in Section 8.5. The error caused by a small cross-polarized target component σ_c is given by (8.5.12):

$$\sigma_\theta = \frac{\theta_3(\Delta_c/\Sigma)}{k_m\sqrt{\sigma/\sigma_c}}$$

where Δ_c is the on-axis response of the Δ pattern to cross-polarized signals. In a typical antenna, with $\Delta_c/\Sigma = -30$ dB, the error was calculated as $0.008\theta_3$ for a cross-polarized target component at 6 dB below the intended polarization.

The cross-polarized jammer attempts to produce at the antenna output a cross-polarized component that is large relative to the component at the intended polarization, driving the error to θ_3 or beyond. This requires that the ratio of cross- to intended polarization emitted by the jammer exceed the ratio Σ/Δ_c of the antenna. For $\Delta_c/\Sigma = -30$ dB, with linear polarization, the jammer must be within 1.8° of orthogonal polarization. In general, for a jammer with cross-polarized power J_c and power in the radar polarization J_i, the error will be

$$\sigma_\theta = \frac{\theta_3(\Delta_c/\Sigma)}{k_m/\sqrt{2(S + J_i)/J_c}} \tag{10.5.8}$$

For any given requirement $J_c/(S + J_i) = 2(\sigma_\theta k_m/\theta_3 \Delta_c/\Sigma)^2 = X$, the departure from orthogonal polarization must be

$$\phi \leqslant \sqrt{(1/X) - S/J} \tag{10.5.9}$$

where J is the total jammer power and $J/S \geqslant X \gg 1$ is assumed.

For example, if errors $\sigma_\theta = 0.2\theta_3$ are adequate to defeat the radar-controlled weapon system, the ratio $J_c/(S + J_i) > 180 = 22$ dB. For perfect cross-polarization, $J/S = 22$ dB is required. For infinite J/S, the polarization must be within 4.3° of the orthogonal polarization. For $J/S = +30$ dB, $\phi < 3.9°$ is required. If the radar polarization is circular or elliptical, the specifications can be derived on the basis of the ratio of the orthogonally polarized jamming power to the power at the radar polarization, J_c/J_i.

In cases where a radar linear polarization is unknown and cannot be measured, the jammer polarization can be rotated to cover all linear polarizations, producing periodic large errors similar to those of the swept scan-frequency jammer (Figure 10.5.1). The duration of the large error now depends on the magnitude of the cross-polarized response of the radar antenna, and the speed of jammer polarization rotation. The fraction of signals subject to error of a given level will be $2\phi/\pi$, where ϕ is the orthogonality tolerance given by (10.5.9). Polarization rotation is inapplicable to elliptically polarized antennas, because the jammer would have to sweep in two polarization parameters to ensure going through the orthogonal polarization.

In order to produce large errors with cross-polarized jamming, the radar receiving polarization must be accurately known at the jammer. The intercept receiver of the jammer can measure the radar transmitted polarization, and for conventional antennas this will give a good indication of the receiving polarization. Even with this knowledge, control of the jammer polarization to within a few degrees, to produce large radar tracking errors, is difficult. The radar ECCM can take one of the following forms:

(1) Reduction of cross-polarized response of the antenna;
(2) Use of polarization-diversity reception to track on the stronger polarization emitted by the jammer;
(3) Use of an arbitrary polarization for reception, unknown to the jammer.

The last technique is easily implemented in most types of antenna, and is inherent in semiactive homing seekers, where the antenna polarization depends on the arbitrary roll angle of the missile.

(f) *Two-Frequency Jamming.* Radar receivers that lack RF preselection, resulting in RF bandwidths greater than the frequency of the first IF, are subject to generation of spurious outputs from two jamming carriers separated by the IF [10.8, pp. 253–256]. This jamming method can be effective against monopulse radars other than those which use amplitude comparison after envelope detection. Relatively high jamming levels are required, however, to produce the spurious outputs at a level which will cause large errors. The method depends on the nonlinearity of the input mixers, as the input signals approach the level of the radar local oscillator. As a result, two-frequency jamming is unlikely to be applied in cases other than homing seekers, which must operate at very short ranges from the SSJ. The radar ECCM is to use RF preselection or amplitude comparison after envelope detection.

(g) *Cross-Eye Jamming.* The cross-eye jammer is a source of enhanced glint error (see Section 3.4). Extreme errors can be generated, if signals from two equal target sources arrive at the radar antenna in phase opposition (Figure 3.4.1). The cross-eye jammer attempts to create this situation, from a single platform, by radiating coherent signals from two separated antennas (e.g., on the wing-tips of an aircraft). If the null of the interferometer pattern created by these signals can be made to lie within the radar antenna, the Σ signal will fade and the Δ signal will be enhanced, generating a large error. The problem in implementing this jammer is that the interferometer null of the antenna pair must be positioned on the radar antenna for maximum effectiveness, and this requires exact control of the relative phase of the transmitted signals, as well as equality in amplitude.

The magnitude of the error, and the tolerance in amplitude and phase to produce a given error, may be estimated from (3.4.4). For perfect phase ($\phi = \pi$) the error for a target span L is

$$\epsilon = L(1 + k)/(1 - k)$$

Thus, to generate a tracking point which is $4L$ from the target center, $k = 0.6 = -2\,\text{dB}$ (or $k = 1.67 = +2\,\text{dB}$). If the two sources are controlled to $k = 1.25 = 1\,\text{dB}$, the allowable phase error from $180°$ is $\pm 7°$ or 0.02λ. For an aircraft wingspan of 15 m, with $\lambda = 0.03$ m, this corresponds to having the null pointed within ± 0.04 mr of the center of the radar antenna. If these tolerances can be held, the radar has no way of avoiding the error. Not only monopulse, but all types of radar angle measurement are affected by this technique.

(h) *Bistatic Jamming.* The bistatic jammer [10.8, pp. 263–266] creates large angle errors or additional false targets from a single jammer platform by illuminating an external object with a directional jamming beam. The radar sees the jamming source at or near the illuminated object, if it lies within the same resolution cell as the target. With enough jamming power and a sufficiently directional jamming antenna, it may be possible to move the tracking point so that the real target is in the sidelobes of the radar antenna pattern. Possible external objects are chaff, towed or expendable decoys, or the surface of the earth. Using the surface, or *ground-bounce jamming*, will be discussed in Section 11.2, as a special case of the natural multipath error in tracking.

For bistatic jamming, the jamming power at the radar receiver can be written

$$J_b = \frac{P_j G_j G_r \lambda^2 \sigma_b}{(4\pi)^3 R_1^2 R_2^2} \qquad (10.5.10)$$

where R_1 is the range from the jammer to the reflecting object, R_2 is the range from that object to the radar, and σ_b is the bistatic cross section of the object. The direct jamming signal, through the sidelobe of the jamming antenna, is

$$J_d = \frac{P_j G_j G_r \lambda^2 F_j^2}{(4\pi)^2 R^2} \qquad (10.5.11)$$

where F_j is the jammer antenna pattern factor in the direction of the radar, and R is the range to the radar. Assuming $R_1 \approx R$, this gives

$$J_b/J_d = \sigma_b/4\pi R_2^2 F_j^2 \qquad (10.5.12)$$

The requirement for effective bistatic jamming is $J_b/J_d > 1$. Objects far from the jammer must be very reflective, or the jammer sidelobes must be very low if distant objects are to be used.

Consider a chaff burst with $\sigma = \sigma_b = 100$ m^2. From (3.6.14) this will require about 0.15 kg of X-band chaff. If this burst is 500 m from the jamming aircraft, the signal ratio will be

$$J_b/J_d = 3 \times 10^{-5}/F_j^2$$

and a jammer sidelobe level of -45 dB will be required to achieve unity ratio. At the same time, the jamming must be sufficient to obscure the echo from the aircraft, which implies a jammer power that is 45 dB greater than for self-screening. Larger bistatic cross sections are clearly desirable, if large errors are to be produced.

The radar ECCM against bistatic jamming is basically better resolution. If the radar main lobe is narrow and the sidelobes are low, the bistatic object must be placed close to the actual target to inject its jamming into the radar. The radar will then produce a jamming strobe or angle indication not far from the target, and will continue to illuminate the target as a seeker homing source. Depending on the direction of arrival of the homing missile, the radar may close on the real target first, transferring its track from the bistatic source to the target as the range ratio between them grows large.

(i) *Multiple Blinking Jammers.* The use of multiple jammer platforms is an effective way of generating erroneous signal inputs to a tracking radar. Multiple noise jammers received simultaneously at the radar, or within the time constant of the tracker, will create an apparent source at the power centroid of the jamming. Slowly blinking sources will cause the tracking point to shift from one jammer to the other, and if the timing is properly adjusted, there may be little or no data on any source before it shifts again. To be effective against a radar with low sidelobes, the jammers must all be located within the main lobe ($\pm\theta_3$ from the axis). Power requirements are the same as for self-screening noise.

The radar ECCM against this type of jamming consists basically of a tracker response which is much faster than required by the system that uses the data. This forces the time-on-target of each individual jammer to be short, permitting the tracker to coast without excessive lag through the interruptions caused by one or two jammers, other than the one selected for tracking. Special processors capable of resolving multiple targets within the beam can also be devised for certain types of jamming.

REFERENCES

[10.1] S.S. Blackman, *Multiple-Target Tracking with Radar Applications*, Artech House, 1986.

[10.2] H.R. Ward, C.A. Fowler, and H.I. Lipson, GCA Radars: Their History and State of Development, *Proc. IEEE* **62**, No. 6, June 1974, pp. 705–716.

[10.3] D.C. Schleher, *Introduction to Electronic Warfare*, Artech House, 1986.

[10.4] L.B. Van Brunt, *Applied ECM*, Vol. 1, EW Communication, 1978.

[10.5] L.B. Van Brunt, *Applied ECM*, Vol. 2, EW Communication, 1982.

[10.6] S.L. Johnston (ed.), *Radar Electronic Counter-Countermeasures*, Artech House, 1979.

[10.7] M.V. Maksimov, *et al., Radar Anti-Jamming Techniques*, Artech House, 1979.

[10.8] A.I. Leonov and K.I. Fomichev, *Monopulse Radar*, Artech House, 1986.

[10.9] D.K. Barton and H.R. Ward, *Handbook of Radar Measurement*, Artech House, 1984.

[10.10] D.K. Barton, Radar Tracking on a Noise Source, Trans. RADC ECCM Symposium, Vol. 1, October 1957; (U.S. Patent 3,196,433; Passive Radar Tracking Apparatus).

[10.11] I. Kanter, Varieties of Average Monopulse Responses to Multiple Targets, *IEEE Trans.* **AES-17,** No. 1, January 1981, pp. 25–28.

Chapter 11

Radar Error Analysis

The measurement performance of a radar system is characterized by the accuracy of its data output under the intended conditions of operation. Accuracy is defined as freedom from error, and so is characterized by the magnitudes and other properties of the errors in the output data. In order to discuss quantitatively the subject of radar error, we shall consider first the methods by which errors are described.

The mathematics of radar error analysis need not be made so difficult that it requires advanced knowledge of statistical theory. In the following discussion, no special background will be assumed, but a general appreciation of the nature of measurement processes will be needed. The treatment will probably not satisfy those who place a high value on rigorous derivations, but the results of the analysis method have proved valuable in many radar system problems. The procedure described has served as the standard method for radar error analysis since it was first introduced in reports on the AN/FPS-16 Instrumentation Radar [11.1], [11.2], and refined in *Radar System Analysis* [11.3].

11.1 DEFINITIONS AND MODELS

In the discussion of radar targets (Chapter 3), statistical methods were used to describe the amplitude and frequency distributions of radar cross section. The same procedures are applicable to description of errors in radar measurement, which have both systematic and random properties. By choosing mathematical models which are close approximations of the actual error components, we may obtain an accurate analysis of the error in a particular system. The information needed to do this consists of test data and theory for each major system element; although the data may be available only for special test conditions, it is possible to extrapolate the results to a wide variety of actual operating conditions.

Definition of Error

The error in a given measurement is defined as the difference between the value indicated by the measuring instrument and the true value of the measured quantity:

$$x = U_{\text{measured}} - U_{\text{true}}$$

The purpose of error analysis is to provide a description of the error which will permit its magnitude to be estimated for any set of operating conditions, without the necessity of running calibrations or tests for all possible combinations of conditions that may be encountered. The error will vary with the time at which the measurement is made, the value of the quantity to be measured, and with environmental conditions.

Errors are commonly divided into systematic (bias) and random (accidental or noise) components. The former are characterized by some degree of predictability, and hence may be partially corrected by a process of calibration applied to the instrument before and after the measurement. In the ideal case, if the error were constant for all conditions, a single number would describe its magnitude, and subtraction of that number from all data would provide error-free data. One calibration would be sufficient to measure the necessary correction. In practice, however, the error will assume some value within a limited range centered about a mean error, as indicated in Figures 11.1.1 and 11.1.2. For any measured set of n error values, we may determine this mean error as

$$\bar{x} = \frac{1}{n} \sum_{i=1}^{i=n} x_i \tag{11.1.1}$$

Other terms used to describe the magnitude of error include the maximum (or peak) error x_m, the peak-to-peak error $x_{p\text{-}p}$, the rms error x_{rms}, and the probable error x_{50}. The rms error is defined by the expression:

$$(x_{rms})^2 = \frac{1}{n} \sum_{i=1}^{i=n} x_i^2 \tag{11.1.2}$$

The probable error is the value which is exceeded in 50% of the readings, and can have any value from zero to just below the peak error.

In those cases where the mean error is not zero, the rms error may be written as the sum of bias and random components, added in rss (root of the sum of the squares) fashion:

$$(x_{\mathrm{rms}})^2 = \bar{x}^2 + \frac{1}{n} \sum_{i=1}^{i=n} (x_i - \bar{x})^2 \qquad (11.1.3)$$

Equations (11.1.1) to (11.1.3) are not limited to only particular types of errors, and may be applied to any arbitrary or measured distribution of error values. Figure 11.1.2, for example, was obtained from the experimental data of Figure 11.1.1 by simply counting the errors occurring within each increment of three scale divisions (0–2, 2–5, *et cetera*), and plotting a smooth curve through the resulting points. The rms values shown were computed from (11.1.3), and the normal distribution curve with the same rms value was drawn to show how well the data could be approximated by this common mathematical function.

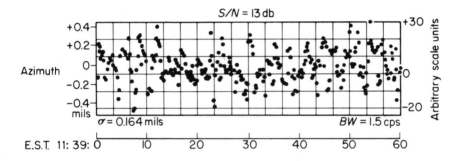

Figure 11.1.1 Typical tracking error *versus* time, derived from boresight telescope.

The square of the standard deviation is the "variance" of the error. Although the term *standard deviation* should be limited to cases where the normal distribution is found, the symbol σ_x, is often applied to designate the rms value of any type of variable error, and will be so used here. The probable error x_{50}, for the special case of the normal distribution, is given by

$$x_{50} = 0.6745\sigma_x \qquad (11.1.4)$$

The normal distribution is often assumed to represent errors of unknown characteristics, and it closely approximates the distribution of many actual errors. In most cases, the peak error observed will correspond to a deviation of $3\sigma_x$ from the mean value, and the peak-to-peak error will be about $6\sigma_x$, when the error appears "noisy" in character. The probability of exceeding the 3σ deviation for the normal distribution is 0.3%, which

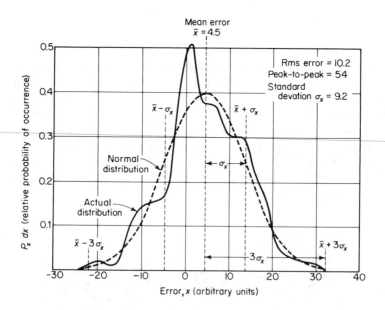

Figure 11.1.2 Distribution of error amplitudes in typical radar track, compared with normal distribution.

implies that one out of 300 independent observations will deviate from the mean by an amount greater than 3σ. When a set of data points is taken over a period of one minute, using a system with a bandwidth of several Hz, we would expect only one or two excursions beyond this value. A deviation of 4σ would be expected only once in 20 or 30 such sets of data, if it followed the normal distribution, and hence we would be likely to measure our peak error near the 3σ point.

The error contributed by a single element in the system may follow a regular pattern, such as those shown in Figure 11.1.3, rather than the random noise pattern. It will be noted that the waveforms are given in order of increasing ratio of peak to rms error, and that these ratios vary from unity for the square wave to three for random noise. True Gaussian noise, of course, would have an infinite "peak" error, but we would have to wait a long time, as mentioned above, to see the 3σ level exceeded.

Time Functions and Frequency Spectra

The variation of error with time and the resulting frequency spectrum of the error are important factors in classifying and describing the performance of a tracking system, and must be known if the effect of smoothing

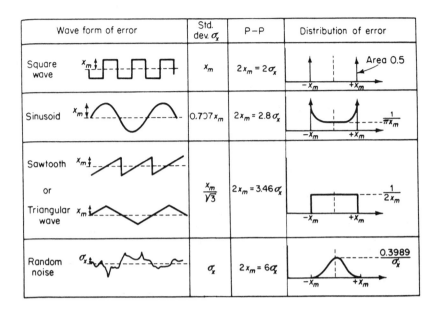

Wave form of error	Std. dev. σ_x	P–P	Distribution of error
Square wave	x_m	$2x_m = 2\sigma_x$	Area 0.5
Sinusoid	$0.707x_m$	$2x_m = 2.8\sigma_x$	$\frac{1}{\pi x_m}$
Sawtooth or Triangular wave	$\dfrac{x_m}{\sqrt{3}}$	$2x_m = 3.46\sigma_x$	$\frac{1}{2x_m}$
Random noise	σ_x	$2x_m = 6\sigma_x$	$\frac{0.3989}{\sigma_x}$

Figure 11.1.3 Relationships between rms, peak, and peak-to-peak errors.

or differentiation is to be determined. The bias error defined above, if it is truly constant for periods of time which are long compared to the calibration and operation times of the system, can be removed by the calibration procedure. Hence, it is of little importance in error analysis. The residual bias error observed in most tests and evaluations of equipment is taken as that portion of error which does not change during the period of an individual test or operation, which may be as short as a few seconds or one minute. When errors are introduced as functions of the measured quantity, the speed at which this quantity varies will determine how much of the error appears as "bias" and how much as "noise." For this reason, no sharp line can be drawn separating bias from noise components in error analysis. An arbitrary time period must be chosen, and those errors which do not change appreciably during this time may be classed as "bias."

The frequency spectrum of the error may be obtained from the observed time function by using one of the harmonic analysis techniques based on the use of the Fourier integral or transform. A typical error spectrum for an angle-tracking system is shown in Figure 11.1.4. True bias is represented by an "impulse function" at zero frequency, with infinitesimal width and an area σ_o^2 equal to $\bar{x}^2$. The apparent bias observed over time intervals shorter than t_o is represented by the area σ_b^2 beneath the spectral density curve between zero frequency and the frequency

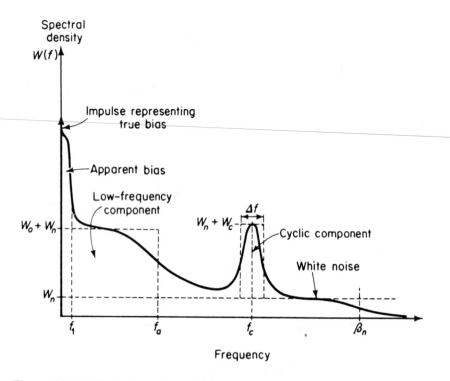

Figure 11.1.4 Typical angle-tracking error spectrum.

$f_1 = 1/(2t_o)$. Above f_1 there appears a low-frequency error component, which can be approximated by a Markoffian spectrum:

$$W_a(f) = W_o \frac{f_a^2}{f_a^2 + f^2} \tag{11.1.5}$$

The variance of this component is given by the value of the integral of W_a between zero and infinity:

$$\sigma_a^2 = \frac{\pi}{2} W_o f_a \tag{11.1.6}$$

This represents the spectrum which results when broadband (white) noise is passed through a single-section low-pass filter consisting of a series resistor and parallel capacitor, where $RC = t_a = 1/(2\pi f_a)$.

Above the frequency f_a in Figure 11.1.4, we see two more error components. The random or white-noise component extends with uniform spectral density W_n to the limit of the observed spectrum, as set by the bandwidth of the measuring device. The variance of the random noise error is

$$\sigma_n^2 = W_n\beta_n \tag{11.1.7}$$

Superimposed upon the white noise is a cyclic component, occupying a narrow band of frequencies centered at f_c. If this component is a pure sinusoid, given by the wave form $x_c = X_c \sin(2\pi f_c t)$, it should appear as an impulse function with area $\sigma_c^2 = X_c^2/2$ at the frequency f_c. In a spectrum obtained from experimental data over a finite period, the same area will be distributed over a narrow band of frequencies, and the error may be approximated by a rectangular spectrum of amplitude W_c and width Δf, chosen so that $\sigma_c^2 = W_c\Delta f$.

The variance of the total error is given by the area under the entire spectrum, or

$$\sigma_x^2 = \int_0^\infty W(f)\, df = \sigma_o^2 + \sigma_b^2 + \sigma_a^2 + \sigma_n^2 + \sigma_c^2 \tag{11.1.8}$$

In the case shown, no error is correlated with any other, and the process of rms addition yields the total variance. It might be expected that two bias components would add directly, rather than in an rms fashion. However, the area σ_b^2 represents a slowly varying error, which will sometimes have the same polarity as the true bias and sometimes oppose it. When evaluated over a large number of intervals, each of duration t_o, the sum $(\sigma_o^2 + \sigma_b^2)^{1/2}$ represents the rms value of the bias observed, and the probable bias will be approximately 67% of this value, in accordance with the normal distribution.

Validity of the Error Model

The statistical description of errors permits us to break down a complex tracking system into many elements, to calculate or to measure the corresponding components of error, and then to combine these in an rss addition process to obtain the overall system error. The underlying assumption is that the several error components are completely uncorrelated with each other, and in practice it has been found safe to ignore possible correlation effects unless there are clear physical links causing the common variation of two or more error components. Two examples taken from analysis of actual radar systems will illustrate the validity of this assumption.

In one case, the radar was investigated to find the extent of the angle-measurement errors resulting from thermal noise in the receiver and from target fluctuation. A lengthy series of digital computer simulations was done, using various combinations of mean *S/N* and modulation, superimposed on the signal to represent target scintillation. After the results were plotted and reviewed, it was found that the entire family of curves resulting from the simulations could be represented accurately by the rss sum of a thermal noise component, varying with mean *S/N*, and an independent scintillation component. Similar results have been obtained by using the data from Swerling's analysis.

In a second case, a tracking pedestal was tested piece by piece with optical procedures, and error curves were determined for each gear mesh, shaft, and major structural member. A detailed analysis showed that the overall pointing error of the antenna axis, evaluated over many points in the hemisphere, was the same as the rss sum of the individual error components. This was true in spite of the fact that each of the errors was a function of the azimuth and elevation angle of the antenna. The several components were characterized by different periods and phase relationships, as referred to the two tracking angles, and they added as though they were completely uncorrelated.

Application of Error Model

In order to apply the error model to a specific radar and environment, the many sources of error must first be identified, for each measured coordinate, and estimates of their amplitude and frequency characteristics must be expressed in terms of the radar parameters and the location of the target within the coverage. Material in Chapters 8 to 10, and in previous chapters, provides the basis for many of these estimates. Before applying these data to error analysis, however, there are two major sources of measurement error which require discussion. These are multipath errors, caused by reflection or forward scatter of the target energy from the surface beneath the target-to-radar path, and clutter errors, caused by backscatter from objects within the surface or volume resolution cell containing the target.

11.2   MULTIPATH REFLECTION ERRORS

The reflection of radar waves from the earth's surface was discussed in Section 6.2. For a flat surface, the amplitude of the reflected wave, relative to that of the direct wave reaching the radar from the same target, was given by the Fresnel reflection coefficient, ρ_0, plotted in Figure 6.2.2 as a function of grazing angle for various surface materials. For rough surfaces, a specular scattering coefficient ρ_s (Figure 6.2.3) multiplies the Fresnel coefficient, reducing the amplitude of the reflected wave. A further reduction may be caused when the surface is covered by vegetation. The total reflection coefficient $\rho = \rho_0\,\rho_s\,\rho_v$ was used to determine the extent to which reflection lobes and nulls would modify the coverage diagram of the radar (Figure 6.2.4) through change in the pattern-propagation factor F in the radar equation.

Surface reflections affect measurements in all four radar coordinates, but the problem is most pronounced in elevation, where the reflections produce the largest extension of the target signal. In analyzing the radar errors, we must consider not only the specular component of reflection, which systematically modifies the pattern-propagation factor and detection coverage, but also the smaller diffuse and diffraction components, which are often negligible in detection analysis. We must also develop models that describe the distributions of multipath sources in four-dimensional radar coordinates. The response of the radar antenna and signal processor to these distributions will determine the errors in output data.

Spatial Distribution of Multipath Sources

The errors caused by multipath reflections can be estimated, if the sources of reflected energy can be located in radar coordinates. Basic radio propagation theory can be combined with experimental data to create models of reflected components from specular reflection, diffuse reflection, and diffraction from obstacles and major irregularities in the surface beneath the target-to-radar path. In detection analysis, the two-way propagation factor is used to predict the signal strength at the radar. In analysis of multipath errors in the angle coordinates, only the return path to the radar is significant, because the transmitting path affects only the amplitude and phase of the signal reflected from the target, and not its angle of arrival at the radar antenna.

(a) *Specular Reflection.* The specular reflection component was described in Section 6.2, where Figure 6.2.1 gave the geometrical relationships between the direct and the reflected paths. Figure 11.2.1 summarizes the relationships for a flat surface. The field amplitudes scattered from the target are A_t for the direct path and A_r for the reflected path, and relative powers are represented by the squares of these amplitudes. At reflection, the field A_r is multiplied by the Fresnel reflection coefficient $\rho_0 < 1$. For a flat surface, the specular scattering coefficient $\rho_s = 1$.

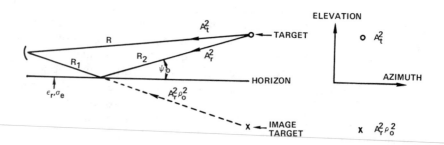

Figure 11.2.1 Specular reflection from a flat surface: (a) rays in the vertical plane containing the direct path; (b) apparent sources in radar azimuth-elevation coordinates.

The radar antenna then receives signals from two discrete sources: (1) the direct target signal with relative power A_t^2, at an elevation angle of

$$\theta_t = \sin^{-1}(h_t - h_r)/R \approx (h_t - h_r)/R \tag{11.2.1}$$

and (2) the specular multipath interference with relative power $A_r^2 \rho_0^2$, at a negative elevation angle of

$$\theta_r = \psi_0 = \sin^{-1}(h_t + h_r)/R \approx (h_t + h_r)/R \tag{11.2.2}$$

Both sources will be at the azimuth of the target, but the time delay of the reflected interference will be greater by δ_0/c, where

$$\delta_0 = R[(\cos\theta_t - \cos\psi) - 1] \approx 2h_t h_r/R \tag{11.2.3}$$

The output of the antenna will depend on the antenna pattern in the elevation plane.

(b) *Diffuse Reflection.* For a completely rough surface, defined by $\rho_s \approx 0$, or $(\sigma_h/\lambda)\sin\psi_0 > 1$, specular reflection will be replaced by diffuse reflection, $\rho_d > 0$. The theory of the *glistening surface* [11.4] can be applied to predict the diffuse power and its spatial distribution. Each surface element dS in Figure 11.2.2 contributes a relative interference power:

$$dI_d = A_r^2 \, \rho_0^2 \, d\rho_d^2 = A_r^2 \, \rho_0^2 \left[\frac{R}{R_1 R_2}\right]^2 \frac{dS}{4\pi\beta_0^2} \exp(\beta^2/\beta_0^2) \tag{11.2.4}$$

where β is the angle between the bisector of the two reflection paths and

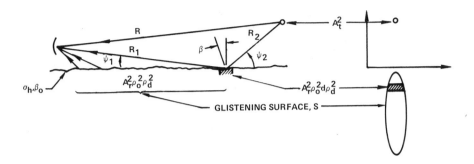

Figure 11.2.2 Diffuse reflection from completely rough surface.

the vertical, and β_0 is $\sqrt{2}$ times the rms slope of the reflecting facets making up the surface.

The glistening surface from which diffuse power arrives at the radar is defined by the elliptical contour in azimuth-elevation space at which the intensity drops to $1/e$ of its maximum value ($\beta = \beta_0$ in (11.2.4)), although sources of decreasing intensity extend outward from this contour. For low grazing angles, at which the upper end of the glistening surface extends to the horizon, considerations of the conservation of energy [11.5] require that the diffuse scattering coefficient $d\rho_d^2$ for each surface element be multiplied by a factor less than unity, which is dependent on the ratio of the grazing angle to the rms slope.

(c) *Partially Diffuse Reflection.* Most actual surface conditions give partially diffuse and partially specular reflection. The relative magnitudes of specular and diffuse scattering coefficients vary over the glistening surface, and as a result the diffuse scattering coefficient for each surface element is multiplied by a roughness factor:

$$F_d^2 = \sqrt{(1 - \rho_{s_1}^2)(1 - \rho_{s_2}^2)} \tag{11.2.5}$$

where ρ_{s_1} and ρ_{s_2} are the specular scattering coefficients for the grazing angles of the two reflection paths of Figure 11.2.3. This factor has the property that a transition to pure specular reflection ($\rho_s \rightarrow 1$) for either part of the reflected path eliminates the possibility of diffuse reflection from the target to that surface element to the radar ($d\rho_d \rightarrow 0$). The magnitude of the specular component reaching the radar depends on the value of the specular scattering coefficient ρ_{s_0} at the specular point, which in turn

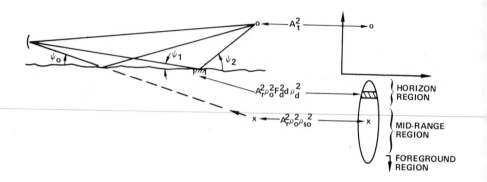

Figure 11.2.3 Reflection from a partially rough surface.

is determined by the grazing angle ψ_0 and the rms surface roughness within the first Fresnel zone around the specular point [11.4, p. 12].

Other differences between the general reflection model [11.5] and the original Beckmann and Spizzichino model [11.4] are given below.

(1) For low grazing angle ψ_1 or ψ_2, the height deviation σ_h of the random surface must be adjusted to account for shadowing of the lower portions of the surface. A ray passing at low grazing angle over an irregular surface will interact only with the upper portions of each irregularity, ignoring the depth of the intervening troughs. All of the energy directed from a source downwards toward the surface will reach the surface, so the shadowing adjustment does not reduce the average reflection coefficient taken over regions larger than the correlation length of the irregularities. However, the surface appears to have a reduced height deviation σ_h' determined empirically to be

$$\sigma_h' = \sigma_h \sqrt[5]{2\psi/\beta_0}, \quad 2\psi/\beta_0 < 1 \tag{11.2.6}$$

where ψ is the lower of the two grazing angles. One effect of this correction is to eliminate the anomaly [11.6, Fig. 2], wherein ρ_s appears not to fall to zero as fast as predicted by (6.2.13).

(2) For very rough surfaces, the diffuse coefficient ρ_d continues to increase to a maximum value $\rho_d = 0.7$, as predicted by theory [11.4, p. 264], rather than leveling off near $\rho_d = 0.3$, as shown experimentally in [11.6, Fig. 4] and [11.4, Fig. 15.3]. It was found [11.7] that the leveling

off of ρ_d was the result of antenna patterns which did not permit all the diffuse reflections to enter the receiver.

(3) The values of slope β_0 used in (11.2.4) of the general model are evaluated for surface facets of size sufficient to make significant contributions to forward scattering at grazing angles of interest. This means that the slopes of small-scale irregularities (e.g., capillary waves on the sea surface, pebbles and small furrows on land) should be excluded from β_0. The mathematical expression for the surface should be smoothed to ignore irregularities for which $\sigma_h/\lambda < 1/8\pi \sin\psi$. As a result, the curves for β_0 *versus* wind speed over the sea [11.4, p. 408], and equivalent data on surface slopes used for estimation of backscatter for land cannot be applied directly to the multipath or forward-scatter case. Instead, the rms height and correlation distance (or length of the sinusoidal sea waves) can be used to calculate β_0 as follows:

$$\beta_0 = 2\sigma_h/L_c \quad \text{(random surfaces)} \tag{11.2.7}$$
$$\beta_0 = \sqrt{2}\,\pi\sigma_h/L \quad \text{(sinusoidal)}$$

where L_c is the correlation distance of the random irregularities, and L is the wavelength of the sinusoidal undulations.

(d) *Effects of Vegetation.* The specular and diffuse components are both subject to attenuation by vegetation on land surfaces. In the generalized multipath model, this can be represented by a vegetation factor ρ_v, which directly multiplies ρ_0 of the underlying surface (Figure 11.2.4). Insufficient data exist to express this factor as a function of grazing angle, wavelength, and type of vegetation, but values from -30 to -10 dB (0.03 $< \rho_v < 0.3$) have been measured. The vegetation also scatters the incident energy more widely than other types of rough surface. This wider scattering appears in two components: (1) as increased backscatter (expressed as reflectivity $\sigma°$ or γ) and wide-angle bistatic scatter, and (2) as a component of forward scatter having a broad angle in both azimuth and elevation about the specular direction. The azimuth distribution is much broader than predicted by the tilted facet model on which the glistening surface theory was based, and this forward-scatter component is larger than predicted from the usual bistatic reflectivity parameter $\sigma_b°$.

(e) *Diffraction.* Surface features that rise to any significant height above the average surface will scatter incident energy by knife-edge diffraction (Figure 11.2.5). For an isolated obstacle, which is fully illuminated from both ends of the reflected path, the magnitude $A(v)$ of the diffraction component is given by Figure 6.3.4 for a diffraction parameter $v > 1$. This

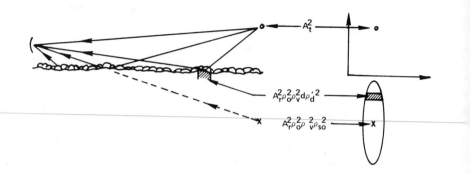

Figure 11.2.4 Effects of vegetation on multipath reflection.

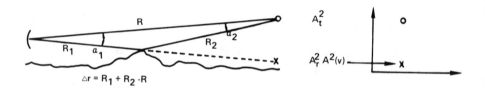

$$\Delta r = R_1 + R_2 - R$$

Figure 11.2.5 Diffraction over an obstacle.

component, when added to the direct signal of unit amplitude, is responsible for the small oscillations in the total diffraction field for paths above the obstacle, $v < 1$ (Figure 6.3.5). However, in analysis of multipath measurement errors, it is used as a separate interference component, as shown in Figure 11.2.5. Multiple diffraction sources can be included in the model, but only rarely will they each meet the criterion for being fully illuminated by the two ray paths.

The criterion for a fully illuminated obstacle is that there must be Fresnel clearance on both paths to the obstacle (Figure 11.2.6): the terrain between the obstacle and both the radar and the target must not project into the ellipses representing paths of length $R_1 + \lambda/2$ and $R_2 + \lambda/2$. If a large area of smooth reflecting surface lies within the Fresnel ellipse,

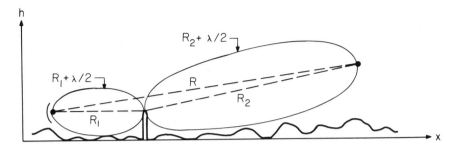

Figure 11.2.6 Conditions for fully illuminated obstacle.

propagation over the corresponding path will be governed by reflection-interference phenomena from that surface [11.8, pp. 39–42]. Penetration of the ellipse by discrete obstacles can be evaluated by using multiple diffraction procedures [11.8, pp. 34–35]. Unfortunately, there is no convenient model for cases in which a random rough surface lies within the ellipse.

Elevation Distribution of Multipath Interference

Having modeled the specular, diffuse, and diffraction components of multipath interference, the power density in the elevation coordinate can be calculated (Figure 11.2.7). Specular and diffraction components appear as delta functions at angles $\theta_0 = -\psi_0$ and θ_d. The diffuse components $d\rho_d^2$ are transformed into a continuous distribution of density η_d (in relative power per unit angle), given by

$$\eta_d = -d\rho_d^2/d\psi_1 \approx (R_1/\psi_1)d\rho_d^2/dR_1 \tag{11.2.8}$$

In radars with low resolution capabilities in range and doppler, the multipath interference components will be received with weights established entirely by the antenna elevation patterns for the Σ and Δ channels, as shown in the figure.

The multipath interference power, relative to the signal power in the Σ channel, will be

$$I_\Delta/S = (1/A_t^2)\left[\int A_r^2(\theta)\rho_0^2(\theta)\eta_d(\theta)\Delta^2(\theta)d\theta\right.$$
$$\left. + A_r^2(\theta_0)\rho_0^2(\theta_0)\Delta^2(\theta_0) + A_r^2(\theta_d)A^2(v)\Delta^2(\theta_d)\right] \tag{11.2.9}$$

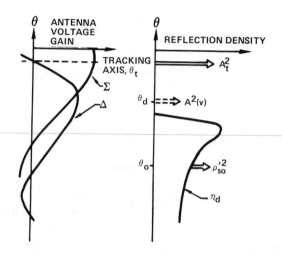

Figure 11.2.7 Elevation distribution and weighting of multipath interference.

This is the reciprocal of the signal-to-interference ratio used in (8.1.9) to determine the elevation error caused by the multipath reflections.

It is usually assumed that the target scatters equally toward the radar and the surface, $A_t = A_r$, so that these terms in (11.2.9) cancel out. The reflection lobe widths of actual targets are, in fact, broader than the angle $2h_r/R$ between these two paths, in a majority of cases, leading to close correlation between the two amplitudes. However, as the radar antenna height and the vertical extent of the target are increased, the two amplitudes may vary independently, with the reflected amplitude becoming large enough to overcome the reduced reflection coefficient ($\rho A_r > A_t$), leading to large multipath errors.

Figure 11.2.7 illustrates some important factors which determine the seriousness of the elevation multipath error problem.

(a) The elevation Δ main lobe extends beyond the Σ main lobe, and hence may receive large multipath reflections, even when these lie outside the main lobe of the Σ pattern. In a typical pattern, the Δ main lobe has its first null at an angle $1.5\theta_3$ below the axis.

(b) The diffuse reflection distribution, typically extending upwards from the specular point to the horizon, is received with higher gain Δ than the specular reflection. If diffuse multipath is to be excluded from the Δ main lobe, the target must be at an elevation angle at least $1.5\theta_3$ above the horizon.

(c) If a radar fence or mask is relied upon to reduce specular and diffuse reflections, it will produce a diffraction source at an angle which is even closer than the reflections to that of the target. In addition, since the vegetation factor ρ_v has no effect on $A(v)$, the resulting diffraction error may be greater than the reflection errors that are eliminated by the mask.

Multidimensional Distributions of Multipath

To evaluate the effects of smoothing, filtering, and frequency diversity or agility on multipath error, a more generalized multipath model is needed. Such a model preserves the time delay and (for moving targets) the doppler coordinates of each multipath component. In time, the minimum delay applies to the specular component, for which the extra path length is given by (11.2.3). Diffuse components of increasing delay are received from regions of the glistening surface on all sides of the specular point. Since the geometry of reflection from tilted facets on the glistening surface restricts the width of that surface, for targets at low elevation angles, it is sufficient to consider all the sources to be in a vertical plane containing the radar-target path. This permits the delay δ for surface elements at ground range x from the radar to be expressed in terms of x and x_t, the ground range to the target:

$$\delta = [h_r(1 - x/x_t) + h_t x/x_t]^2/2x(1 - x/x_t) \qquad (11.2.10)$$

For low-angle tracking ($\theta_t < \beta_0$), the maximum delays at the ends of the glistening surface are, near the radar, $\delta \approx \beta_0 h_r$, and near the target $\delta \approx \beta_0 h_t$.

In the usual low-angle tracking case, with the radar antenna at low altitude, radar range resolution is not sufficient to exclude the major portions of multipath reflection. For example, with $h_r = 10$ m and the target at an elevation angle $\theta_t \approx h_t/R = 0.01$ radian, the excess path length for specular reflection is only 0.2 m, or 0.67 ns in time delay. For $\beta_0 = 0.05$ radian, the diffuse components nearest the radar have an excess path $\delta = 0.5$ m, but at that point they are at a depression angle of $2\beta_0 = 0.1$ radian from the antenna, and will generally lie in the sidelobes of the antenna pattern. Diffuse components from the regions closest to the target may be resolvable on the basis of range delay.

Knowledge of the distribution of multipath reflections in range permits the error in range measurement to be estimated. In this case, the impulse response for the round-trip path of both the direct and reflected signals must be considered, and this is the self-convolution of the one-way path impulse response (Figure 11.2.8). When this response is convolved with the radar waveform (including the carrier term), the phase of the reflected components may be either positive or negative, depending on the phase contribution at the delay δ_0. Convolution of the path impulse response with the sinusoidal carrier, followed by integration over the waveform duration, gives the propagation factor F^2 of (6.2.7).

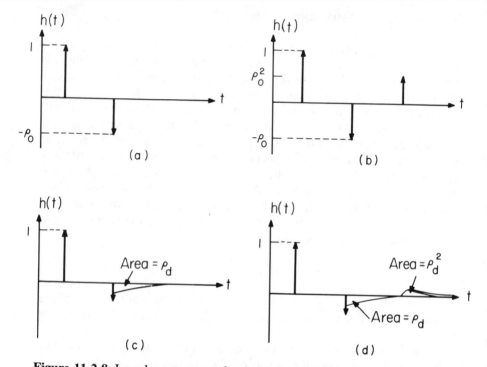

Figure 11.2.8 Impulse response of radar channels with specular and diffuse multipath: (a) one-way path with specular reflection; (b) two-way path with specular reflection; (c) one-way path with diffuse reflection; (d) two-way path with diffuse reflection.

The doppler shift relative to the doppler on the direct path, for a target flying horizontally, is

$$\Delta f_d = (v_t/\lambda)(\cos\theta_t - \cos\psi_2) \approx (v_t/2\lambda)(\psi_2^2 - \theta_t^2) \tag{11.2.11}$$

For the specular reflection, where $\psi_2 = \psi_0$, this gives

$$\Delta f_d = \delta_0 v_t/\lambda R \tag{11.2.12}$$

Again, for the usual low-angle geometry, the doppler shift is seldom sufficient to permit exclusion of major multipath components. In the previous example, with $\delta_0 = 0.2$ m, the relative doppler shift of the specular component for $v_t = 300$ m/s, $R = 10$ km, $\lambda = 0.03$ m is only $\Delta f_d = 0.2$ Hz. The resulting sinusoidal error, having a period of 5 s, cannot be removed from the data by smoothing, unless very long smoothing times are used, bringing with them large lag errors for even small target maneuvers. The diffuse components nearest the target will have the largest doppler shift, permitting them to be excluded by filtering in the receiver or smoothing of the output data. These may be the same components which are resolvable in range, however.

Elevation Multipath Error

The general expression for elevation multipath error can be found by substitution of (11.2.9) into (8.1.10), replacing the general resolution cell z_3 by θ_e and the slope k_z by the applicable angle error slope (e.g., k_m for monopulse radar):

$$\sigma_\theta = \frac{\theta_e}{k_m\sqrt{2(S/I_\Delta)n_e}} \qquad (11.2.13)$$

For conical-scan and other tracking techniques, the multipath errors are essentially the same as for monopulse, but additional components may appear when the multipath reflections fluctuate at or near the scan frequency.

The number of independent samples, n_e, averaged in the data output is then found by considering the spectrum of the multipath doppler frequency Δf_d, given by (11.2.11). As noted above, that spectrum often involves such low frequencies that $n_e \approx 1$, and the output error is then given directly by (11.2.13). The equation is valid only for $S/I_\Delta \gg 1$, and does not apply to the special case of strong specular multipath. That special case, and others for which simplified equations may be used, can be summarized as follows. The generalized reflection coefficient ρ will be used to denote the product of Fresnel coefficient, specular or diffuse coefficient, and vegetation coefficient in each case.

(a) *Strong Specular Multipath.* When the specular reflection coefficient $\rho_0\rho_s\rho_v > 0.7$, and the target is separated by less than $1.4\theta_e$ from its specular image, a two-element target is created (Section 3.4). The average tracking point, in theory, lies at the stronger (direct) target position, but, for $k = \rho A_r/A_t \to 1$, Figure 3.4.1 shows that the tracking point will lie at the horizon most of the time, rising steeply above the real target when the relative phase nears 180°. The resulting tracking angle is plotted in Figure 11.2.9, as a function of true target angle, for surfaces with varying degrees of roughness. For the smooth surface, $\rho \rightarrow 1$ and the target data are lost as θ_t reaches $0.7\theta_e$. Below this angle, the tracking point is either at the horizon or an angle $0.7\theta_e$, which bears no relation to the target elevation. Over surfaces of greater roughness, the loss of target data is delayed until $\rho > 0.7$. If the instantaneous reflected signal exceeds the direct signal, the excursions from the horizon will be to $-0.7\theta_e$. The radar error cannot be expressed meaningfully as an rms value under these circumstances.

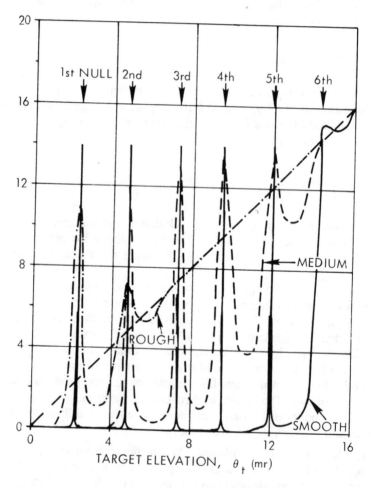

Figure 11.2.9 Tracking error with strong specular reflection.

This phenomenon, known as *nose-diving,* can result in loss of track in all coordinates, as the radar antenna may overshoot during the fades that generate the large peak errors, completely breaking the tracking loop. Several means of preventing loss of track for strong specular reflections have been devised, and will be discussed below.

(b) *Weak Specular Reflection.* When $\rho < 0.7$, (11.2.13) may be applied, with $I_\Delta/S = [\rho k_m(\theta_t + \psi_0)]^2$, resulting in

$$\sigma_\theta = \rho(\theta_t + \psi_0)/\sqrt{2} = \sqrt{2}\,\rho h_t/R \qquad (11.2.14)$$

The error will oscillate slowly (at frequency Δf_d given by (11.2.12)) about the true target position.

(c) *Specular Reflection in Sidelobes.* At target elevations above about $0.7\theta_e$, the specular image will lie in the sidelobes of the elevation Δ pattern, giving $S/I_\Delta \ll 1$. For a ratio G_{se} of main lobe Σ gain to average Δ sidelobe gain, the error becomes

$$\sigma_\theta = \frac{\theta_e\rho}{k_m\sqrt{2G_{se}}} \qquad (11.2.15)$$

This expression can be used to establish the required Δ-channel sidelobe ratio for a given tracking accuracy. For example, if $\rho = 0.7$ is expected, and an rms error $\sigma_\theta = 0.01\theta_e$ is required, with $k_m = 1.4$, $G_{se} > 1250 = 31$ dB must be obtained at angles equal to about twice the target elevation angle.

(d) *Diffuse Reflection in Main Lobe.* When tracking targets at elevation angles below $1.5\theta_e$, the specular reflection from near the horizon will enter the Δ-channel main lobe. The diffuse reflection density η_d must be weighted by $\Delta(\theta)$ and integrated over the glistening surface, according to (11.2.13). There is no general approximation for this case, but a calculator program based on the material in [11.5] has been published [11.9], from which data such as Figure 11.2.10 can be obtained. In this figure, the abscissa represents the approximate target elevation, normalized to the beamwidth, $\theta_t/\theta_e = h_t/R\theta_e$, and the error has also been normalized to the beamwidth. Errors near $0.05\theta_e$ are shown for vertical polarization over water, at target elevation angles from about $0.7\theta_e$ down to the angle at which nose-diving invalidates the radar data. Over land surfaces, and for beamwidths narrower than the 40 mr illustrated, the diffuse errors would tend to increase toward the curve for $\rho_0 = 1$, because the Fresnel reflection coefficient would be higher than for water surfaces at the 40 mr beamwidth.

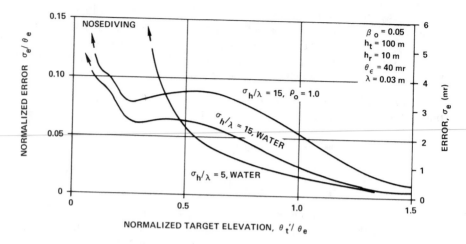

Figure 11.2.10 Typical error from diffuse reflection.

(e) *Diffraction Edge within Main Lobe.* A fence or other obstacle can be used for low-sited radars over land to block most multipath reflections. If the diffraction from such an obstacle is the dominant multipath component, and if it lies within the region of linear Δ slope, the error can be estimated from

$$\sigma_\theta = A(\theta_d)(\theta_t - \theta_d)/\sqrt{2} \tag{11.2.16}$$

where the diffraction amplitude $A(\theta_d)$ is found from Figure 6.3.4 for an edge at elevation angle θ_d.

As an example, for the geometry shown in Figure 11.2.11 with an X-band radar, the diffraction parameter $v = 1.08$ and $A(\theta_d) = -15$ dB $= 0.18$. The angle between the target and the edge is 4 mr, and this will be assumed to place it within the region of linear Δ slope. The error will be

$$\sigma_\theta = 0.18 \times 4/1.4 = 0.5 \text{ mr}$$

Whether this represents an improvement over the error that would have been encountered without the fence depends on the beamwidth of the radar and the reflection conditions of the terrain.

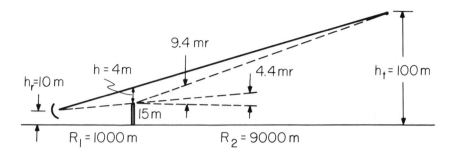

Figure 11.2.11 Typical radar fence diffraction geometry.

Azimuth Multipath Error

Azimuth error is caused when the multipath components arrive from outside the vertical plane containing the direct path to the target. This can happen when specular reflections arrive from a tilted surface, in which case the azimuth error is calculated simply as the sine of the tilt angle times the elevation error that will result from a level surface. A more common source of azimuth error is the spread of the diffuse components across the width of the glistening surface. This spread, measured between the $1/e$ points of the surface, is

$$\theta_{da} = 4\beta_0(\theta_t + \psi_1) \tag{11.2.17}$$

The elevation beamwidth of the antenna sets a limit near $\theta_e/2$ on the angle $\theta_t + \psi_1$ from which reflections will enter the azimuth Δ pattern with significant gain. If the target elevation is above $\theta_e/2$, there will be negligible azimuth multipath from the glistening surface. For targets near $\theta_t = \theta_e/2$, the azimuth error will be

$$\sigma_\theta = \rho\beta_0\theta_e \tag{11.2.18}$$

where ρ is the diffuse reflection coefficient for that portion of the glistening surface which is near enough to the horizon to enter the mainlobe of the elevation pattern. In a typical case of a rough surface, with $\beta_0 = 0.05$ radian and $\rho_d = 0.3$, this gives

$$\sigma_\theta = 0.015\theta_e$$

The actual azimuth errors when tracking targets below $\theta_e/2$ appear to be larger than predicted by the theory of the glistening surface [11.10]. Over land surfaces, this can be explained by the broadened forward scatter lobe from vegetation (Figure 11.2.4). Similar diffraction effects may arise from rough sea surfaces, generating horizontal gradients in the field at the radar antenna and increasing the azimuth error beyond the predictions of glistening surface theory.

Reduction of Multipath Errors

Many attempts have been made to devise antenna patterns and signal processing procedures that will reduce multipath error in elevation measurement. Some of these are discussed in [11.5] and [11.7]. A few general comments can be made on these techniques, as follows.

(a) *High Resolution in Angle.* The most powerful technique for reducing multipath error is to narrow the elevation beamwidth. The error scales directly to the beamwidth, in most cases, and except for strong specular reflections, the error in elevation should not exceed

$$\sigma_\theta \approx 0.05\theta_e$$

Thus, if the beamwidth does not exceed 20 times the allowable error, no special error reduction procedures are necessary.

(b) *High Resolution in Range.* The specular reflection arrives with an excess path length:

$$\delta_0 = 2h_r h_t/R \approx 2h_r\theta_t$$

and a great part of the diffuse reflected power lies very close to this delay. Only in cases where the radar antenna can be located at considerable height, and when the target is at or above this height, can range resolution be expected to solve the multipath problem. For example, if the radar and target are both at $h = 10$ m, the excess path length for a target at a range as short as $R = 3$ km is 0.07 m, and the time delay in 0.2 ns. While systems having this time resolution (bandwidth of 5 GHz) are not impossible, they achieve this resolution at the expense of considerable difficulty in solving the other problems of tracking and measurement. There is little point in removing the effects of multipath unless the resulting system accuracy is better than that of a conventional system with multipath error.

(c) *Frequency Diversity and Agility*. The use of frequency diversity and agility offers some reduction in multipath error, by obtaining in a short time several independent samples of the error for averaging. The required frequency shift to obtain independence is reciprocally related to the delay of the multipath reflections relative to the direct signal. It is easier to operate with successive pulses at frequencies $1/\delta$ than to generate and process signals with this bandwidth. The total number of independent samples available for averaging is on the order of

$$n_e = 2\delta B$$

where B is the bandwidth over which the transmissions are spread. Radars operating within the conventional microwave bands have difficulty in achieving sufficient bandwidth for multipath reduction.

(d) *Doppler Resolution and Frequency Diversity*. Calculations have been made for the doppler difference between the direct and reflected components, showing from (11.2.12) that the difference is often less than one hertz. This renders it impossible to resolve on the basis of doppler shift, and data smoothing is similarly ineffective unless long averages are taken. Even when the radar is a homing seeker, approaching the target at supersonic speed, the doppler difference for the specular component is typically only a few tens of hertz at ranges beyond 1000 m. Smoothing of the data can be effective in such cases.

(e) *Off-Axis Tracking*. An old technique for avoiding the nose-diving phenomenon is to lock the radar antenna at about $\theta_b = 0.75\theta_e$, monitoring the elevation error signal for targets below that angle. The results are shown in Figure 11.2.12, for the same radar and target parameters used in Figure 11.2.10 for null tracking. The errors are reduced only slightly, below the lock angle, but oscillation of the antenna and loss of track are prevented. Data below $\theta_t = \theta_e/6$ are of doubtful value.

(f) *Multiple-Target Estimators*. Many techniques have been proposed on the basis of maximum-likelihood and related procedures for estimating the positions of both the target and its image. These procedures show promise in cases where the reflections are purely specular, but fail when the diffuse reflections are significant (which is the usual case). Because the diffuse reflections cannot be modeled as white noise, the mathematics of the maximum-likelihood estimator are difficult to apply to the real world of multipath. This has not prevented the many attempts to apply the theory, but at this writing there has been universal lack of success in devising a workable procedure for multipath reduction by using this method.

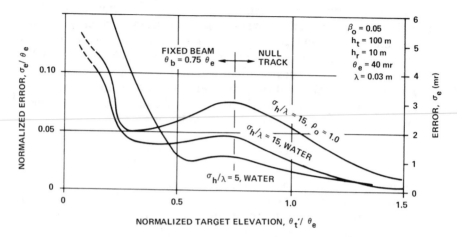

Figure 11.2.12 Typical multipath error for off-axis monopulse tracker.

(g) *Polarization Methods.* At low grazing angles (below the Brewster angle in Figure 6.2.2), the phase angle of reflection is the same for vertical and horizontal reflection components, and no reversal of sense occurs when a circularly polarized wave is reflected from the surface. The only practical use of polarization in reducing multipath reflections is when the radar can look down at the target (and the surface) with a vertically polarized wave at angles near the Brewster angle. The Fresnel reflection coefficient then drops substantially, proportionately reducing the multipath error. This approach is valid for airborne radars and missile seekers using vertical polarization, and ensures freedom from strong specular multipath.

(h) *Millimeter-Wave Radar.* As the radar frequency is increased, the surfaces become relatively rougher, so that the millimeter-wave radar seldom sees a specular reflection from natural surfaces. When used with broadband frequency agility and narrow beamwidths, millimeter-wave radar offers substantial reduction in multipath error effects, at those ranges where propagation permits the radar to detect and to track the targets.

Surface-Bounce Jamming

The surface-bounce jammer exploits the multipath error process to prevent the radar from obtaining adequate data on the jamming platform. In effect, the jammer uses steep gradient in its antenna pattern to ensure a ratio $A_r/A_t \gg 1$ and $A_r\rho/A_t > 1$, producing a multipath component which is the dominant source of tracking signal. To the extent that this is successful, in the geometry of the radar-target path, the radar will be denied elevation angle data on the target.

All the analysis of multipath errors and multipath reduction techniques, given above, can be applied to surface-bounce jamming by adjusting the ratio A_r/A_t to represent the properties of the jamming antenna. It must be assumed that the jamming masks the echo signal from the target, even with the reduced value of jammer A_t along the direct path.

11.3 ERRORS FROM CLUTTER

The calculation of clutter-to-signal ratios for the Σ channel of a radar have been covered in Section 1.5. In order to evaluate the measurement errors caused by clutter, the ratio must be calculated for the Δ channels as well, and substituted in (8.1.10) for each coordinate. The errors in angle tracking will generally be of greatest concern.

Difference-Channel Clutter Power

Clutter that is broadly distributed about the radar beam will produce zero mean voltage in the Δ channel. The clutter power C_Δ in the Δ channel will be less than C in the Σ channel by an extra beamshape loss $L'_p = 1.4$ dB, because the transmitting pattern is the same as the Σ pattern and is poorly matched to the Δ receiving pattern. Hence, for broadly distributed rain or chaff, the equations for C/S can be used with $C_\Delta/S = 0.7C/S$.

The clutter error can be written (for monopulse):

$$\sigma_\theta = \frac{\theta_3}{k_m\sqrt{2(S/C_\Delta)n_c}} \tag{11.3.1}$$

Surface clutter appears in the azimuth Δ channel with this same 1.4 dB additional beamshape loss, but the situation in the elevation Δ channel is quite different, unless the beam is pointing at the horizon from a low antenna. In any given range cell, the elevation angle to the surface can be calculated and used to find the gains $\Sigma(\theta)$ and $\Delta(\theta)$ for clutter in that cell. The gain $\Delta(\theta)$ will generally be higher, because the mainlobe of that pattern extends $1.5\theta_e$ from the beam axis. If the Σ-channel clutter has been calculated, the Δ-channel clutter power will be $[\Delta(\theta)/\Sigma(\theta)]^2$ times that value.

Spectrum of the Clutter Error Signal

The clutter error signal is detected by mixing with the target Σ signal, and hence its spectrum will contain frequencies representing the differences in doppler between clutter scatterers and the target, aliased by the PRF of the radar. If doppler processing is used prior to the error detector, to reduce the clutter power C_Δ at the detector, the spectrum of the residue may be broadened significantly by radar instabilities.

The number of independent samples of clutter error, n_c, can fluctuate over broad limits, as the relative velocity of target and clutter varies over the blind speed. For clutter with a narrow spectrum differing by f_d from the target, $n_c = 1 + t_o(f_d - if_r)$ may be used, where i is the integer giving a frequency difference less than $f_r/2$. In a low-PRF radar, unless PRF shifting is used to keep the target and aliased clutter frequencies from approaching each other, there will be periods when the difference between these frequencies passes through zero, giving $n_c = 1$. This will lead to a large output error, as the smoothing effect vanishes during the time that the frequency difference is less than the tracker bandwidth or $1/2t_o$. Where the single-pulse C/S is not small, and averaging of many pulses has been relied on to keep the error small, track may be lost during these periods of small frequency difference.

Tracking at Low S/C Ratio

Noncoherent trackers operating at angles near the horizon will have a clutter error which consists of first- and second-order components, the first representing the cross-product of C_Δ with S, and the second being the cross-product of C_Δ with C. This is similar to the case for low S/N, except that there is correlation between the clutter in the two channels, leading to both tracking bias and a larger, unsmoothed second-order noise component. The equation for elevation error can be written

$$\sigma_\theta = (\theta_e/k_m)\{[C_\Delta/2S][(1/n_c) + (C/S)]\}^{1/2} \tag{11.3.2}$$

For $C/S \ll 1/n_c$, this reduces to (11.3.1), while for $C/S \gg 1/n_c$, it becomes

$$\sigma_\theta = (\theta_e/k_m S) \sqrt{C_\Delta C/2} \tag{11.3.3}$$

Thus, the noncoherent radar cannot track accurately at low elevation angles over clutter, unless the *single-pulse* C/S is small, regardless of how many pulses are averaged in the tracking filter.

The use of doppler processing in the Σ channel, to reduce C/S, reduces the second-order error. Processing in the Δ channel, to reduce C_Δ/S, reduces both components of error.

Tracking in Rain and Chaff

The clutter amplitudes in Σ and Δ channels will be uncorrelated, if the clutter is distributed uniformly over the beam. Hence, the clutter error, instead of (11.3.2), will be given by

$$\sigma_\theta = (\theta_e/k_m)\{[C_\Delta/2Sn_c][1 + (C/S)]\}^{1/2} \tag{11.3.4}$$

The averaging over n_c samples will now reduce both first- and second-order error components. However, the average tracking point will be quite sensitive to asymmetries in the clutter distribution, so high accuracy cannot be expected.

Range and Doppler Tracking in Clutter

Clutter will usually be found as broadly distributed in time delay with respect to the signal and the range gates. Hence, the equations of Sections 9.1 and 9.2 for thermal noise can be used, with S/C replacing S/N and n_c replacing n.

In doppler tracking, the clutter spectrum will normally be excluded from the tracking filter, and only residue from system instabilities will compete with the target. In this case, the output S/C replaces nE_1/N_0 in (9.3.9).

11.4 ANGLE ERROR ANALYSIS

Classification of Angle Errors

In addition to separation and classification by their amplitude distributions, waveforms, and spectral properties, errors may be further classified by source and the degree of their dependence upon target characteristics, propagation conditions, or parameters of the tracker itself. This classification will serve to clarify the operating conditions under which a given level of performance can be achieved, and should also simplify the problem of devising tests to verify the results of theoretical analysis. In testing and evaluating angle trackers, an additional source of error is introduced by the reference instrumentation against which the tracker is compared. The apparent errors from this source must be isolated from the actual errors of the tracker under test. Another classification of errors is according to the point of entry into the tracking system. Those errors which are due to the tracking axis leaving the target may be termed *tracking errors,* whereas the errors in reading the position of the axis and providing numerical values relative to the fixed reference system may be called *translation errors.* This classification is of particular importance when performance tests are to be run by using an optical reference (e.g., boresight telescope), which may yield direct readings of tracking error, but no data on translation error.

Using these two classifications and the frequency spectrum classification (bias *versus* noise), we may list the error sources in a typical radar angle-tracking system as shown in Table 11.1. We shall discuss certain of these errors in more detail below, and give the relationships by which the error may be predicted from the known parameters of the radar and target situation.

In the class of radar-dependent noise errors, we have listed thermal noise, multipath, clutter, and jamming, as well as several errors that depend on servo and mechanical design of the antenna and pedestal. Most of these errors have been discussed in previous sections, and detailed analyses of the antenna errors caused by wind was covered in [11.3, pp. 331–335]. Some further notes on errors caused by jamming are in order.

Angle Errors from Jamming

Jamming that originates outside the target can cause angle errors in target tracking, even when the jamming is not strong enough to mask the target. Noise jamming is the most common type, and the error can be estimated by using the equations derived in Sections 8.2 and 8.4 for thermal noise. The ratio S/N is replaced by S/J_Δ, where J_Δ is the jamming power entering the Δ channel. Many antenna designs have Δ sidelobes much higher than the Σ sidelobes, and this larger response to jamming must be considered in the calculation of jamming errors.

Jamming that originates at the target has the effect of providing a beacon for tracking. In this case, S/N is replaced by J/N, usually reducing the thermal noise error to a negligible level. However, specialized deception jamming may introduce new error components, as discussed in Section 10.5.

Table 11.1 Sources of Angle Error

Class of Error	Bias Components	Noise Components
Radar-dependent tracking errors	Boresight axis setting and drift	Thermal noise
	Torque caused by wind and gravity	Multipath
		Clutter jamming
		Torque caused by wind gusts
	Servo imbalance and drift	Servo noise
		Deflection of antenna caused by acceleration
Radar-dependent translation errors	Pedestal leveling	Bearing wobble
	Azimuth alignment	Data gear nonlinearity and backlash
	Orthogonality of axes	
	Pedestal flexure caused by gravity force	Data takeoff nonlinearity and granularity
	Pedestal flexure caused by solar heating	Pedestal deflection caused by acceleration
		Phase shifter error
Target-dependent tracking errors	Dynamic lag	Glint
		Dynamic lag variation
		Scintillation or beacon modulation
Propagation errors	Average refraction of troposphere	Irregularities in refraction of troposphere
	Average refraction of ionosphere	Irregularities in refraction of ionosphere
Apparent or instrumentation errors	Stability of telescope or reference instrument	Vibration or jitter in reference instrument
	Stability of film base or emulsion	Film transport jitter
	Optical parallax	Reading error
		Granularity error
		Variation in parallax

Radar-Dependent Bias Error

In modern radars, the servo design can reduce fixed imbalance in the servo, and effects of static imbalance in the antenna, to a negligible level. The remaining radar-dependent bias errors are those which shift the electrical axis relative to the mechanical axis established by the pedestal and the mechanical mounting of the antenna.

Collimation and Drift Errors in the Electrical Axis

If the electrical axis of the radar is stable over long periods of time, the accuracy with which it can be collimated is dependent primarily upon the care and patience which are exercised in calibration. Comparisons of the electrical axis with boresight telescope observations on visible targets can be made over a period of time and over a range of operating conditions such that noise components of error are averaged to very low values. This process is especially accurate if photographic readings are taken from the telescope while a point-source target is being tracked at relatively high elevation angle, where multipath and propagation errors are minimized. The residual errors are caused by drift components, which change too rapidly to be removed by calibration. These can be the result of variation in several operating parameters of the radar, and of environmental factors such as uneven heating of the radar components. In a complete error analysis, the variation in position of the axis must be determined as a function of the following:

(1) Frequency of operation within the radar band;
(2) Tuning of the system (center IF frequency);
(3) Phase or gain variations in the receiver;
(4) Signal strength;
(5) Temperature or intensity of solar (thermal) radiation.

When these effects are determined, it will be possible to devise calibration and collimation procedures, and to estimate the errors remaining in the radar output at various times after calibration. For example, consider the experimental curve of Figure 11.4.1, which represents the shift in position of the null point in the RF error pattern of the AN/FPS-16 as its operating frequency is varied over a 10% band. No RF tuning elements are included in the antenna system, so the error can be reduced during operation only by collimation at the frequency to be used (it can be reduced in data processing by use of a calibration curve). If it is assumed that many different frequencies are to be used in the period between collimations, and that no calibrations are to be applied to correct for the shift, the error can be expressed as the rms value of the curve shown, or about 0.05 mil.

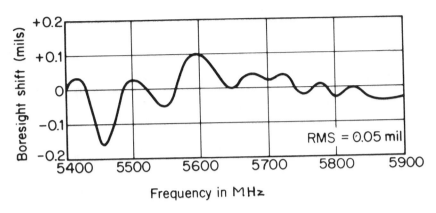

Figure 11.4.1 Typical boresight shift *versus* IF tuning.

This presumes that the frequency chosen for collimation gives an error near zero. When a different frequency is used for operation, a bias error will appear in elevation and traverse angle. The azimuth error Δ_a will be equal to the traverse bias Δ_{tr} at low elevation angles, but will increase as the target rises in elevation.

$$\Delta_a = \Delta_{tr} \sec E \tag{11.4.1}$$

Although the azimuth error becomes infinite at zenith, the linear error in target position does not change.

    A curve showing the sensitivity of a precision monopulse radar to IF excursions (tuning error) is given in Figure 11.4.2. Assuming that collimation is done with the system properly tuned, the shift in the tracking axis may be estimated for any operating condition which leads to tuning error by assigning to the tuning error a probability distribution and calculating the rms value of the corresponding error from this figure. The tuning error may arise from doppler shift, drift in the transmitting frequency of a beacon, or uncorrected drifts in the radar oscillators. Let us assume that the tuning error in a particular case is normally distributed with a standard deviation of 1/3 MHz, and that the wide receiver bandwidth is used in the radar. For simplicity, we shall represent the tuning error distribution by a series of rectangular segments, and we shall calculate the rms error based on the boresight shifts occurring at the center of each rectangle. A typical calculation based on the elevation error curve (Figure 11.4.2 b) takes the form of Table 11.2.

    The rms boresight error for this case is 0.026 mil. The simplified procedure leads to a slight overestimation of the error, but is well within

Table 11.2 Calculation of Tuning Error Distribution

Tuning Error (MHz)	Boresight Error (x)	Probability (P)	Variance Px^2
−1.0 to −0.67	−0.063	0.023	0.00009
−0.67 to −0.33	−0.37	0.136	0.00019
−0.33 to 0	−0.012	0.341	0.00005
0 to 0.33	0.012	0.341	0.00005
0.33 to 0.67	0.037	0.136	0.00019
0.67 to 1.0	0.063	0.023	0.00009
		1.000	$0.00066 = \sigma^2$

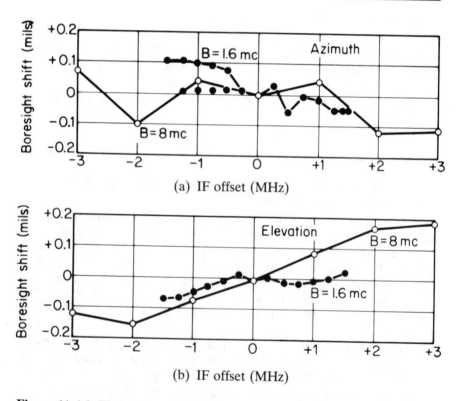

(a) IF offset (MHz)

(b) IF offset (MHz)

Figure 11.4.2 Typical boresight shift *versus* IF tuning.

the accuracy required for error analysis. Results of the type shown above may be used to establish realistic limits for tuning accuracy of the receiver and transmitter system, to guide design of automatic-frequency-control circuits, or to set the allowable deviations in IF amplifier phase shift and linearity over the center portion of the passband.

The boresight errors caused by faulty implementation of the monopulse networks and receivers were discussed in Section 8.5.

Radar-Dependent Translation Errors

The ability of the radar to keep its tracking axis pointed at the target is only one part of the tracking problem. It is also necessary to translate this axis position into output data in a form usable by computers and similar devices, and properly related to a known coordinate system. A list of error sources which affect this translation was given in Table 11.1. The techniques for evaluating and expressing these error components are similar to those described above, but there is little general theory which can be applied in the absence of a specific example. Most of the "bias" errors listed will be functions of the angles at which the radar is pointed, whereas the noise components will vary in a more rapid or random fashion.

An example will serve to illustrate the analysis of errors of this type. Figure 11.4.3 shows a plot of error in the readings of a high-accuracy digital encoder, measured over both short intervals of time and angle and at a number of such intervals around a complete circle. Two types of error are evident: one is a rapid, noise-like variation in the indicated position when the encoder is fixed at any point; the second is a systematic error which rises and falls slowly as the encoder is rotated, and which is almost exactly repeatable with any single encoder unit. The distribution of total error is shown in Figure 11.4.4, along with the distributions of the two separate components. Since the two errors are uncorrelated, it is possible to express the rms value of the total error as the square root of the sum of the two variances. The spectrum of the systematic component is proportional to the rate at which the target moves in angle, with spectral components at harmonics of the rotation rate of the encoder. The noise component has been found to approximate white noise over the bandwidth determined by the sampling rate of the encoder. In other words, successive samples have uncorrelated noise components, even at readout rates of 50 or 100 samples per second. When tracking a nonmaneuvering target, it would be possible, in principle, to reduce the noise by taking an average (or smoothing) over a period containing many samples, and to reduce the systematic portion by careful calibration against some more accurate standard. This would leave a residual error considerably less than the rms value shown for the two components. With a device of this accuracy, however, it is difficult to find and to apply a suitable external standard, and to reduce

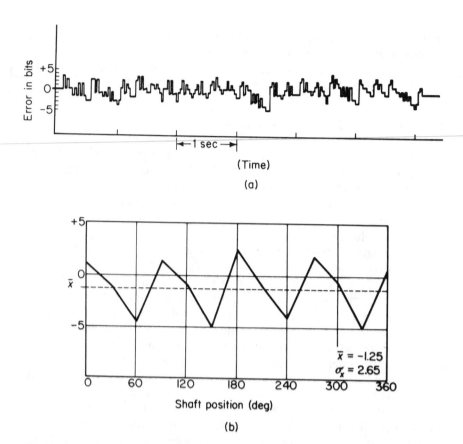

Figure 11.4.3 Digital encoder error (19-bit encoder, least bit = 0.012 mils): (a) noise error with encoder shaft fixed; (b) systematic error measured over one shaft rotation.

the other sources of error to the point where the additional accuracy provided by calibration would be useful.

Target Error and Nonradar Components

The error components listed as "target dependent" have been discussed in Chapters 3 and 8. Propagation errors are covered in some detail in Chapter 6. Apparent or test-instrumentation errors are listed in Table 11.1 as a reminder that the instruments used to evaluate the radar error are not perfect, and that allowance must be made for their errors in using

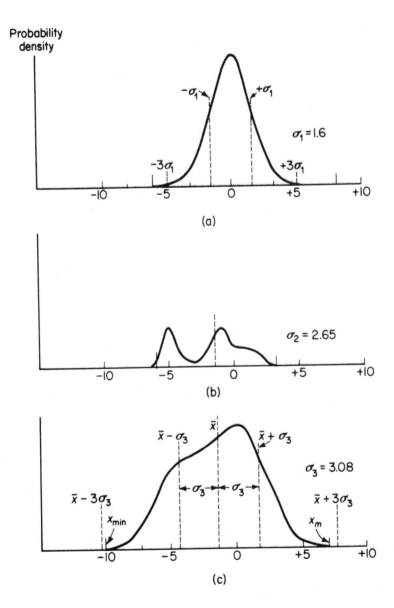

Figure 11.4.4 Distributions for encoder error: (a) distribution of noise about short-term average; (b) distribution of systematic error about mean value; (c) distribution of total error about mean value.

experimental data which has been gathered with their aid. In error analysis of an angle tracker, all the actual error components, along with the radar-dependent errors previously described, must be added in an appropriate fashion (usually rms), to find the total error. An example will be given to illustrate this process.

Example of Angle Error Analysis in Mechanical Tracker

To illustrate how the theory is applied to arrive at estimates of total angle error of a tracking radar, the example of the AN/FPS-16, used in [11.3], will be repeated here. Although this is one of the early monopulse radars, it remains in wide use and represents one of the most accurate tracking devices employed in test-range instrumentation. Its major parameters are listed in Table 11.3.

Errors of Fixed rms Value

A number of the errors listed in Table 11.1 may be described as "fixed" in the sense that their rms values are nearly independent of the conditions under which the radar is operated. This is especially true of the radar-dependent translation errors (principally mechanical in nature), but it may apply also to some of the other errors over a broad range of conditions. As a first step in analysis of total error, we may list these fixed error values, as in Table 11.4, to set the minimum angle error attainable on any target. In some cases, the values shown represent the result of operating experience (e.g., collimation and alignment error), whereas others are based on typical measured values or those calculated in the preceding paragraphs. The boresight axis drift was calculated on the assumption that the system is calibrated within 50 MHz of the operating frequency, tuned within 1/3 MHz, and that the differential phase shifts between receiver channels are within 5° rms. The tropospheric refraction terms are representative of typical conditions where the target is about 5° above the horizon.

Errors of Variable rms Value

In order to proceed further with analysis of error, specific operating conditions must be assumed. For our example, we shall assume that the target is a satellite in orbit at 185 km altitude, a corner reflector being used to provide extended radar tracking range. Radar operation at 1 MW peak power, 1 μs pulsewidth, and 340 pulses per second will be assumed,

Table 11.3 Characteristics of Instrumentation Radar AN/FPS-16

Frequency	5400 to 5900 MHz
Power output	1.0 MW peak (5480 MHz)
	250 kW peak (5450–5825 MHz)
Pulsewidths	0.25, 0.5, and 1.0 μs
Repetition rates	160 to 1707 Hz
Pulse codes	Up to 5 pulses
Duty cycle	0.001 at 1.0 MW, 0.0016 at 250 kW
Antenna size	3.7-m diameter reflector
Antenna gain	44.5 dB
Beamwidth	1.1°
Monopulse feed type	4-horn, amplitude comparison
Receiver noise factor	11 dB maximum
Receiver bandwidths	1.6 and 8.0 MHz
Coverage	Azimuth 360°
	Elevation $-10°$ to 85°
	Range 0.5 to 365 km
Maximum rates	Azimuth 48°/s
	Elevation 37°/s
	Range 9300 m/s
Servo bandwidths (β_n)	0.5 to 6 Hz
Velocity error constant	300 (angle), 2500 (range)
Acquisition scans	Adjustable circle and sector
Data outputs	Digital, synchro, and potentiometer
Displays	Dual A-scope, dials, digital
Detection range (1.0 m²)	375 km
Accurate tracking range (1.0 m²)	187 km
Angle error (bias)	0.1 mil rms
(noise)	0.1 mil rms
Range error (total)	10 m rms
Input power	75 kV, 208/115 V, 60 Hz
Installation	Fixed building

with the target moving at 7.5 km/s along an orbit which crosses a point 75 km in ground range from the radar. We shall evaluate the errors near crossover ($R = 200$ km) and at a point near the maximum range of the ranging system ($R = 375$ km).

As a starting point, signal-to-noise ratios are determined by using the radar range equation, and angular rates and accelerations found from Tables 10.4 and 10.5. The optimum servo bandwidths will then be calculated from (10.2.5). The results of these steps are shown in Table 11.5.

Only the error components caused by multipath, glint, and scintillation remain to be estimated.

At this high elevation angle, the multipath error will drop to very low values, and may be ignored. When a corner reflector is used as a target, the glint term is reduced to zero, and any scintillation error is eliminated by the use of fast AGC in the monopulse radar. As a result, the total angle error may be found by combining the "fixed" error level from Table 11.4 with the thermal-noise and lag terms from Table 11.5. It should be noted that some of the azimuth errors are larger near crossover because of the high elevation angle, which causes the traverse error to be multiplied by the secant of elevation to give a larger azimuth error. The error in the position of the target is not increased, because the azimuth error is multiplied by the ground range, rather than by slant range, to find the linear error in target position. The rms values of error in both angular coordinates are summarized at the end of Table 11.5, which also shows the error converted to target position in feet. The most serious bias error components are those due to azimuth lag and to fixed error in elevation. The largest noise components are from thermal noise. Figure 11.4.5 shows how the error varies with time during the track from 370 km to crossover, and also shows how thermal noise and total error would increase at long range if servo bandwidth were not reduced to the optimum value.

Example of Error Analysis for Phased Array

Consider now a hypothetical phased array tracker, tracking multiple targets within an azimuth sector of 90°, from the horizon to 30° elevation. The broadside beamwidth will be assumed equal to that of the AN/FPS-16, implying a total number of radiating elements, from (4.3.7), given by

$$T = 3.14/\theta_a\theta_e = 3.14/(0.02 \times 0.02) = 7850$$

Use of three-bit phase shifters will be assumed. Replacing the radar-dependent error budget of Table 11.4 will be three major components:

Boresight axis collimation:	0.05 mr (bias)	total bias = 0.064
Boresight axis drift:	0.04 mr (bias)	
Phase-shifter quantization:	0.04 mr (noise)	(Eq. (4.3.21))

Replacing the radar-dependent translation error budget of Table 11.4 will be a single bias error for initial alignment of the array to the reference system:

Level and north alignment: 0.02 mr (bias)
Resultant total fixed error: 0.067 mr (bias)
0.04 mr (noise)

This is an improvement of about 20% relative to a mechanical tracker.

Table 11.4 Errors of Fixed rms Value

	Angle Error (in mils rms)	
Error source	Bias	Noise
Boresight axis collimation	0.025	—
Boresight axis drift	0.04	—
Wind forces (25 m/s)	0.02	0.012
Servo noise and unbalance	0.01	0.02
Subtotal: Radar-dependent tracking errors	0.052	0.023
Leveling and north alignment	0.015	—
Orthogonality of axes	0.02	—
Mechanical deflections	0.01	—
Thermal distortion	0.01	—
Bearing wobble	—	0.005
Data gear error	—	0.03
Digital encoder error	—	0.025
Subtotal: Radar-dependent translation errors	0.023	0.04
Tropospheric refraction	0.05	0.03
Total "fixed" error	0.078	0.054

The tracking errors resulting from thermal noise and dynamic lag will depend on the assumptions made for radar power. If the average power and energy per pulse is increased by 10 dB relative to the mechanical tracker, the energy available per target will be the same. For equal receiver noise temperatures and loss budgets, the tracking accuracies will be the same for regions near the array broadside, and the array will have the major advantage of giving data on 10 targets. At $f_r = 340$ Hz, each target receives 34 pulses per second at $S/N = 40$ dB per pulse, giving a single-sample precision of about 0.1 mr. The array can support tracking loop bandwidths up to about 15 Hz. As targets become farther from broadside, the performance of the array declines as a result of the broader beam and

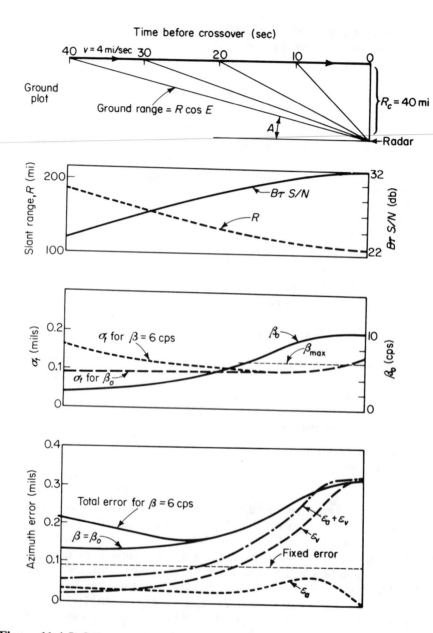

Figure 11.4.5 Calculation of error for crossing target.

Table 11.5 Thermal-Noise and Lag Error Calculation

		Azimuth		Elevation	
		$R = 200$ (km)	$R = 370$ (km)	$R = 200$ (km)	$R = 370$ (km)
S/N	(dB)	30	20	30	20
Angle sector	(°)	60–90	12–18	60–70	25–35
Angle rate (max)	(rad/s)	0.1	0.006	0.017	0.011
Angle acceleration (max)	(rad/s²)	0.0065	0.0004	-0.003	0.0004
Effective target acceleration	(g)	54	14	-63	150
Optimum servo bandwidth (β_o)	(Hz)	6.5	1.9	7.2	2.0
Thermal-noise error* (σ_t)	(mil)	0.09	0.078	0.043	0.078
Velocity-lag error (rms) (ϵ_v)	(mil)	0.30	0.02	0.057	0.037
Acceleration-lag error† (ϵ_a)	(mil)	0.05	0.04	-0.033	0.04
Total "variable" error	(mil)	0.32	0.1	0.1	0.11
Fixed error from Table 11.4	(mil)	0.095	0.095	0.095	0.095
Total error	(mil)	0.33	0.138	0.138	0.145
Position error of target	(m)	27	42	28	53

*Calculated for bandwidth not exceeding 6 Hz (max), and includes secant term.
†Although the elevation lags tend to cancel before crossover, they add after crossover, and the totals shown include the sum of the two lags.

decreased gain. The beamwidth becomes 28 mr and S/N drops to 35 dB, giving a single-sample thermal noise error of 0.22 mr. The phased array radar can redistribute its energy to reduce this error, at the expense of greater error on the other targets.

Other errors, such as glint and multipath, will be the same for the phased array as for mechanical tracking systems.

11.5 RANGE ERROR ANALYSIS

The procedures used in describing and combining error components in range measurement systems are similar to those used for angle measurement. The mathematical descriptions were discussed in Section 11.1. Table 11.6 lists some of the errors frequently encountered in range tracking systems. The form of the listing is similar to Table 11.1 for angle errors, and the definitions of the several classes of error will not be repeated. The more significant error sources are discussed briefly below.

Table 11.6 Sources of Range Error

Class of Error	Bias Components	Noise Components
Radar-dependent tracking error	Zero range setting Range discriminator shift Receiver delay	Thermal noise Multipath Clutter Jamming variation in receiver delay
Radar-dependent translation error	Range oscillator (velocity of light)	Range-doppler coupling Internal jitter Data encoding Range oscillator stability
Target-dependent tracking error	Dynamic lag Beacon delay	Dynamic lag Glint Scintillation Beacon jitter
Propagation error	Average tropospheric refraction Average ionospheric refraction	Variation in tropospheric refraction Variation in ionospheric refraction

Velocity of Light

The velocity of light in a vacuum is

$$c = 2.997925 \times 10^8 \text{ m/s}$$

and the variation in this velocity, in the earth's atmosphere, is about 300 parts per million. The accuracy with which atmospheric refraction errors can be corrected is discussed in Section 6.4, where it is concluded that measurements can be corrected to within one meter, if the refraction index at the surface is measured. The significant errors in measurement of range then result from the estimation of time delay between transmission and reception of the radar waveform.

Measurement of Time Delay

The two steps in measurement of time delay (Section 9.1) are the location of the centroid or other definable point on the noisy received waveform, and the measurement of delay between this point and the time when it was transmitted. Errors in estimation of the centroid or leading edge have been discussed. Errors in measuring the delay to this defined point can be made very small in modern radar systems. Digital counters, operating at frequencies approaching 1 GHz or interpolating to 1 ns between longer clock periods, can measure the time delay to the center of the range gate to 1 ns, equivalent to a range increment of 0.3 m. The major sources of error are then in the ability of the gate to track the target, the range glint of the target itself, and the variable delay through the radar receiver.

The receiver delay results primarily from the use of cascaded filters in the IF amplifier stages. The delay of each stage is approximately the reciprocal of its bandwidth B_1, and if m stages of equal bandwidth are used the total bandwidth is

$$B = B_1/\sqrt{m} \tag{11.5.1}$$

Thus, the total delay for $m = 10$ is about

$$\delta_t = 3/B \tag{11.5.2}$$

or about twice the pulsewidth for a typical system without pulse compression. The receiver delay varies as a function of temperature, signal strength, and tuning of the receiver. Very accurate tuning is usually needed to keep the delay variation within $\delta_t/50 = 1/15B$. When pulse compression is used, the delay is dominated by the compression filter, and is somewhat greater than the width of the transmitted pulse.

Range-Doppler Coupling

In FM pulse compression systems, the dispersive compression line introduces a systematic shift of delay with doppler shift. For a rectangular pulse of width τ and bandwidth B, the delay variation with doppler shift is

$$\delta_t = \tau f_d/B \tag{11.5.3}$$

with the sign of the shift determined by the upward or downward slope of the transmitted waveform. For example, with $\tau = 100$ μs and $B = 1$ MHz, a doppler shift $f_d = 1$ kHz will cause a delay change $\delta = 0.1$ μs, giving a range error of 15 m. This is far greater than other errors commonly encountered in range measurement.

There are two ways in which the range shift can be corrected:

(a) If range rate is measured over a train of pulses, the doppler shift can be determined and a correction made, based on (11.5.3).
(b) The range measurement can be considered to have been taken at a time shifted by Δt from that of the actual measurement [11.11], where

$$\Delta t = f\tau/B \tag{11.5.4}$$

Here, f is the carrier frequency. For example, if the carrier frequency of our previous 100-μs, 1-MHz pulse were 3000 MHz, the time displacement of the range data would be 0.3 s. A measurement displaced by Δt will be accurate for any target doppler shift, regardless of how great the slope τ/B. Receiver tuning is much more critical in pulse compression radars, as an error in tuning will cause the same effect as a doppler shift (11.5.4), and this will not be corrected by assuming a time displacement of the range data. Since amplifier-type transmitters must be used for pulse compression, it is relatively easy to lock the receiving oscillators to those used in developing the transmitted carrier, avoiding all tuning error. This leaves only the shifts caused by temperature and other slow changes in the networks.

11.6 DOPPLER ERROR ANALYSIS

The principal components of error in doppler measurement are listed in Table 11.7. The contributions of thermal noise, multipath, clutter, and jamming have been discussed in Sections 9.3, 10.5, 11.2, and 11.3. These, plus the target and propagation effects, will dominate the error in a well designed radar system.

Table 11.7 Sources of Doppler Error

Class of Error	Bias Components	Noise Components
Radar-dependent tracking errors	Discriminator zero setting and drift Gradient of receiver delay	Thermal noise Multipath Clutter Jamming variation in receiver delay
Radar-dependent translation errors	Transmitting oscillator frequency	VCO frequency measurement Radar frequency stability
Target-dependent tracking errors	Dynamic lag	Dynamic lag Target rotation (glint) Target modulation
Propagation error	Gradient of atmospheric refraction	Fluctuation in atmospheric refraction

A particular problem in doppler tracking is the presence of modulated scatterers on the target (e.g., rotating propellers and turbines of jet engines) [11.11]. These scatterers can produce offset lines in the spectrum, which can be aliased by the PRF to any part of the target spectrum, causing false readings of doppler that cannot be characterized by an rms error. Only by inspecting the entire target spectrum over some period of time can the radar operator be sure that the tracking loop is locked on the doppler of the target body, rather than on one of these offset lines.

The error in doppler velocity caused by atmospheric refraction is usually not great, but the direction to which this velocity is projected will be in error by the elevation refraction error (and other angle errors). Whether this error in the projection of the vector is important will depend on the use to be made of the data. In general, the nonradial components of velocity will be known only to the accuracy of differentiated $R\theta$, which is dominated in any case by the angle errors.

REFERENCES

[11.1] D.K. Barton, Final Report, Instrumentation Radar AN/FPS-16 (XN-1), RCA, Moorestown, NJ, under Contract DA36-034-ORD-151, 1957.

[11.2] D.K. Barton, Section 5, Final Report, Instrumentation Radar AN/FPS-16 (XN-2), RCA, Moorestown, NJ, under Contract BuAer NOas 55-869c, 1959.

[11.3] D.K. Barton, *Radar System Analysis,* Prentice-Hall, 1964; Artech House, 1976.

[11.4] P. Beckmann and A. Spizzichino, *The Scattering of Electromagnetic Waves from Rough Surfaces,* Pergamon Press, 1963; Artech House, 1987.

[11.5] D.K. Barton, Low-Altitude Tracking over Rough Surfaces, I: Theoretical Predictions, *IEEE Eascon-79 Record,* pp. 224–234.

[11.6] C.I. Beard, Coherent and Incoherent Scattering of Microwaves from the Ocean, *IRE Trans.* **AP-9,** No. 5, September 1961, pp. 470–483; reprinted in D.K. Barton (ed.), *Radars,* Vol. 4, *Radar Resolution and Multipath Effects,* Artech House, 1975.

[11.7] D.K. Barton, Low-Angle Radar Tracking, *Proc. IEEE* **62,** No. 6, June 1974, pp. 687–704; reprinted in D.K. Barton (ed.), *Radars,* Vol. 4, *Radar Resolution and Multipath,* Artech House, 1975.

[11.8] M.L. Meeks, *Radar Propagation at Low Altitudes,* Artech House, 1982.

[11.9] W.A. Skillman, *Radar Calculations Using the TI-59 Programmable Calculator,* Artech House, 1983.

[11.10] P.E. Cornwell and J. Lancaster, Low-Altitude Tracking over Rough Surfaces, II: Experimental and Model Comparisons, *IEEE Eascon-79 Record,* pp. 235–248.

[11.11] R. Hynes and R.E. Gardner, Doppler Spectra of S-Band and X-Band Signals, *IEEE Eascon 1967 Record,* November 1967, pp. 356–365.

Appendix A

TI-59 Calculator Program: Propagation Factor

(a) *Spherical earth geometry (Figure A.1)*. A line-of-sight path exists when the range R between terminals is less than the sum R_h of the two horizon ranges:

$$R_h = \sqrt{2ka} \left(\sqrt{h_1} + \sqrt{h_2}\right) > R \tag{A.1}$$

Only diffracted energy passes between the terminals for $R \geq R_h$ (blocked path). For line-of-sight paths, the range to the specular reflection point is found by using the method of Fishback [A.1], who presented the solution for the approximate cubic equation applicable to ground or airborne terminals:

$$d_1 = R/2 - p \cos\left(\frac{\phi + \pi}{3}\right) \tag{A.2}$$

$$p^2 = [4ka(h_1 + h_2) + R^2]/3 \tag{A.3}$$

$$\cos\phi = 2ka(h_2 - h_1)R/p^3 \tag{A.4}$$

If the spherical earth divergence factor is neglected (as is done here to avoid unnecessary complication and errors in interpolation for near grazing paths), the reflection-interference factor may now be calculated on the basis of flat-earth reflection from the plane tangent at d_1, with reduced terminal heights given by

$$h'_1 = h_1 - d_1/2ka \tag{A.5}$$

$$h'_2 = (R - d_1)h'_1/d_1 \tag{A.6}$$

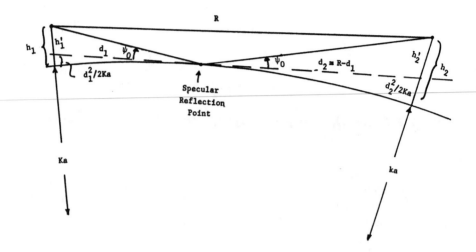

Figure A.1 Geometry of spherical earth reflection.

The grazing angle ψ_0 and excess length δ_0 of the reflection path are

$$\psi_0 = \tan^{-1}(h_1'/d_1) \tag{A.7}$$

$$\delta_0 = 2h_1'h_2'/R \tag{A.8}$$

This completes the geometric description of the line-of-sight case. During calculation, values of d_1, h_1', and h_2' are printed for later reference.

(b) *Diffraction factor.* The method of Burroughs and Atwood [A.2] is used to calculate the loss of the first diffraction mode relative to free-space propagation. This result is accurate for paths blocked by the smooth spherical earth, and is useful for extrapolation on line-of-sight paths where the reflection path excess length $\delta_0 < \lambda/4$. To avoid confusion, the diffraction factor F_d is set to unity (0 dB) for $\delta_0 \geq \lambda/4$.

In the Burroughts and Atwood procedure, the loss is expressed as the product of four components:

$$F_d = 2A_1 F_s H_{L1} H_{L2} \tag{A.9}$$

where $2A_1$ = flat-earth factor; F_s = curved-earth factor; H_{L1}, H_{L2} = height gain factors. The flat earth factor is found from

$$A_1 = \frac{1}{p'R} \quad \text{for} \quad \lambda > 3 \text{ m}, \quad R > \frac{50}{p'} \tag{A.10}$$

$$p' = \frac{2\pi}{\lambda} \frac{[(\epsilon_r - 1)^2 + (60\sigma_e\lambda)^2]^{1/2}}{[\epsilon_r^2 + (60\sigma_e\lambda)^2]^b} \tag{A.11}$$

where λ = wavelength, ϵ_r = relative dielectric constant, σ_e = conductivity, b = 0 for horizontal polarization, b = 1 for vertical polarization, R = path length.

The curved earth (shadowing) factor is

$$F_s = 2.507(sR)^{3/2} \exp(-1.607sR) \tag{A.12}$$
$$s = 4.43 \times 10^{-5}\lambda^{-1/3}$$

for standard atmosphere, R and λ in m, leading to

$$F_s = 7.396 \times 10^{-7}\lambda^{-1/2}R^{3/2} \exp(-7.12 \times 10^{-5}R\lambda^{-1/3}) \tag{A.13}$$

The height gain factor is unity for antenna heights between zero and h_{min}, where

$$h_{min} = \sqrt{\lambda/2\pi p'} = \frac{\lambda}{2\pi} \frac{[\epsilon_r^2 + (60\sigma_e\lambda)^2]^{b/2}}{[(\epsilon_r - 1)^2 + (60\sigma_e\lambda)^2]^{1/4}} \tag{A.14}$$

Above this height, it increases linearly to $h = h_c'$, and then more rapidly

$$H_L = 1, \qquad 0 \le h \le h_{min} \tag{A.15}$$

$$H_L = gh/h_{min}, \quad h_{min} \le h \tag{A.16}$$

$$g = 1, \qquad h < h_c = 30\lambda^{2/3} \tag{A.17}$$

$$g = 0.1356(h/2h_c)^{-0.904}10^{0.948\sqrt{h/2h_c}}, \quad h \ge h_c \tag{A.18}$$

thus, setting

$$h'' = h_{min}, \quad h < h_{min} \tag{A.19}$$
$$h'' = h, \qquad h \ge h_{min}$$

we can write

$$H_L = gh''/h_{min}$$

(Note that the Burroughs and Atwood height gain factor differs from those defined in Kerr [A.4] and Rice *et al.* [A.3]. Thus, the four terms in (A.9) may be combined to give

$$F_d = \frac{2}{p'R} \times 7.396 \times 10^{-7}\lambda^{-1/2}R^{3/2} \exp(-7.12 \times 10^{-5}R\lambda^{-1/3})$$

$$\times (g_1 h_1''\sqrt{2\pi p'/\lambda}) \times (g_2 h_2''\sqrt{2\pi p'/\lambda})$$

$$= 9.29 \times 10^{-6}\lambda^{-3/2}R^{1/2} \exp(-7.12 \times 10^{-5}R\lambda^{-1/3})g_2 h_2'' \qquad (A.20)$$

$$20 \log F_d = 100.6 - 30 \log\lambda + 10 \log R - R/1617\sqrt[3]{\lambda}$$

$$+ 20 \log h_1'' + 20 \log h_2'' + 20 \log g_1 + 20 \log g_2 \qquad (A.21)$$

$$20 \log g = 0, \quad h < h_c$$

$$= -17.4 - 18.1 \log(h/2h_c) + 18.96\sqrt{h/2h_c}, \quad h \geq h_c \qquad (A.22)$$

Equations (A.20) and (A.21) are seen to consist of eight terms, including the constant.

The program evaluates each of the eight components of F_d in (A.21), accumulating the result in Register 19 for eventual printout. To save running time, values of h_{min} and $2h_c$ are calculated and stored in the initialization program.

A comparison of the Burroughs and Atwood method with that described by Rice, Longley, and Norton [A.3] shows agreement, except for a constant factor of 3 dB (-103.5 in place of -100.6 used in (A.21)). Approximations limit the applicability of (A.21) to the following region:

$$\lambda \leq 3 \text{ m}$$

$$R > 50/p' = 100\pi h_{min}^2/\lambda$$

$$F_d < -6 \text{ dB (limit of diffraction region)}$$

Values of F_d greater than -6 dB are calculated, as long as $\delta_0 < \lambda/4$, to help interpolation in the region between valid diffraction and reflection data (see Figure 6.3.2).

(c) *Reflection factor.* With the assumption of grazing angles ψ_0 much smaller than antenna elevation beamwidths, the pattern-propagation factor defined by Kerr in [A.4] is reduced simply to the reflection factor F_r:

$$F_r^2 = 1 + \rho^2 + 2\rho \cos(\alpha + 2\pi\delta_0/\lambda) \qquad (A.23)$$

where ρ is the magnitude of the surface reflection coefficient, α is its phase angle, and δ_0 is the excess reflection path length previously computed. The surface reflection coefficient is the product of two factors (excluding divergence factor as noted earlier):

$$\rho = |\rho_0|\rho_s \qquad (A.24)$$

where the Fresnel reflection coefficient for the surface material is [A.5]:

$$\rho_0 = \frac{\sin\psi_0 - \sqrt{\epsilon' - \cos^2\psi/\epsilon'^b}}{\sin\psi_0 + \sqrt{\epsilon'\cos^2\psi/\epsilon'^b}} = |\rho_0|e^{j\alpha} \qquad (A.25)$$

The complex dielectric constant ϵ' is

$$\epsilon' = \epsilon_r - j60\sigma_e\lambda \qquad (A.26)$$

where ϵ_r is the relative dielectric constant and σ_e is the conductivity of the surface. The exponent b in (A.25) is 0 for horizontally polarized and 1 for vertically polarized waves. Values of ϵ_r and σ_e are given in Table A.1 for different surfaces.

The second component of ρ is the specular scattering factor for rough surfaces

$$\rho_s^2 = \exp[-(4\pi\sigma_h/\lambda)^2 \sin^2\psi_0] \qquad (A.27)$$

where σ_h is the rms deviation from the mean surface.

The program first evaluates ρ_s, and for very rough surfaces sets $F_r = 1$ (or 0 dB), skipping the remaining calculations. Otherwise, ρ_s is printed, and ρ_0 is calculated by using (A.25) and the programs in the Master Module. Phase angle α is held in storage, $|\rho_0|$ is printed and used in (A.24) and (A.23) to arrive at F_r, which is converted to decibels and stored.

(d) *Results.* Upon completion of calculations, values of δ_0, ψ_0, F_r, and F_d are in the last four registers, and are printed out to conclude the run. Appearance of $F_r = -200$ dB indicates a blocked path. Appearance of $F_d = 0$ dB indicates a clear path, $\delta_0 \geq \lambda/4$. Values of δ_0 and ψ_0 can be used as aids to interpretation of the results. Figure 6.3.2 is an example of curves prepared from computed results, starting with the sample problem (the input parameters of which are stored on Card 4, appearing when the input cuing sequence is used). Interpolation between F_r and F_d is shown in the intermediate region, $0 < \delta_0 < \lambda/4$. To define the first reflection null at $\lambda = 0.03$ m, $R = 23$ km, the values of δ_0 are inspected for $R = 25$ km and 20 km, and small intermediate steps are chosen to obtain $\delta_0 \approx \lambda$. The same process could be used to locate other nulls for $\delta_0 = 2\lambda$, *et cetera*.

(e) *Initialization.* Because most repetitive calculations are made with varying R or h_2, the values of h_{min} and $2h_c$ are calculated and stored in the initialization subroutine. Whenever a new value of λ polarization or surface parameter is inserted, this initialization subroutine must be repeated.

(f) *Input Data.* Key E' starts the full input procedure, printing and displaying the prestored value of each input with the identification of the parameters and a question mark (see sample problem). If the value is correct, R/S goes to the next input parameter. If change is required, the new value is entered and R/S stores it before bringing up the next input. Following inputs of h_1, h_2, or R, the new value is printed. Following input of λ, the initialization subroutine is called, leading to a printout of all input parameters.

After a run, if only a change in h_1, h_2, or R is needed, Key A', B', or C' will make the change, and a new run may be started with Key A or by C' R/S.

REFERENCES

[A.1] W.T. Fishback, Methods for Calculating Field Strength with Standard Refraction, in Kerr [A.4].

[A.2] C.R. Burrows and S.S. Attwood, *Radio Wave Propagation*, Academic Press, 1949, pp. 404–408.

[A.3] P.L. Rice, A.G. Longley, and K.A. Norton, Transmission Loss Predictions for Tropospheric Communication Circuits, National Bureau of Standards, *Technical Note 101*, May 7, 1965, Revised June 1967, AD 687820, AD 687821.

[A.4] D.E. Kerr, *Propagation of Short Radio Waves*, Vol. 13 of MIT Rad. Lab. Series, McGraw-Hill, 1951, p. 38.

[A.5] W.T. Fishback, in Kerr [A.4], p. 99.

List of Symbols for Appendix A

A_1	Flat-earth diffraction factor
α	Phase of Fresnel reflection coefficient
b	Polarization parameter
d_1	Range from terminal 1 to point of specular reflection
δ	Excess path length of specular reflection
ϵ'	Complex dielectric constant
ϵ_r	Relative dielectric constant of surface
F_d	Diffraction factor
F_r	Reflection factor
F_s	Curved-earth diffraction factor
h_1, h_2	Heights of terminals 1 and 2
h_1', h_2'	Heights of terminals 1 and 2 above target plane
h_c	Height above which extra height gain applies
$h_{\min}$	Height above which height gain applies
h_1'', h_2''	Effective heights in height gain equation
h_{L1}, H_{L2}	Height gain factor
ka	Effective earth radius
λ	Wavelength of transmission
p	Intermediate step in calculating d_1
ϕ	Intermediate step in calculating d_1
p'	Intermediate step in diffraction calculation
ρ_s	Specular scattering factor (for rough surface)
ψ	Grazing angle of specular reflection
R	Range distance between terminals
R_h	Line-of-sight range
ρ	Magnitude of combined reflection coefficient
ρ_0	Fresnel reflection of coefficient
s	Intermediate step in diffraction calculation
σ_e	Conductivity of surface
σ_h	rms surface height deviation

Table A.1 Values of Surface Parameters

	ϵ_r	σ_e (mho/m)
"Good" soil	25	0.02
Average soil	15	0.005
"Poor" soil	3	0.001
Snow, ice	3	0.001
Fresh water ($\lambda = 1$ m)	81	1
Fresh water ($\lambda = .03$ m)	65	10
Salt Water ($\lambda = 1$ m)	81	5
Salt Water ($\lambda = .03$ m)	65	15

For any nominally flat surface, σ_h may vary from 0.1 to 1 m. For rolling and rough terrain, values up to 50 m are encountered, but spherical earth calculations are applicable only for terminal heights h_1 or $h_2 > 10\,\sigma_h$.

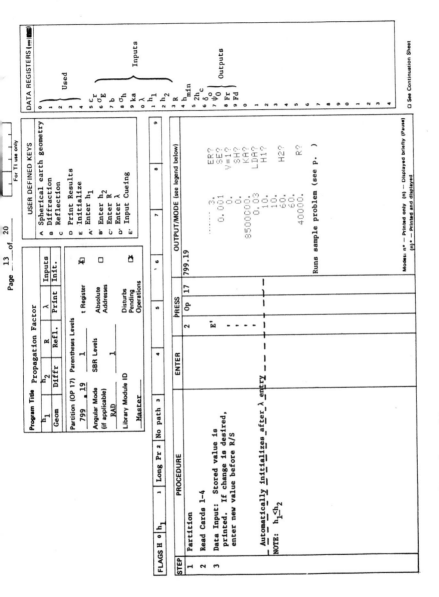

Example A.1 Sample problem.

Sample Problem

Statement of Example

      Uses input data stored on Card 4. If input cueing is used, initialization follows λ entry, and program runs after R entry, with long-form print of all inputs (as below).

      After this sample run at R = 40 km, a new range R = 50 km is introduced (see over). This results in a blocked path (F_r = -200dB). With F_d = -23.4dB. A further increase to 60 km reduces F_d to -42.5dB, at which point ranges are reduced to 30 km, 25km, and 20 km. Plotting of these data (Fig. 2) permits connection of diffraction with reflection results, and shows need for more data near the reflection null.

☐ **See Continuation Sheet**

ENTER	PRESS	OUTPUT/MODE (see legend below)			COMMENT
		40000.			R input
		10630.6739			d_1, ⎫ spherical
		3.352280727			h_1', ⎬ Earth geometry
		9.261334393			h_2' ⎭ and
		1.			ρ_s ⎫ reflection
		.9995541409			ρ_0 ⎬ coefficients
		3.	05	ε_r	printed if
		0.001	06	σ_e	line of sight
		0.	07	b	exists
		0.	08	σ_h	
		8500000.	09	Ka	
		0.03	10	λ	
		10.	11	h_1	
		60.	12	h_2	
		40000.	13	R	
		.0033761855	14	h_{min} ⎫	computed at
		5.792936308	15	$2h_c$ ⎬	initialization
		.0015523296	16	δ_0	reflected path delay
		.0003153404	17	ψ_0	grazing angle
		-9.799371328	18	F_r(dB)	reflection
		-4.479388599	19	F_d(dB)	diffraction
					Run time ≈45 sec.

Modes: n* — Printed only (n) — Displayed Briefly (Pause)
(n)* — Printed and displayed

■ **Over**

Example A.1 (cont'd)

ENTER	PRESS	OUTPUT/MODE (see legend below)		COMMENT
	C'	40000.	R?	
50000	R/S	50000.		
		0.	16	
		0.	17	Blocked path
		-200.	18	
		-23.4131824	19	
	C'	50000.	R?	
60000	R/S	60000.		
		0.	16	
		0.	17	
		-200.	18	Try short ranges
		-42.52426387	19	
	C'	60000.	R?	
30000	R/S	30000.		
		6468.254072		
		7.538922898		
		27.42687845		
		1.		
		.9983530558		
		.0137846081	16	
		.0011655262	17	
		5.942901616	18	Clear Path
		0.	19	
	C'	30000.	R?	
25000	R/S	25000.		
		4827.163444		
		8.629323123		
		36.06215678		
		1.		
		.9974750666		Near first null
		.0248953603	16	at $\delta_0 = \lambda$
		.0017876573	17	
		0.151849899	18	
		0.	19	
	C'	25000.	R?	
20000	R/S	20000.		
		3487.456102		
		9.284567644		
		43.96093494		
		1.		
		.9962420699		
		.0408158274	16	Other side of null;
		.0026622688	17	use small increments
		5.142335393	18	in R to define curve
		0.	19	of Fig. 2.

Example A.1 (cont'd)

LOC	CODE	KEY	COMMENTS	LOC	CODE	KEY	COMMENTS	LOC	CODE	KEY	COMMENTS
000	98	ADV		055	13	13	R	110	55	÷	
001	91	R/S		056	33	X²		111	43	RCL	k
002	76	LBL	Spherical	057	54	)		112	09	09	
003	11	A	earth	058	55	÷		113	94	+/-	
004	29	CP	geometry	059	03	3		114	85	+	
005	43	RCL	Test for	060	54	)		115	43	RCL	
006	07	07	horizontal	061	34	ΓX		116	11	11	
007	94	+/-	polarization	062	42	STO	P	117	54	)	
008	77	GE		063	00	00		118	42	STO	h₁
009	86	STF		064	45	YX	Eq.(4)	119	02	02	printed
010	61	GTO		065	03	3		120	99	PRT	
011	57	ENG		066	54	)		121	55	÷	
012	76	LBL	Horizontal	067	35	1/X		122	43	RCL	d₁
013	86	STF	polarization	068	65	×		123	00	00	
014	86	STF		069	43	RCL	R	124	54	)	
015	00	00		070	13	13		125	42	STO	h₁/d₁
016	76	LBL	Line of	071	65	×		126	01	01	
017	57	ENG	sight	072	02	2		127	22	INV	
018	43	RCL	geometry	073	65	×		128	30	TAN	
019	13	13	R	074	53	(		129	42	STO	ψ₀
020	75	-		075	43	RCL	h₂	130	17	17	
021	53	(		076	12	12		131	43	RCL	R
022	02	2		077	75	-		132	13	13	
023	65	×		078	43	RCL	h₁	133	75	-	
024	43	RCL		079	11	11		134	43	RCL	d₁
025	09	09		080	54	)		135	00	00	
026	54	)	2Ka	081	65	×		136	54	)	
027	34	ΓX		082	43	RCL	ka	137	49	PRD	h₂
028	65	×		083	09	09		138	01	01	
029	53	(		084	54	)		139	43	RCL	
030	43	RCL	h₁	085	22	INV		140	02	02	h₁
031	11	11		086	39	COS	φ	141	65	×	
032	34	ΓX		087	85	+	Eq. (2)	142	43	RCL	h₂
033	85	+		088	89	π		143	01	01	
034	43	RCL	h₂	089	54	)		144	99	PRT	Printed
035	12	12		090	55	÷		145	65	×	
036	34	ΓX	Rₕ	091	03	3		146	02	2	
037	54	)	Test for	092	54	)		147	55	÷	
038	54	)	blocked	093	39	COS		148	43	RCL	R
039	77	GE	path	094	65	×		149	13	13	
040	23	LNX	Eq.(3)	095	43	RCL	P	150	54	)	
041	43	RCL		096	00	00		151	42	STO	δ₀
042	09	09		097	94	+/-		152	16	16	
043	65	×		098	85	+		153	61	GTO	
044	04	4	4ka	099	43	RCL	R	154	12	B	
045	65	×		100	13	13		155	76	LBL	Blocked
046	53	(		101	55	÷		156	23	LNX	path
047	43	RCL	h₁	102	02	2		157	02	2	
048	11	11		103	54	)		158	00	0	
049	85	+		104	99	PRT		159	00	0	
050	43	RCL	h₂	105	42	STO	d₁				
051	12	12		106	00	00	printed				
052	54	)		107	33	X²					
053	85	+		108	55	÷					
054	43	RCL		109	02	2					

MERGED CODES
62 ☐ ☐ 72 STO ☐ 83 GTO ☐
63 ☐ ☐ 73 RCL ☐ 84 ☐ ☐
64 ☐ ☐ 74 SUM ☐ 92 INV SBR

Program Listing

LOC	CODE	KEY	COMMENTS	LOC	CODE	KEY	COMMENTS	LOC	CODE	KEY	COMMENTS
160	94	+/-	Set F	215	00	00		270	76	LBL	Height
161	42	STO	=-200^{r}dB	216	86	STF	h_1 terms	271	33	x²	gain
162	18	18	.	217	01	01	finished	272	43	RCL	h_1 or h_2
163	00	0		218	75	-		273	00	00	
164	42	STO	$\delta_0=\psi_0=0$	219	43	RCL	h_{min}	274	28	LOG	
165	16	16		220	14	14		275	65	×	
166	42	STO		221	54	)		276	02	2	
167	17	17		222	77	GE	Test for	277	00	0	
168	86	STF	Skip Refl	223	33	x²	$h_2 \geq h_{min}$	278	54	)	
169	03	03		224	43	RCL		279	44	SUM	F_{d5} or F_{d6}
170	76	LBL	Diffraction	225	14	14	$h''_2=h_{min}$	280	19	19	
171	12	B		226	28	LOG		281	43	RCL	h_1 or h_2
172	43	RCL	δ_0	227	65	×		282	00	00	
173	16	16		228	02	2		283	75	-	
174	75	-		229	00	0		284	43	RCL	
175	43	RCL		230	54	)		285	15	15	
176	10	10		231	44	SUM	F_{d6}	286	55	÷	
177	55	÷		232	19	19		287	02	2	
178	04	4	λ/4	233	76	LBL	Range	288	54	)	h_c
179	54	)		234	35	1/X	terms	289	77	GE	Test for
180	77	GE	Test for	235	43	RCL	R	290	34	⌐X	$h>h_c$
181	30	TAN	clear path	236	13	13		291	87	IFF	Height gain
182	25	CLR	$\delta_0 \geq \lambda/4$	237	28	LOG		292	01	01	finished
183	01	1		238	65	×		293	35	1/X	
184	00	0	Eq. (21)	239	01	1		294	61	GTO	Do h_2
185	00	0		240	00	0	F_{d3}	295	38	SIN	terms
186	93	.		241	75	-		296	76	LBL	Extra height
187	06	6		242	43	RCL	λ	297	34	⌐X	gain
188	94	+/-		243	10	10		298	43	RCL	
189	42	STO	F_{d1}	244	28	LOG		299	15	15	$2h_c$
190	19	19		245	65	×		300	22	INV	
191	43	RCL	h_1	246	03	3		301	49	PRD	
192	11	11		247	00	0		302	00	00	$h/2h_c$
193	42	STO		248	75	-	F_{d2}	303	43	RCL	
194	00	00		249	43	RCL		304	00	00	
195	75	-		250	13	13	R	305	28	LOG	
196	43	RCL	h_{min}	251	55	÷		306	65	×	
197	14	14		252	01	1		307	01	1	
198	54	)		253	06	6		308	08	8	
199	77	GE	Test for	254	01	1		309	93	.	
200	33	x²	$h_1 \geq h_{min}$	255	07	7		310	01	1	
201	43	RCL		256	55	÷		311	75	-	
202	14	14	$h_1 = h_{min}$	257	43	RCL	λ	312	43	RCL	
203	28	LOG		258	10	10		313	00	00	
204	65	×		259	22	INV		314	34	⌐X	
205	02	2		260	45	YX		315	65	×	
206	00	0		261	03	3	F_{d4}	316	01	1	
207	54	)		262	54	)		317	08	8	
208	44	SUM	F_{d5}	263	44	SUM	F_{d5}	318	93	.	$F_{d2}+F_{d3}+F_{d4}+F_{d5}$
209	19	19		264	19	19		319	09	9	
210	76	LBL	h_2 terms	265	22	INV	Diffr.				
211	38	SIN		266	86	STF	finished				
212	43	RCL	h_2	267	01	01					
213	12	12		268	61	GTO	Reflection				
214	42	STO		269	13	C					

MERGED CODES		
62 ▪▪ ▪▪	72 STO ▪▪	83 GTO ▪▪
63 ▪▪ ▪▪	73 RCL ▪▪	84 ▪▪ ▪▪
64 ▪▪ ▪▪	74 SUM ▪▪	92 INV SBR

Program Listing (cont'd)

LOC	CODE	KEY	COMMENTS	LOC	CODE	KEY	COMMENTS	LOC	CODE	KEY	COMMENTS
320	06	6		375	17	17	ψ_σ	430	75	−	
321	85	+		376	39	COS		431	43	RCL	Re√X/Y
322	01	1		377	33	X²		432	00	00	
323	07	7		378	94	+/−		433	54	)	
324	93	.		379	85	+		434	36	PGM	Re N
325	04	4		380	43	RCL	ε_r	435	04	04	
326	54	)		381	05	05		436	11	A	
327	22	INV	F_{d7} or F_{d8}	382	54	)		437	43	RCL	
328	44	SUM		383	36	PGM	Re X	438	18	18	Im√X/Y
329	19	19		384	05	05		439	94	+/−	
330	87	IFF	Height gain	385	11	A		440	36	PGM	Im N
331	01	01	finished	386	43	RCL	λ	441	04	04	
332	35	1/X		387	10	10		442	11	A	
333	61	GTO	Do h_2 terms	388	65	×		443	43	RCL	ψ
334	38	SIN		389	43	RCL	σ_ε	444	17	17	
335	76	LBL	Clear path	390	06	06		445	38	SIN	
336	30	TAN		391	65	×		446	85	+	
337	00	0		392	06	6		447	43	RCL	Re√X/Y
338	42	STO	F_d = 0dB	393	00	0		448	00	00	
339	19	19		394	54	)		449	54	)	
340	76	LBL	Reflection	395	94	+/−		450	36	PGM	Re D
341	13	C		396	42	STO	$-60\lambda\sigma_e$	451	04	04	
342	87	IFF	Skip if no	397	00	00		452	16	A'	
343	03	03	line of sig	398	36	PGM	Im X = $\sqrt{\varepsilon'-\omega_0^2 k}$	453	43	RCL	
344	14	D		399	05	05		454	18	18	Im √X/Y
345	09	9	Eq. (27)	400	11	A		455	36	PGM	
346	32	X:T		401	36	PGM	√X	456	04	04	Im D
347	43	RCL		402	05	05		457	16	A'	
348	17	17	ψ_a	403	14	D		458	36	PGM	
349	38	SIN		404	87	IFF	Skip for	459	04	04	ρo
350	65	×		405	00	00	H pol.	460	18	C'	
351	43	RCL		406	42	STO		461	32	X:T	
352	08	08	σ_h	407	43	RCL	ε_r	462	22	INV	
353	65	×		408	05	05		463	37	P/R	
354	04	4		409	36	PGM	Re Y = $R_e \varepsilon'$	464	42	STO	
355	65	×		410	04	04		465	00	00	∝
356	89	π		411	16	A'		466	32	X:T	
357	55	÷		412	43	RCL	$-60\lambda\sigma_e$	467	99	PRT	$\lvert \rho_0 \rvert$
358	43	RCL	λ	413	00	00		468	49	PRD	printed
359	10	10		414	36	PGM	Im Y = Im ε'	469	07	07	Eq. (23)
360	54	)		415	04	04		470	43	RCL	ρ
361	33	X²		416	16	A'		471	00	00	α
362	77	GE	ρ_s = 0	417	36	PGM	√X/Y	472	85	+	
363	39	COS		418	04	04		473	02	2	
364	94	+/−		419	18	C'		474	65	×	
365	22	INV		420	76	LBL	H pol.	475	89	π	
366	23	LNX		421	42	STO		476	65	×	
367	34	√X		422	42	STO		477	43	RCL	δ
368	42	STO	ρ_s	423	00	00	Re√X/Y	478	16	16	
369	07	07	printed	424	32	X:T		479	55	÷	
370	99	PRT		425	42	STO					
371	29	CP		426	18	18	I_m √X/Y				
372	76	LBL	Fresnel Co-	427	43	RCL	ψ_0				
373	43	RCL	efficient,	428	17	17					
374	43	RCL	Eq. (25)	429	38	SIN					

MERGED CODES

62			72 STO		83 GTO	
63			73 RCL		84	
64			74 SUM		92 INV	SBR

Program Listing (cont'd)

LOC	CODE	KEY	COMMENTS	LOC	CODE	KEY	COMMENTS	LOC	CODE	KEY	COMMENTS
480	43	RCL	λ	535	05	5		590	49	PRD	h_{min}
481	10	10		536	22	INV		591	14	14	
482	54	)		537	90	LST		592	06	6	
483	39	COS		538	81	RST	Exit	593	00	0	Eq.(17)
484	65	×		539	76	LBL	Initialize	594	65	×	
485	43	RCL		540	15	E		595	43	RCL	
486	07	07	ρ	541	70	RAD		596	10	10	
487	65	×		542	29	CP	Eq. (14)	597	22	INV	
488	02	2		543	43	RCL	λ	598	45	YX	
489	85	+		544	10	10		599	01	1	
490	43	RCL		545	55	÷		600	93	.	
491	07	07		546	02	2		601	05	5	
492	33	x²	ρ^2	547	55	÷		602	54	)	
493	85	+		548	89	π		603	42	STO	
494	01	1		549	54	)		604	15	15	$2h_c$
495	54	)		550	42	STO		605	92	RTN	
496	28	LOG	F_r^2	551	14	14	$\lambda/2\pi$	606	76	LBL	
497	65	×		552	43	RCL		607	10	E'	
498	01	1		553	05	05	ε_r	608	86	STF	Long
499	00	0		554	33	x²		609	02	02	print
500	54	)		555	85	+		610	01	1	
501	42	STO		556	53	(		611	07	7	E
502	18	18	F_r(dB)	557	06	6		612	03	3	
503	87	IFF	H pol	558	00	0		613	05	5	R
504	00	00		559	65	×		614	07	7	
505	52	EE		560	43	RCL		615	01	1	?
506	01	1	V pol.	561	06	06	σ_e	616	69	OP	
507	42	STO		562	65	×		617	04	04	
508	07	07	restore 1	563	43	RCL	λ	618	43	RCL	ε_r
509	61	GTO	Print	564	10	10		619	05	05	
510	14	D	results	565	54	)		620	69	OP	
511	76	LBL	Rough	566	33	x²		621	06	06	
512	39	COS	Surface	567	42	STO	$(60\lambda\sigma_e)^2$	622	91	R/S	Enter
513	00	0		568	00	00		623	42	STO	new ε_r
514	42	STO		569	54	)		624	05	05	
515	18	18	F_r = 0dB	570	34	ΓX		625	03	3	S
516	61	GTO		571	45	YX		626	06	6	
517	14	D		572	43	RCL		627	01	1	E
518	76	LBL		573	07	07	b	628	07	7	
519	52	EE	restore 0	574	54	)		629	07	7	?
520	00	0	for H pol	575	49	PRD		630	01	1	
521	42	STO		576	14	14		631	69	OP	
522	07	07		577	43	RCL		632	04	04	
523	76	LBL	print	578	05	05	εr	633	43	RCL	σ_e
524	14	D	results	579	75	-		634	06	06	
525	87	IFF	Long	580	01	1		635	69	OP	
526	02	02	print	581	54	)		636	06	06	
527	44	SUM		582	33	x²		637	91	R/S	Enter
528	01	1		583	85	+		638	42	STO	new σ_e
529	06	6	Short	584	43	RCL		639	06	06	
530	22	INV	print	585	00	00					
531	90	LST		586	54	)					
532	81	RST	Exit	587	34	ΓX					
533	76	LBL	Long	588	34	ΓX					
534	44	SUM	Print	589	22	INV					

MERGED CODES					
62 ▢▢	72 STO ▢	83 GTO ▢
63 ▢▢	73 RCL ▢	84 ▢ ▢
64 ▢▢	74 SUM ▢	92 INV SBR

Program Listing (cont'd)

LOC	CODE	KEY	COMMENTS	LOC	CODE	KEY	COMMENTS	LOC	CODE	KEY	COMMENTS
640	04	4		695	07	7	?	750	69	OP	
641	02	2	v	696	01	1		751	04	04	
642	06	6	=	697	69	OP		752	43	RCL	
643	04	4		698	04	04	λ	753	13	13	R
644	00	0	1	699	43	RCL		754	69	OP	
645	02	2		700	10	10		755	06	06	
646	07	7	?	701	69	OP		756	91	R/S	Enter new
647	01	1		702	06	06		757	42	STO	R
648	69	OP		703	91	R/S	Enter	758	13	13	
649	04	04		704	42	STO	new λ	759	99	PRT	
650	43	RCL	b	705	10	10		760	61	GTO	Start
651	07	07		706	71	SBR		761	11	A	
652	69	OP		707	15	E					
653	06	06	Enter	708	76	LBL	h$_1$				
654	91	R/S		709	16	A'		Time saver: Use $\rho = 1$,		
655	42	STO	b=1(v pol)	710	02	2		inserting 9 steps 345-53:		
656	07	07	b=0(H pol)	711	03	3	H				
657	03	3	S	712	00	0		340	76	LBL	
658	06	6		713	02	2	1	341	13	C	
659	02	2	H	714	07	7		342	87	IFF	
660	03	3		715	01	1	?	343	03	03	
661	07	7		716	69	OP		344	14	D	
662	01	1	?	717	04	04		345	01	1	
663	69	OP		718	43	RCL		346	42	STO	
664	04	04		719	11	11	h$_1$	347	07	07	
665	43	RCL		720	69	OP		348	89	π	
666	08	08	σ$_h$	721	06	06		349	42	STO	
667	69	OP		722	91	R/S	Enter	350	00	00	
668	06	06		723	42	STO	new h$_1$	351	61	GTO	
669	91	R/S	Enter	724	11	11		352	04	04	
670	42	STO	new σ$_h$	725	99	PRT		353	70	70	
671	08	08	K	726	76	LBL	h$_2$				
672	02	2		727	17	B'					
673	06	6		728	02	2					
674	01	1		729	03	3	H				
675	03	3	A	730	00	0	2				
676	07	7		731	03	3					
677	01	1	?	732	07	7					
678	69	OP		733	01	1	?				
679	04	04		734	69	OP					
680	43	RCL	ka	735	04	04					
681	09	09		736	43	RCL					
682	69	OP		737	12	12	h$_2$				
683	06	06		738	69	OP					
684	91	R/S	Enter	739	06	06					
685	42	STO	new ka	740	91	R/S	Enter new				
686	09	09		741	42	STO	h$_2$				
687	76	LBL	λ	742	12	12					
688	19	D'		743	99	PRT					
689	02	2		744	76	LBL	R				
690	07	7	L	745	18	C'					
691	01	1	D	746	03	3					
692	06	6		747	05	5					
693	01	1	A	748	07	7	R				
694	03	3		749	01	1	?				

MERGED CODES

62 ▩ ▩   72 STO ▩   83 GTO ▩
63 ▩ ▩   73 RCL ▩   84 ▩ ▩
64 ▩ ▩   74 SUM ▩   92 INV PRT

Program Listing (cont'd)

Appendix B
List of Symbols

The page number is the location of the first use. Where the symbol is defined later, that page is also given.

Symbol	Meaning	Page
A	Area of target	121
	Area of antenna aperture	150
	Azimuth angle	459
A'	Effective area of antenna aperture, for blockage	158
$A(f)$	Signal voltage spectrum	17
$A(f_d, R_c)$	Correlation factor for oscillator phase noise	265
$A(v)$	Diffracted field strength	517
A_b	Area of antenna blockage	158
A_p	Area of flat plate	101
A_r	Target field strength toward reflecting surface	514
A_t	Target field strength toward radar	514
$A_0(f)$	Spectrum of fine line	70
$A_1(f)$	Line structure of spectrum	70
$A_2(f)$	Spectral envelope	70
A_b	Effective aperture area of beacon or intercept antenna	12
A_c	Area within resolution cell	42
A_m	Azimuth search sector	24
A_r	Effective aperture area of receiving antenna	10
a	Radius of the earth	41
	Radius of sphere	101
	Side length of corner reflector	103
	Weibull distribution spread parameter	129
	Constant in estimation of refraction error	305
a_t	Target acceleration	459

Symbol	Meaning	Page
$a_x(t)$	Output waveform	223
$a(t)$	Signal waveform	17
a_0	Peak of output waveform	223
$a_0(t)$	Received signal envelope	70
$a_1(t)$	Pulse repetition rate function	70
a_1	Radius of disk	101
	Precomparator amplitude error in monopulse	413
$a_2(t)$	Waveform of individual pulse	70
a_2	Postcomparator amplitude error in monopulse	413
B_a	Half-power bandwidth of array	187
	Signal band limit	433
B_f	Bandwidth of doppler filter	20
B_h	Filter band limit	435
B_j	Noise bandwidth of jammer	33
B_n	Equivalent noise bandwidth of filter	14
B_s	Scan-frequency sweep bandwidth	494
B_v	Video bandwidth	79
B_w	Wide bandwidth in Dicke-fix CFAR	89
B_1	Width of input filter	217
	Bandwidth of single IF amplifier stage	549
B_3	Width of output filter	217
B_{3a}	Half-power width of signal spectrum	381
b	Constant in estimation of refraction error	305
C	Clutter power	38
C_a	Detector loss for monopulse and AGC	467
C_B	Blake's bandwidth matching factor	20
C_d	Detector loss for conical scan	467
C_e	Dimensional constant	211
C_v	Dimensional constant	211
C_x	Detector loss	64
CA	Clutter attenuation	244
C/S	Clutter-to-signal power ratio	39
C_0	Effective clutter spectral density	48
C_Δ	Clutter power in difference channel	532
c	Velocity of light	20
c_g	Velocity of wave in waveguide	168
D	Detectability factor for ideal processing	21
	Diameter of target	122
	Divergence factor	291
D_b	Detectability factor with binary integration	87
D_c	Detectability factor for synchronous detection	63

Symbol	Meaning	Page
$D_e(n,n_e)$	Detectability factor with diversity	84
D_u	Duty factor of transmission	32
D_x	Detectability factor with losses	19
D_{xc}	Clutter detectability factor with losses	40, 95
D_0	Detectability factor for steady target	26, 62
$D_0(1)$	Detectability factor of single sample from steady target	20
$D_0(n)$	Detectability factor for each of n pulses	19
D_1	Detectability factor for Swerling Case 1 target	81
d	Cathode-ray-tube sweep speed	79
	Diameter of antenna blockage	160
	Element spacing in array	165
dP	Differential probability	59
d_1	Distance from radar to obstacle	299
	Precomparator error in monopulse	413
d_2	Distance from target to obstacle	299
	Postcomparator error in monopulse	413
E	Total signal energy	17
	Elevation angle	303
E_f	Signal energy in coherent integration time	45
E_n	Instantaneous noise envelope voltage	59
E_s	Peak voltage of sinusoidal signal	61
E_t	Threshold voltage	58
E_1	Single-pulse signal energy	17
E_Δ	Normalized error output	409
e	Voltage from intended target polarization	416
e_c	Voltage from cross-polarization	416
e_n	Instantaneous noise voltage	410
F	Pattern-propagation factor	18
F_a	Fraction of active elements in thinned array	173
	Fraction of clear area in delay-doppler plane	260
F_c	Pattern-propagation factor to clutter	38
F_d	Roughness factor	515
F_j	Pattern-propagation factor from jammer to radar	33
F_n	Receiver noise factor	14
F_r	Receiving pattern-propagation factor	21
F_t	Transmitting pattern-propagation factor	21
f	Frequency	15
$f(\theta, \phi)$	Antenna voltage pattern relative to peak gain	150
$f(z)$	Voltage response in arbitrary coordinate	378

Symbol	Meaning	Page
f'	Voltage pattern after blockage	159
f_a	Array pattern factor	165
	Frequency of Markoffian error component	510
f_b	Voltage of blockage pattern	158
f_c	Intermediate frequency	58
	Correlation frequency of target	86
	Critical frequency of ionosphere	309
	Frequency of cyclic error component	510
f_d	Doppler frequency shift	143
Δf_d	Relative doppler shift of multipath component	522
f_e	Element pattern factor	165
f_r	Pulse repetition frequency	19
	Voltage pattern of receiving antenna	145
f_s	Signal frequency	58
	Scan frequency	383
f_t	Voltage pattern of transmitting antenna	145
f_0	Center frequency of carrier	15
Δf	Diversity system bandwidth	86
f_1	Width of spectrum of apparent bias error	510
G	Power gain	101
$G(f)$	Glint spectral density	120
G_b	Beacon receiving antenna gain	12
G_c	Coherent integration gain	272
G_{cs}	Gain of cosecant-squared antenna	162
$G_d(n_e)$	Diversity gain for n_e target samples	85
G_f	Gain of thinned array	174
$G_i(n)$	Integration gain for n pulses	62, 71
G_j	Jammer transmitting gain	33
G_m	Gain of antenna on beam axis	145
G_n	Monopulse null depth	413
G_r	Radar receiving antenna gain	10
G_s	Antenna sidelobe gain	157
G_{se}	Difference-channel main-lobe/sidelobe gain ratio	406
G_{sr}	Sum-channel main-lobe/sidelobe gain ratio	406
G_t	Radar transmitting antenna gain	10
G_0	Gain for uniformly illuminated aperture	150
$g(x,y)$	Illumination function of aperture	151
$g'(x)$	Illumination function collapsed onto x axis	185
$g_d(x)$	Illumination function of difference channel	400
$H(f)$	Filter frequency response	15
$H_1(f)$	Frequency response of input filter	218

Symbol	Meaning	Page
	Frequency response of single-delay canceler	237
$H_2(f)$	Frequency response of double-delay canceler	238
$H_3(f)$	Frequency response of output filter	218
h	Height of antenna aperture	150
	Scale height of atmospheric refractivity	304
$h(t)$	Filter impulse response	17
$h_a(t)$	Impulse response of array	185
h_d	Duct height	309
h_m	Maximum search altitude	26
h_r	Antenna height above surface	21
h_r'	Effective antenna height above clutter	128
h_t	Target altitude	46
I	In-phase component in receiver	58
I_m	MTI or doppler improvement factor	40
I_q	Improvement factor limit caused by quantization	248
I_r	Power density at radar receiving antenna	10
I_s	Power density at target	10
	Improvement factor limit caused by PRF stagger	247
I_x	Improvement factor of x-delay canceler	248
I_0	Interference spectral density	49
I_1	Improvement factor of single-delay canceler	245
I_2	Improvement factor of double-delay canceler	245
I_3	Improvement factor of triple-delay canceler	245
J	Jamming power at radar receiver	490
J_b	Bistatic jamming power	502
J_c	Jamming power in cross-polarization	499
J_d	Direct jamming signal from jammer sidelobe	502
J_i	Jamming power in intended radar polarization	499
J_0	Jamming power density at receiver	33
K	Number of degrees of freedom in chi-square distribution	85
	Relative difference slope	401
K_a	Acceleration error constant	463
K_B	Beaufort wind scale number	125
K_f	Frequency measurement slope	441
K_h	Ratio of input to output pulsewidth	430
K_r	Difference slope ratio	402
K_v	Velocity error constant	463
K_3	Jerk error constant	463
K_θ	Beamwidth constant	153
K_θ'	Beamwidth constant in synthetic aperture radar	362

Symbol	Meaning	Page
k	Boltzmann's constant	13
	Effective earth radius factor	41
	Constant used to adjust number of CFAR reference cells	89
	$2P/L$	102
	Ratio of amplitudes for two-element target	116
k_m	Monopulse error slope	401
k'_m	Monopulse error slope with implementation errors	413
k_p	Sector scanning error slope	393
k_s	Conical scan error slope	387
k_{sh}	Wind shear coefficient	136
k_t	Time measurement slope	429
k_z	Error slope in z coordinate	379
k_α	Attenuation coefficient of atmosphere	279
L	Target length	102
	Length of sinusoidal surface irregularity	517
L_c	Collapsing loss	31, 77
	Correlation length of surface roughness	517
L_{cd}	Clutter distribution loss	92
L_{cs}	Cosecant-squared pattern loss	162
L_d	Scan distribution loss	31
L_{ea}	Azimuth straddling loss	269
L_{ec}	Eclipsing loss	270
L_{ef}	Filter straddling loss	269
L_{eg}	Transient gating loss	270
L_{er}	Range straddling loss	251
L_{ev}	Velocity response loss	270
L_f	Target fluctuation loss	31, 81
L_g	Threshold or CFAR loss	90
L_i	Integration loss	30, 71
L_k	Crossover loss in conical scan	387
L_m	Receiver matching loss	17
L_{mf}	Doppler filter matching loss	67
L_{mti}	MTI loss	250
L_n	Beam pattern constant	24, 155
L_p	Beam shape loss	18
L_q	Quantization loss	171
L_r	Receiving line loss	14
	Target radial length	86
L_s	Total search loss	24
L_t	Transmitting line loss	18

Symbol	Meaning	Page
L_x	Miscellaneous processing loss	20
	Target cross-range span	86
L_z	Loss caused by limited array bandwidth	187
L_α	Atmospheric attenuation	18
$L_{\alpha c}$	Two-way atmospheric attenuation to clutter	38
$L_{\alpha j}$	One-way atmospheric attenuation from jammer	33
$L_{\alpha m}$	Actual atmospheric attenuation at R_m	286
$L_{\alpha 1}$	First estimate of atmospheric attenuation	21
$L_{\alpha 2}$	Second estimate of atmospheric attenuation	21
L_ϕ	Loss caused by phase errors	183
$\mathscr{L}$	rms aperture width	381
$\mathscr{L}_0$	rms width of rectangular aperture	382
M	Receiver matching factor	20, 78
m	Binary detection threshold	69, 73
	Number of extra noise samples in collapsing	77
	Number of reference cells in CFAR system	89
	Number of bits in phase shifter	171
	Number of chips in phase-coded pulse	221
	Number of pulses in MTI cancelation process	250
	Number of bursts with PRF diversity	260
	Number of bits in A/D converter	268
	Number of acquisition scans	455
N	Noise power	14
	Refractivity of atmosphere	304
N_d	Refractivity at top of duct	309
N_e	Electron density of ionosphere	309
N_i	Refractivity of ionosphere	309
N_s	Refractivity at earth's surface	304
N_0	Noise spectral density	13
n	Number of pulses integrated	19
	Refractive index of atmosphere	303
	Number of error samples evaluated	506
n'	Number of independent outputs from doppler filter	20
	Number of pulses integrated in reduced bandwidth	468
n_a	Number of elements in row	149
n_b	Number of pulses used in sector scanning	393
n_c	Number of pulses in doppler processing burst	257
	Number of independent samples of clutter	318
n_d	Number of detection opportunities	483
n_e	Number of elements in column	149
	Number of independent target samples	84

Symbol	*Meaning*	*Page*
	Number of independent error samples	381
n_f	Number of doppler channels	76
n_s	Number of search beam positions	458
n_t	Number of range gates	76
P	Atmospheric pressure	304
P_a	Probability of acquisition	453
P_{av}	Transmitter average power	17
P_c	Cumulative probability of detection	88, 260
P_{ca}	Cumulative probability of acquisition	455
P_d	Detection probability	21
P_{fa}	False-alarm probability	21
P'_{fa}	False-alarm probability after collapsing	77
P_i	Power intercepted by target	10
P_j	Jammer transmitter power	33
P_{ja}	Average jammer power of repeater	490
P_p	Probability of target suppression by clutter peak	252
P_r	Probability that target lies within circle	455
P_t	Transmitter peak power	10
P_v	Probability that target lies within scan volume	453
P_x	Probability that target lies within interval X	455
P_2	Probability that target lies within rectangle	455
p	Partial pressure of water vapor	304
p_d	Detection probability for single pulse	74
p_{fa}	False-alarm probability for single pulse	74
Q	Quadrature component in receiver	58
$Q(E)$	Probability integral	63
R	Target range	10
R'	Maximum range without atmospheric attenuation	21
R_b	Range at which beam reaches surface	265
R_{bt}	Burnthrough range with jamming	35
R_c	Range to clutter	38
R_h	Horizon range	41
R_{ht}	Line-of-sight range to target	46
R_j	Jammer range	33
R_m	Maximum detection range	19
R_u	Unambiguous range	51, 234
R_0	Maximum range in free space	20
R_1	Transition range for clutter propagation	124
	Transition range for phase noise correlation	265
	First iteration for maximum range	21
R_2	Second iteration for maximum range	286

Symbol	Meaning	Page
$\mathcal{R}$	Doubled energy ratio $2E/N_0$	381
r	Rainfall rate	133
S	Signal power at radar receiver	10
$S(f)$	Power spectrum	115
S_i	Signal power at beacon	12
S_r	Signal power from beacon	12
SCV	Subclutter visibility	244
S/C	Signal-to-clutter power ratio	40
S/I	Signal-to-interference ratio	380
S/I_Δ	Signal-to-interference ratio in difference channel	380
S/J	Signal-to-jamming power ratio	94
S/N	Signal-to-noise power ratio	17
$(S/N)_m$	Signal-to-noise ratio on beam axis	387
s	Cathode-ray-tube spot diameter	79
	Slot spacing along waveguide	168
T	Total number of elements in array	149
	Temperature of atmosphere	304
T_a	Effective antenna temperature	14
T'_a	Sky temperature	14
T_e	Effective receiver temperature	14
T_f	Number of active elements in thinned array	173
T_g	Ground component of antenna temperature	21
T_j	Effective noise temperature of jamming	34
T_p	Thermal temperature of loss medium	288
T_r	Effective receiving line temperature	14
T_s	Effective system input temperature	13
T_{ta}	Antenna thermal temperature	21
T_{tg}	Ground thermal temperature	21
T_{tr}	Thermal temperature of receiving line	21
t_a	Acquisition time	451
t_c	Correlation time of target	86
t'_c	Correlation time of envelope-detected target	388
t_d	Range delay	143
t_{dc}	Range delay to farthest clutter	261
t_f	Coherent integration time	20
t'_f	Coherent processing time after transient gating	261
t_{fa}	False-alarm time	60, 76
t_g	Duration of transient gate	271
t_j	Duration of jamming transient	494
t_o	Signal observation (integration) time	17
t_r	Pulse repetition interval	70

Symbol	Meaning	Page
t_s	Search frame time	24
	Jammer sweep time	494
t_t	Settling time of doppler filter	271
t_v	Video integration time during scan	79
$U(f)$	Frequency response of Urkowitz filter	236
V_c	Volume of resolution cell	38
V_0	Blake's visibility factor	20
v	Instantaneous IF noise voltage	59
	Diffraction parameter	299
v_b	Blind speed	234
v_m	Maximum target velocity	339
v_p	Platform velocity	245
v_r	Target radial velocity	439
v_t	Target forward velocity	459
v_w	Wind speed	249
W	Weight of chaff	134
$W(f)$	Power spectrum of target scintillation	388
	Power spectrum of error	510
$W_n(f)$	Spectrum of phase noise in oscillator	265
W_0	Zero-frequency power spectral density	388
w	Width of antenna aperture	150
X	Ratio of cross-polarized jamming to useful signal	500
x	Instantaneous noise power	59
	CFAR ratio	90
	Horizontal coordinate in aperture plane	146
	Error value	506
	Ground distance to reflecting surface element	521
x_t	Ground range to target	521
x_0	Range to specular reflection point	290
y	Vertical coordinate in aperture plane	146
z	Coordinate normal to aperture plane	146
	Normalized clutter spectral spread	244
	Arbitrary measurement coordinate	378
z_a	Normalized spread caused by antenna scanning	244
z_i	Location of interference source in z coordinate	380
z_k	Offset of individual response in z coordinate	379
z_3	Half-power width of response in z coordinate	379
α	Half-angle of cone	103
	Weibull distribution amplitude parameter	129
	Angle to wind vector	136
	Phase angle of reflected wave	291

Symbol	Meaning	Page
	rms time duration of signal	381
α_0	rms duration of rectangular signal envelope	382
β	Angle from forward scatter, in slant plane	122
	Tilt of surface facet	125
	rms bandwidth of signal	381
β_a	rms bandwidth of band-limited rectangular pulse	433
β_n	Servo noise bandwidth (one-sided)	389
β_{no}	Servo bandwidth for high S/N	468
β_0	$\sqrt{2}$ times rms surface slope	125
	rms bandwidth of rectangular spectrum	382
γ	Surface clutter reflectivity factor	42, 124
	PRI stager ratio	248
Δ	Lobe width	102
Δ	Difference-channel voltage	199
Δ_a	Synthetic aperture azimuth resolution	361
	Azimuth error channel output	409
Δ_c	Width of clutter spectrum	257
	Difference-channel voltage from cross-polarized component	415
ΔE	Elevation refraction error	303
	Elevation sector	316
Δ_e	Elevation error channel output	409
Δ_i	Interference signal in difference channel	380
Δ_t	Target signal in difference channel	380
ΔR	Range refraction error	306
	Range resolution	39
Δ_r	Eclipsed range interval	260
Δ_v	Width of velocity notch at -6 dB level	257
Δ_x	Cross-range resolution	359
$\Delta\theta$	Elevation error	353
δ	Normalized glint error	117
$\delta(\cdot)$	Delta function	70
δ_t	Time delay in IF amplifier	549
δ_0	Delay of specular reflection relative to direct signal	290
ϵ	Error	501
ϵ_a	Azimuth lag error	463
	Acceleration lag error	471
ϵ_c	Complex dielectric constant of surface material	292
ϵ_r	Relative dielectric constant of surface material	292
ϵ_θ	Beam-steering error	171
η_a	Aperture efficiency	152

Symbol	Meaning	Page
η_b	Efficiency factor for blockage	162
η_d	Elevation density of diffuse multipath	519
η_f	Filter efficiency factor	429
η_i	Illumination efficiency	162
η_s	Efficiency factor for spillover	162
η_t	Efficiency factor for surface tolerance	162
	Efficiency of power generation	180
η_v	Volume clutter reflectivity	38, 133
η_x	Efficiency of illumination in horizontal coordinate	152
η_y	Efficiency of illumination in vertical coordinate	152
θ	Elevation angle	26
	Aspect angle of target	102
	Angle from forward scatter, in azimuth plane	122
	Angle from antenna beam axis	146
	Angle between ray paths over obstacle	299
θ_a	Azimuth half-power beamwidth	24
θ_d	Maximum ray elevation trapped in duct	309
θ_{da}	Azimuth width of glistening surface	527
θ_e	Elevation half-power beamwidth	24
θ_k	Squint angle of beam	385
θ_m	Effective maximum search elevation	24
θ'_m	Effective search elevation for time budget	27
θ_n	Elevation angle between reflection nulls	291
θ_s	Angle of sidelobe	159
θ_t	Target elevation	290
θ_0	Minimum search elevation	24
θ_1	Maximum search elevation at R_m	26
θ_2	Maximum search elevation of cosecant-squared pattern	26
θ_3	Half-power antenna beamwidth	153
λ	Radar wavelength	11
λ_c	Critical wavelength of waveguide	168
λ_g	Wavelength in waveguide	168
λ_{g0}	Wavelength in waveguide for broadside beam	168
λ_0	Wavelength for broadside beam	168
μ	Parameter in Student distribution for glint	116
ν	Decision rate	76
ρ	Collapsing ratio	77
	Reflection coefficient of surface	291
$\rho(t)$	Correlation function	114
ρ_d	Diffuse reflection coefficient	514

Symbol	Meaning	Page
ρ_s	Specular scattering coefficient	291
ρ_v	Coefficient for vegetative absorption	291
ρ_0	Fresnel reflection coefficient	291
Σ	Sum channel voltage	199
Σ_c	Sum channel voltage from cross-polarized component	415
σ	Radar cross section of target	10, 100
σ_a	rms amplitude error in array	188
σ_b	Bistatic cross section	12
σ_c	Radar cross section of clutter	38
	Cross-polarized target cross section	416
σ_c'	Effective cross section of ambiguous clutter	324
σ_e	Conductivity of surface material	292
	rms elevation error	353
σ_f	rms frequency error	118
	Forward scatter cross section	121
	rms spread of signal spectrum	427
σ_g	rms glint	116
σ_h	rms surface roughness	124
σ_h'	Reduced surface roughness with shadowing	516
σ_r	rms range error	118
σ_s	rms scintillation error	388
σ_t	rms width of impulse response	187
	rms error in time delay	381
σ_v	rms velocity spread	136
σ_{va}	Velocity spread caused by antenna scanning	245
σ_{vi}	Velocity spread caused by radar instability	245
σ_{vp}	Velocity spread caused by platform motion	245
σ_{vs}	Velocity spread caused by wind shear	245
σ_{vt}	Velocity spread caused by internal motion	245
σ_x	rms error in target designation	452
	rms error	507
σ_y	Standard deviation of log-normal distribution	114
σ_z	rms error of measurement in z coordinate	380
σ_θ	rms angle error	118
σ^0	Surface clutter reflectivity	42, 124
σ_b^0	Bistatic reflectivity of surface	517
σ_f^0	Forward scatter clutter reflectivity	125
σ_2	Thermal noise component from off-axis position	421
σ_{50}	Median cross section	111

Symbol	Meaning	Page
σ_ϕ	rms phase shift error	171
τ	Transmitted pulsewidth	17
τ_g	Gate width	79
τ_n	Effective (processed) pulsewidth	38
τ_o	Half-power pulsewidth from matched filter	426
τ_x	Output half-power pulsewidth	187
τ_3	Half-power pulsewidth	145
ϕ	Phase angle	116
	Angle from forward scatter, in elevation plane	122
	Angle in antenna coordinates	146
ϕ_1	Precomparator phase error in monopulse	413
ϕ_2	Postcomparator phase error in monopulse	413
X	Response function	143
X_0	Matched-filter response function, or ambiguity function	209
ψ	Grazing angle	39
ψ_b	Beam solid angle	24
ψ_c	Critical grazing angle	124
ψ_s	Search solid angle	24
ψ_1	Grazing angle from radar to multipath reflecting element	515
ψ_2	Grazing angle from target to multipath reflecting element	515
ω_a	Antenna azimuth scan rate	32
	Rotation rate of target	86
ω_e	Antenna elevation scan rate	79
ω_x	Rotation rate of target	394

Index